NanoScience and Technology

Springer
Berlin
Heidelberg
New York
Hong Kong
London
Milan
Paris
Tokyo

Physics and Astronomy ONLINE LIBRARY
springeronline.com

NanoScience and Technology

Series Editors: P. Avouris   K. von Klitzing   H. Sakaki   R. Wiesendanger

The series NanoScience and Technology is focused on the fascinating nano-world, mesoscopic physics, analysis with atomic resolution, nano and quantum-effect devices, nanomechanics and atomic-scale processes. All the basic aspects and technology-oriented developments in this emerging discipline are covered by comprehensive and timely books. The series constitutes a survey of the relevant special topics, which are presented by leading experts in the field. These books will appeal to researchers, engineers, and advanced students.

Sliding Friction
Physical Principles and Applications
By B.N.J. Persson
2nd Edition

Scanning Probe Microscopy
Analytical Methods
Editor: R. Wiesendanger

Mesoscopic Physics and Electronics
Editors: T. Ando, Y. Arakawa, K. Furuya,
S. Komiyama,
and H. Nakashima

Biological Micro- and Nanotribology
Nature's Solutions
By M. Scherge and S.N. Gorb

**Semiconductor Spintronics
and Quantum Computation**
Editors: D.D. Awschalom, N. Samarth,
D. Loss

Semiconductor Quantum Dots
Physics, Spectroscopy and Applications
Editors: Y. Masumoto and T. Takagahara

Nano-Optoelectronics
Concepts, Physics and Devices
Editor: M. Grundmann

Noncontact Atomic Force Microscopy
Editors: S. Morita, R. Wiesendanger,
E. Meyer

Nanoelectrodynamics
Electrons and Electromagnetic Fields
in Nanometer-Scale Structures
Editor: H. Nejo

Single Organic Nanoparticles
Editors: H. Masuhara, H. Nakanishi,
K. Sasaki

Epitaxy of Nanostructures
By V.A. Shchukin, N.N. Ledentsov and D.
Bimberg

Nanostructures
Theory and Modeling
By C. Delerue and M. Lannoo

**Nanoscale Characterisation
of Ferroelectric Materials**
Scanning Probe Microscopy Approach
Editors: M. Alexe and A. Gruverman

C. Delerue
M. Lannoo

Nanostructures

Theory and Modeling

With 149 Figures

 Springer

Dr. Christophe Delerue
IEMN
Département ISEN
41 boulevard Vauban
59046 Lille, France

Dr. Michel Lannoo
CNRS
3 rue Michel-Ange
75794 Paris, Cedex 16, France

Series Editors:
Professor Dr. Phaedon Avouris
IBM Research Division, Nanometer Scale Science & Technology
Thomas J. Watson Research Center, P.O. Box 218
Yorktown Heights, NY 10598, USA

Professor Dr., Dres. h. c. Klaus von Klitzing
Max-Planck-Institut für Festkörperforschung, Heisenbergstrasse 1
70569 Stuttgart, Germany

Professor Hiroyuki Sakaki
University of Tokyo, Institute of Industrial Science, 4-6-1 Komaba, Meguro-ku
Tokyo 153-8505, Japan

Professor Dr. Roland Wiesendanger
Institut für Angewandte Physik, Universität Hamburg, Jungiusstrasse 11
20355 Hamburg, Germany

ISSN 1434-4904
ISBN 3-540-20694-9 Springer-Verlag Berlin Heidelberg New York

Library of Congress Cataloging-in-Publication Data.

Delerue, C.
Nanostructures: theory and modeling/C. Delerue, M. Lannoo.
p.cm. – (Nanoscience and technology, ISSN 1434-4904)/
Includes bibliographical references and index.
ISBN 3-540-20694-9 (acid-free paper)
1. Nanostructures. I. Lannoo, M. (Michel), 1942– . II. Title. III. Series.
QC176.8.N35D44   2004   530.4'1–dc22   2003067388

This work is subject to copyright. All rights are reserved, whether the whole or part of the material is concerned, specifically the rights of translation, reprinting, reuse of illustrations, recitation, broadcasting, reproduction on microfilm or in any other way, and storage in data banks. Duplication of this publication or parts thereof is permitted only under the provisions of the German Copyright Law of September 9, 1965, in its current version, and permission for use must always be obtained from Springer-Verlag. Violations are liable for prosecution under the German Copyright Law.

Springer-Verlag is a part of Springer Science+Business Media.

springeronline.com

© Springer-Verlag Berlin Heidelberg 2004
Printed in Germany

The use of general descriptive names, registered names, trademarks, etc. in this publication does not imply, even in the absence of a specific statement, that such names are exempt from the relevant protective laws and regulations and therefore free for general use.

Typesetting: Data conversion by the authors using a Springer TeX macro package
Final processing by Frank Herweg, Leutershausen
Cover design: *design& production*, Heidelberg

Printed on acid-free paper   SPIN: 10963623   57/tr 3141 - 5 4 3 2 1 0

To Catherine and Ginette

Preface

This book is an introduction to the theory of nanostructures. Its main objectives are twofold: to provide basic concepts for the physics of nano-objects and to review theoretical methods allowing the predictive simulation of nano-devices. It covers many important features of nanostructures: electronic structure, dielectric properties, optical transitions and electronic transport. Each topic is accompanied by a review of important experimental results in this field. We have tried to make the book accessible to inexperienced readers and it only requires basic knowledge in quantum mechanics and in solid state physics. Whenever possible, each concept is introduced on the basis of simple models giving rise to analytical results. But we also provide the reader with the more elaborate theoretical tools required for simulations on computers. Therefore, this book is intended not only for the students beginning in this field but also for more experienced researchers.

The context of the book is the rapid expansion of nano-technologies resulting from important research efforts in a wide range of disciplines such as physics, biology and chemistry. If much work is presently focusing on the elaboration, the manipulation and the study of individual nano-objects, a major challenge for nano-science is to assemble these objects to make new materials and new devices, opening the door to new technologies. In this context, as the systems become more and more complex, and because probing the matter at the nanoscale remains a challenge, theory and simulation play an essential role in the development of these technologies. A large number of simulation tools are already available in science and technology but most of them are not adapted to the nano-world because, at this scale, quantum mechanical descriptions are usually necessary, and atomistic approaches become increasingly important. Thus, one main objective of the book is to review recent progress in this domain. We show that ab initio approaches provide accurate methods to study small systems ($\lesssim$100–1000 atoms). New concepts allow us to investigate these systems not only in their ground state, but also in their excited states and out of equilibrium. The domain of application of ab initio methods also becomes wider thanks to the decreasing size of the systems, to the increasing power of the computers and to novel algorithms. But these developments are by far not sufficient enough to cover all the needs, in particular when the number of atoms in the systems becomes large ($\gtrsim$100–1000).

Thus, most of the problems in nano-science must be investigated using semi-empirical approaches, and ab initio calculations are used to test or to calibrate the semi-empirical methods in limiting cases. Therefore, an important part of the book is devoted to semi-empirical approaches. In particular, we present recent improvements which greatly enhance their predictive power.

Due to the huge existing literature in this field, we have limited our bibliography to what we believe are the most basic papers. It is also clear that we have not covered all the aspects. For example, we have omitted nano-magnetism which merits a book of its own.

The book is divided into eight chapters. Chapter 1 gives a general overview of the basic theoretical methods which allow an understanding of the electronic properties from condensed matter to molecules and atoms. We present ab initio descriptions of the electronic systems in their ground state, in particular those based on the density functional theory, and we review recent approaches dealing with one-particle and two-particle excitations. Then, semi-empirical methods are introduced, from the simple effective mass approach to more elaborate theories such as tight binding and pseudopotential methods. Chapter 2 provides a general introduction to quantum confined semiconductor systems, from two to zero dimensions. We compare different computational techniques and we discuss their advantages and their limits. The theoretical predictions for quantum confinement effects are reviewed.

Chapter 3 deals with the dielectric properties of nano-objects. Microscopic methods based on electronic structure calculations are presented. Screening properties in semiconductor nanostructures are analyzed using both macroscopic and microscopic approaches. The concept of local dielectric constant is introduced and we conclude by discussing the possibility of using the macroscopic theory of dielectrics in nano-systems. We also point out the importance of surface polarization charges at dielectric interfaces for Coulomb interactions in nanostructures.

In Chapter 4, we focus on the description of quasi-particles and excitons, starting from the simpler methods based on the effective mass theory and progressing to more complex approaches treating dynamic electronic correlations. Chapter 5 discusses the optical properties of nanostructures. It begins with the basic theory of the optical transitions, concentrating on problems specific to nano-objects and including the influence of the electron–phonon coupling on the optical line-shape. The optical properties of semiconductor nanocrystals are then reviewed, both for interband and intraband transitions. Chapter 6 is devoted to hydrogenic impurities and point defects in nanostructures. In view of the importance of surfaces in small systems, surface dangling bond defects are discussed in detail. The chapter closes with study of self-trapped excitons showing that their existence is favored by confinement effects.

Non-radiative processes and relaxation mechanisms are considered in Chap. 7. The effect of the quantum confinement on the multi-phonon capture on point defects is studied. We present theoretical formulations of the Auger recombination in nanostructures and we discuss the importance of this mechanism by reviewing the experimental evidence. Then we address the problem of the relaxation of hot carriers in zero-dimensional objects. In strongly confined systems, phonon-assisted relaxation is slow due to the phonon bottleneck effect, but we explain why this effect is difficult to observe due to competitive relaxation mechanisms.

Chapter 8 discusses non-equilibrium transport in nanostructures. We introduce theoretical methods used to simulate current–voltage characteristics. We start with the regime of weak coupling between the nano-device and the electron leads, introducing the so-called orthodox theory. Situations of stronger coupling are investigated using the scattering theory in the independent particle approximation. Electron–electron interactions are then considered in mean-field approaches. The limits of these methods are analyzed at the end of the chapter.

Finally, we are greatly indebted to G. Allan for a long and fruitful collaboration. We are grateful to all our colleagues and students for discussions and for their contributions. We acknowledge support from the "Centre National de la Recherche Scientifique" (CNRS) and from the "Institut Supérieur d'Electronique et du Numérique" (ISEN).

Lille, Paris, *C. Delerue*
December 2003 *M. Lannoo*

Table of Contents

1 General Basis for Computations and Theoretical Models . 1
  1.1 Ab initio One-Particle Theories for the Ground State 1
      1.1.1 Non-interacting N Electron System 2
      1.1.2 The Hartree Approximation 3
      1.1.3 The Hartree–Fock Approximation 5
      1.1.4 Correlations and Exchange–Correlation Hole 6
      1.1.5 Local Density Approaches 10
  1.2 Quasi-particles and Excitons 13
      1.2.1 One-Particle Eigenvalues 14
      1.2.2 The Exchange–Correlation Hole and Static Screening . 15
      1.2.3 Dynamically Screened Interactions 16
      1.2.4 The GW Approximation 21
      1.2.5 Excitons 26
      1.2.6 Towards a More Quantitative Theory 28
      1.2.7 Time-Dependent Density Functional Theory (TDDFT) 29
  1.3 Semi-empirical Methods 31
      1.3.1 The Empirical Tight Binding Method 32
      1.3.2 The Empirical Pseudopotential Method 39
      1.3.3 The $\mathbf{k}\cdot\mathbf{p}$ Description and Effective Masses 43

2 Quantum Confined Systems 47
  2.1 Quantum Confinement and Its Consequences 47
      2.1.1 Idealized Quantum Wells 47
      2.1.2 Idealized Quantum Wires 51
      2.1.3 Idealized Cubic Quantum Dots 52
      2.1.4 Artificial Atoms: Case of Spherical Wells 53
      2.1.5 Electronic Structure from Bulk to Quantum Dots 54
  2.2 Computational Techniques 56
      2.2.1 $\mathbf{k}\cdot\mathbf{p}$ Method and Envelope Function Approximation . . 57
      2.2.2 Tight Binding and Empirical Pseudopotential Methods 60
      2.2.3 Density Functional Theory 63
  2.3 Comparison Between Different Methods 64
  2.4 Energy Gap of Semiconductor Nanocrystals 67
  2.5 Confined States in Semiconductor Nanocrystals 69

        2.5.1   Electron States in Direct Gap Semiconductors 69
        2.5.2   Electron States in Indirect Gap Semiconductors 70
        2.5.3   Hole States 72
   2.6 Confinement in Disordered and Amorphous Systems 74

3 **Dielectric Properties**..................................... 77
   3.1 Macroscopic Approach: The Classical Electrostatic Theory... 78
        3.1.1   Bases of the Macroscopic Electrostatic Theory
                of Dielectrics..................................... 78
        3.1.2   Coulomb Interactions in a Dielectric Quantum Well... 80
        3.1.3   Coulomb Interactions in Dielectric Quantum Dots 83
   3.2 Quantum Mechanics of Carriers in Dielectrics:
        Simplified Treatments 84
        3.2.1   Dielectric Effects in Single-Particle Problems......... 84
        3.2.2   Dielectric Effects in Many-Particle Problems 86
   3.3 Microscopic Calculations of Screening Properties 90
        3.3.1   General Formulation in Linear-Response Theory 90
        3.3.2   Random-Phase Approximation 91
        3.3.3   Beyond the Random-Phase Approximation 93
        3.3.4   From Microscopic to Macroscopic Dielectric Function
                for the Bulk Crystal 93
   3.4 Concept of Dielectric Constant for Nanostructures 94
        3.4.1   The Importance of Surface Polarization Charges...... 94
        3.4.2   Dielectric Screening in Quantum Wells 95
        3.4.3   Dielectric Screening in Quantum Dots............... 96
        3.4.4   General Arguments on the Dielectric Response
                in Nanostructures 97
        3.4.5   Conclusions...................................... 100
   3.5 Charging of a Nanostructure 100
        3.5.1   Case of a Quantum Dot 100
        3.5.2   Case of a Quantum Well 103

4 **Quasi-particles and Excitons** 105
   4.1 Basic Considerations 105
   4.2 Excitons in the Envelope Function Approximation 108
        4.2.1   Theory of Bulk Excitons 108
        4.2.2   Excitons in Quantum Wells........................ 109
        4.2.3   Exciton Binding Energy in Limiting Situations 110
        4.2.4   The Influence of Dielectric Mismatch 111
   4.3 Excitons in More Refined Semi-empirical Approaches 112
        4.3.1   General Discussion 112
        4.3.2   Excitons in Nanocrystals
                of Direct Gap Semiconductors...................... 114
        4.3.3   Excitons in Si Nanocrystals 116

		4.3.4	Screening of the Electron–Hole Interaction and Configuration Interaction 120

- 4.4 Quantitative Treatment of Quasi-particles 121
  - 4.4.1 General Arguments 122
  - 4.4.2 Tight Binding GW Calculations 123
  - 4.4.3 Conclusions.................................... 126
- 4.5 Quantitative Treatment of Excitons 129
  - 4.5.1 Numerical Calculations 129
  - 4.5.2 Interpretation of the Results 131
  - 4.5.3 Comparison with Experiments 132
- 4.6 Charging Effects and Multi-excitons 133
  - 4.6.1 Charging Effects: Single Particle Tunneling Through Semiconductor Quantum Dots 133
  - 4.6.2 Multi-excitons 138
- 4.7 Conclusion 140

5 Optical Properties and Radiative Processes 141

- 5.1 General Formulation 141
  - 5.1.1 Optical Absorption and Stimulated Emission 141
  - 5.1.2 Luminescence 148
  - 5.1.3 Nanostructures in Optical Cavities and Photonic Crystals 149
  - 5.1.4 Calculation of the Optical Matrix Elements 150
- 5.2 Electron–Phonon Coupling and Optical Line-Shape 151
  - 5.2.1 Normal Coordinates 152
  - 5.2.2 Calculation of Phonons in Nanostructures 153
  - 5.2.3 Configuration Coordinate Diagram 154
  - 5.2.4 General Expression for the Optical Transition Probabilities 156
  - 5.2.5 Calculation of the Coupling Parameters 161
  - 5.2.6 Fröhlich Coupling: Optical Modes 162
  - 5.2.7 Coupling to Acoustic Modes 169
  - 5.2.8 The Importance of Non-adiabatic Transitions 172
- 5.3 Optical Properties of Heterostructures and Nanostructures of Direct Gap Semiconductors.......................... 174
  - 5.3.1 Interband Transitions 175
  - 5.3.2 Intraband Transitions 181
  - 5.3.3 The Importance of Electron–Phonon Coupling 184
- 5.4 Optical Properties of Si and Ge Nanocrystals 185
  - 5.4.1 Interband Transitions 186
  - 5.4.2 Intraband Transitions 191

XIV Table of Contents

6 Defects and Impurities 195
  6.1 Hydrogenic Donors 195
    6.1.1 Envelope Function Approximation 195
    6.1.2 Tight Binding Self-Consistent Treatment 197
  6.2 Deep Level Defects in Nanostructures 200
  6.3 Surface Defects: Si Dangling Bonds 205
    6.3.1 Review of the Properties of Si Dangling Bonds 205
    6.3.2 Si Dangling Bonds at the Surface of Crystallites 207
    6.3.3 Dangling Bond Defects
          in III–V and II–VI Semiconductor Nanocrystals 208
  6.4 Surface Defects: Self-Trapped Excitons 210
  6.5 Oxygen Related Defects at Si–SiO$_2$ Interfaces 214

7 Non-radiative and Relaxation Processes 219
  7.1 Multi-phonon Capture at Point Defects 219
  7.2 Auger Recombination 225
    7.2.1 Theoretical Calculation 225
    7.2.2 Experimental Evidence for Auger Recombination 230
  7.3 Hot Carrier Relaxation: Existence of a Phonon Bottleneck ... 233

8 Transport 235
  8.1 Description of the Systems and of the Boundary Conditions.. 236
  8.2 Weak Coupling Limit 237
    8.2.1 Perturbation Theory 237
    8.2.2 Orthodox Theory of Tunneling 239
  8.3 Beyond Perturbation Theory 245
    8.3.1 Elastic Scattering Formalism 245
    8.3.2 Calculation of the Green's Functions 249
  8.4 Electron–Electron Interactions Beyond the Orthodox Theory. 253
    8.4.1 Self-Consistent Mean-Field Calculations 253
    8.4.2 The Self-Consistent Potential Profile 255
    8.4.3 The Coulomb Blockade Effect 258
  8.5 Transport in Networks of Nanostructures 263
    8.5.1 Tunneling Between Nanostructures 263
    8.5.2 Hopping Conductivity 266
    8.5.3 Coherent Potential Approximation 268
    8.5.4 Example of a Network of Silicon Nanocrystals 270

A Matrix Elements of the Renormalizing Potential 273

B Macroscopic Averages in Maxwell's Equations 277

C Polarization Correction 279

References 281

Index 299

1 General Basis for Computations and Theoretical Models

This chapter describes theoretical concepts and tools used to calculate the electronic structure of materials. We first present ab initio methods which are able to describe the systems in their ground state, in particular those based on the density functional theory. Introducing the concept of quasi-particles, we show that excitations in the systems can be accurately described as excitations of single particles provided that electron–electron interactions are renormalized by the coupling to long-range electronic oscillations, i.e. to plasmons. We then review the main semi-empirical methods used to study the electronic structure of nanostructures.

1.1 Ab initio One-Particle Theories for the Ground State

This section is an attempt to summarize the basic methods which have allowed an understanding of a wide range of electronic properties not only in condensed matter but also in molecules. The basic difficulty is due to the inter-electronic repulsions which prevent from finding any tractable solution to the general N electron problem. One is then bound to find approximate solutions. Historically most of these have tried to reduce this problem to a set of one-particle Schrödinger equations. Of course such a procedure is not exact and one must find the best one-particle wave functions via a minimization procedure based on the variational principle. This one is however valid for the ground state of the system and can only be applied exceptionally to excited states for which the total wave function is orthogonal to the ground state.

The general solution of the N electron system must be antisymmetric under all permutations of pairs of electron coordinates. We start by applying the constraint to the case of N non interacting electrons. We review on that basis the Hartree and Hartree–Fock approximations and give a qualitative discussion of correlation effects. We then pay special attention to the so-called density functional theories of which the most popular one is the local density approximation (LDA). These have the advantage of leading to a set of well-defined one-particle equations, much simpler to solve than in Hartree–Fock theory, and to provide at the same time fairly accurate predictions for

1.1.1 Non-interacting N Electron System

We start by discussing a hypothetical system of independent electrons for which the Hamiltonian can be written

$$H = \sum_{i=1}^{N} h(\boldsymbol{x}_i) , \tag{1.1}$$

where $\boldsymbol{x}_i$ contains both space and spin coordinates ($\boldsymbol{x}_i = \boldsymbol{r}_i, \xi$). Each individual Hamiltonian $h(\boldsymbol{x}_i)$ is identical and has the same set of solutions:

$$h(\boldsymbol{x})u_k(\boldsymbol{x}) = \varepsilon_k u_k(\boldsymbol{x}) . \tag{1.2}$$

For such a simple situation the eigenstate ψ of H with energy E can be obtained as a simple product of one-electron states (also called spin–orbitals)

$$\psi = \prod_{k=1}^{N} u_k(\boldsymbol{x}_k) , \tag{1.3}$$

its energy being obtained as the sum of the corresponding eigenvalues:

$$E = \sum_{k=1}^{N} \varepsilon_k . \tag{1.4}$$

Although these solutions are mathematically exact they are not acceptable for the N electron system since ψ given by (1.3) is not antisymmetric. The way to solve this difficulty is to realize that any other simple product ψ_{kl} obtained from ψ by a simple permutation of $\boldsymbol{x}_k$ and $\boldsymbol{x}_l$ has the same energy E and is thus degenerate with ψ. The problem is thus to find the linear combination ψ_{AS} of ψ and all ψ_{kl} that is antisymmetric under all permutation $\boldsymbol{x}_k \leftrightarrow \boldsymbol{x}_l$. This turns out to be a determinant called the Slater determinant defined by:

$$\psi_{\text{AS}} = \frac{1}{\sqrt{N!}} \begin{vmatrix} u_1(\boldsymbol{x}_1) & \cdots & u_1(\boldsymbol{x}_N) \\ u_2(\boldsymbol{x}_1) & \cdots & u_2(\boldsymbol{x}_N) \\ \vdots & & \vdots \\ u_N(\boldsymbol{x}_1) & \cdots & u_N(\boldsymbol{x}_N) \end{vmatrix} . \tag{1.5}$$

This determinant still has the energy given by (1.4). The ground state of the system is thus obtained by choosing for ψ_{AS} the N one-particle states u_i which have the lowest eigenvalues ε_k. However in doing this one must take care of the fact that the Slater determinant ψ_{AS} vanishes when two u_k are taken identical. This is the Pauli exclusion principle according to which two electrons cannot be in the same quantum state. If $h(\boldsymbol{x})$ is spin independent the spin orbitals can be factorized as a product of a space part $u_k(\mathbf{r})$ and a spin part $\chi_\sigma(\xi)$

$$u_{k\sigma}(\mathbf{x}) = u_k(\mathbf{r})\chi_\sigma(\xi) , \qquad (1.6)$$

where $\mathbf{r}$ is the position vector, ξ the spin variable and $\sigma =\uparrow$ or $\downarrow$. In such a case the Pauli principle states that two electrons can be in the same orbital state if they have opposite spin. The ground state of the system is thus obtained by filling all lowest one-electron states with two electrons with opposite spin per state.

1.1.2 The Hartree Approximation

The full Hamiltonian of the interacting N electron system is

$$H = \sum_k h(\mathbf{x}_k) + \frac{1}{2}\sum_{kk'} v(\mathbf{r}_k,\mathbf{r}_{k'}) + V_{\text{NN}} , \qquad (1.7)$$

where the one-electron part h is the sum of the kinetic energy and the Coulomb interaction with the nuclei, v is the electron–electron interaction

$$v(\mathbf{r}_k,\mathbf{r}_{k'}) = \frac{e^2}{|\mathbf{r}_k - \mathbf{r}_{k'}|} , \qquad (1.8)$$

and V_{NN} is the Coulomb energy due to the interaction between the nuclei (throughout this chapter we use electrostatic units, i.e. $4\pi\epsilon_0 = 1$). It is of course the existence of the terms (1.8) which prevents from factorizing H and getting a simple solution as in the case of independent electrons. A first step towards an approximate solution to this complex problem came from the intuitive idea of Hartree [1–3] who considered that each electron could be treated separately as moving in the field of the nuclei plus the average electrostatic field due to the other electrons. This corresponds to writing an individual Schrödinger equation

$$\left[h(\mathbf{x}_k) + \sum_{k\neq k'} \int v(\mathbf{r}_k,\mathbf{r}_{k'})|u_{k'}(\mathbf{x}_{k'})|^2 d\mathbf{x}_{k'}\right] u_k(\mathbf{x}_k) = \varepsilon_k u_k(\mathbf{x}_k) \qquad (1.9)$$

for each of the N electrons of the system. To connect with the following we rewrite this equation in a more standard form

$$\left[h(\mathbf{x}) + V_{\text{H}}(\mathbf{x}) - \Sigma_k^{\text{SI}}(\mathbf{x})\right] u_k(\mathbf{x}) = \varepsilon_k u_k(\mathbf{x}) \qquad (1.10)$$

which is obtained by adding and subtracting the term $k = k'$ in (1.9). $V_{\text{H}}(\mathbf{x})$ is the so-called Hartree potential, i.e. the electrostatic potential due to the total density $n(\mathbf{x})$ (including the term $k = k'$):

$$V_{\text{H}}(\mathbf{x}) = \int v(\mathbf{r},\mathbf{r}')n(\mathbf{x}')d\mathbf{x}' ,$$
$$n(\mathbf{x}) = \sum_l n_l |u_l(\mathbf{x})|^2 . \qquad (1.11)$$

The n_l introduced in the definition of $n(\boldsymbol{x})$ are the occupation numbers, $n_l = 1$ if there is an electron in u_l, $n_l = 0$ in the opposite case. The last term Σ_k^{SI} is the self-interaction correction, removing the unphysical term $k = k'$ introduced in the definition of V_{H}:

$$\Sigma_k^{\text{SI}}(\boldsymbol{x}) = \int v(\boldsymbol{r}, \boldsymbol{r}') |u_k(\boldsymbol{x}')|^2 \mathrm{d}\boldsymbol{x}' \ . \tag{1.12}$$

The Hartree equations coupled with a spherical averaging of the potential in (1.10) have provided a quite accurate picture of the electronic structure of isolated atoms. They are a basis for understanding the periodic table of the elements and also produce good electron densities $n(\boldsymbol{x})$ as compared with those obtained experimentally from X-ray scattering.

The Hartree equations have been put on firm theoretical grounds by use of the variational principle [2, 4]. For this one takes as trial wave function ψ the simplest form one can obtain for independent electrons, without taking account of the antisymmetry. This one is thus the simple product of spin orbitals given by (1.3). The optimized ψ belonging to this family of wave functions must minimize the energy given by the expectation value of H for this wave function. This is equivalent to solving

$$\langle \delta\psi | E - H | \psi \rangle = 0 \ , \tag{1.13}$$

where $\delta\psi$ is an infinitesimally small variation of ψ. If one now varies each u_k separately in (1.13) by δu_k one directly gets the set of equations (1.9) or (1.10).

The Hartree method then succeeds in reducing approximately the N electron problem to a set of N one-particle equations. However the price to pay is that the potential energy in each equation (1.10,1.11) contains the unknown quantity $n(\boldsymbol{x}') - |u_k(\boldsymbol{x}')|^2$. One must then solve these equations iteratively introducing at the start some guess functions for the $|u_k|^2$ in the potential energy, solve the equations, re-inject the solutions for the $|u_k|^2$ (or some weighted averages) into the potential energy and so on (Fig. 1.1). The pro-

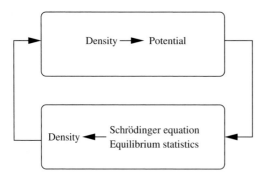

Fig. 1.1. The electron density and the potential must be calculated self-consistently taking into account the occupation of the levels

1.1.3 The Hartree–Fock Approximation

This is less intuitive than the Hartree method and must be directly introduced from a variational treatment. The starting point is similar except that instead of choosing for ψ a simple product function one now makes use of Slater determinant of the form (1.5) in which the spin orbitals are assumed orthonormal. The total energy $E = \langle \psi | H | \psi \rangle$ of such a determinant can be shown [2, 3, 5, 6] to be given by

$$E = \sum_k n_k \langle k|h|k\rangle + \frac{1}{2}\sum_{k,l} n_k n_l (\langle kl|v|kl\rangle - \langle kl|v|lk\rangle) + V_{\text{NN}} \quad (1.14)$$

with:

$$\langle k|h|k\rangle = \int u_k^*(\boldsymbol{x}) h(\boldsymbol{x}) u_k(\boldsymbol{x}) \mathrm{d}\boldsymbol{x} \, ,$$

$$\langle ij|v|kl\rangle = \int u_i^*(\boldsymbol{x}) u_j^*(\boldsymbol{x}') v(\boldsymbol{r}, \boldsymbol{r}') u_k(\boldsymbol{x}) u_l(\boldsymbol{x}') \mathrm{d}\boldsymbol{x} \mathrm{d}\boldsymbol{x}' \, . \quad (1.15)$$

We want to minimize E with respect to the u_k under the constraint that these remain orthonormal, i.e. $\int u_k^*(\boldsymbol{x}) u_l(\boldsymbol{x}) \mathrm{d}\boldsymbol{x} = \delta_{kl}$. This can be achieved via the method of Lagrange multipliers. If we apply a first order change δu_k^* this requires that the quantity $\delta E - \sum_{kl} \lambda_{kl} \int \delta u_k^*(\boldsymbol{x}) u_l(\boldsymbol{x}) \mathrm{d}\boldsymbol{x} = 0, \forall \delta u_k^*$. This leads to the set of one-particle equations:

$$\left[h(\boldsymbol{x}) + \sum_l n_l \int v(\boldsymbol{r}, \boldsymbol{r}') |u_l(\boldsymbol{x}')|^2 \mathrm{d}\boldsymbol{x}' \right] u_k(\boldsymbol{x})$$

$$- \sum_l n_l \left[\int v(\boldsymbol{r}, \boldsymbol{r}') u_l^*(\boldsymbol{x}') u_k(\boldsymbol{x}') \mathrm{d}\boldsymbol{x}' \right] u_l(\boldsymbol{x}) = \sum_l \lambda_{kl} u_l(\boldsymbol{x}) \, . \quad (1.16)$$

This can be simplified by noticing that a unitary transformation applied to the Slater determinant does not modify it apart from a phase factor and thus does not change the structure of the equations. It is thus possible to rewrite (1.16) under diagonal from, i.e. with:

$$\lambda_{kl} = \varepsilon_k \delta_{kl} \, . \quad (1.17)$$

For obvious reasons, the last term on the left hand side of (1.16) is called the exchange term, the second one being the Hartree potential V_{H}. We now rewrite (1.16) using (1.17) under a form which will be generalized in the following:

$$[h(\boldsymbol{x}) + V_{\text{H}}(\boldsymbol{x})] u_k(\boldsymbol{x}) + \int \Sigma_{\text{x}}(\boldsymbol{x}, \boldsymbol{x}') u_k(\boldsymbol{x}') \mathrm{d}\boldsymbol{x}' = \varepsilon_k u_k(\boldsymbol{x}) \, , \quad (1.18)$$

Σ_x corresponding to the non-local exchange potential:

$$\Sigma_\mathrm{x}(\boldsymbol{x},\boldsymbol{x}') = -v(\boldsymbol{r},\boldsymbol{r}') \sum_l n_l u_l(\boldsymbol{x}) u_l^*(\boldsymbol{x}') \; . \tag{1.19}$$

The $l = k$ term in (1.19) when injected into (1.18) directly corresponds to the self-interaction Σ_k^SI of (1.12). The Hartree–Fock (HF) procedure thus reproduces the Hartree equations plus corrective exchange terms for $l \neq k$.

When the spin orbitals are factorized as in (1.6) one can perform the integration over the spin variables directly in the HF equations. In that case the result is that the integrations over $\boldsymbol{x}'$ can be replaced by integrations over $\boldsymbol{r}'$ at the condition of multiplying V_H by a factor 2 for spin degeneracy while the exchange term remains unchanged since opposite spins give a vanishing contribution to (1.16).

While the HF approximation improves over the Hartree one, especially for magnetic properties, it does not provide an accurate enough technique for the ground state properties as well as the excitation energies. This is due to correlation effects which are important in both cases as will be discussed in the following. Furthermore HF leads to heavy calculations due to the non-local character of the exchange term.

1.1.4 Correlations and Exchange–Correlation Hole

By definition correlation effects are the contributions not included in the HF approximation. Conceptually the simplest way to include them is to use the method of configuration interaction (CI). The principle of the CI technique is to expand the eigenstates of the interacting N electron system on the basis of the Slater determinants built from an infinite set of orthonormal one-particle spin orbitals :

$$\psi = \sum_n c_n \psi_{\mathrm{SD},n} \; . \tag{1.20}$$

Quite naturally the starting point in such an expansion could be the ground state HF determinant, the others being built by substitution of excited HF spin orbitals. However the CI technique is quite heavy and does not converge rapidly so that it can be applied only to small molecules (typically 10 atoms maximum). This means that it cannot be applied to solids. We thus now discuss the only case where practically exact results have been obtained for infinite systems, i.e. the free electron gas.

The free electron gas is an idealized model of simple metals in which the nuclear charges are smeared out to produce a uniform positive background charge density. This one is fully compensated by the uniform neutralizing electron density. This produces a constant zero potential in all space. The solutions of the one-particle Hartree equations are

$$u_{\boldsymbol{k}}(\boldsymbol{r}) = \frac{1}{\sqrt{V}} e^{i\boldsymbol{k}\cdot\boldsymbol{r}} ,$$

$$\varepsilon_{\boldsymbol{k}} = \frac{\hbar^2}{2m_0} |\boldsymbol{k}|^2 , \tag{1.21}$$

where V is the volume of the system and m_0 is the electron mass.

In the ground state, these one-particle states are filled with two electrons of opposite spin up to the Fermi level $\varepsilon_{\rm F} = \hbar^2 k_{\rm F}^2/(2m_0)$ where $k_{\rm F}$ is the Fermi momentum. One can then express the electron density n as:

$$n = \frac{k_{\rm F}^3}{3\pi^2} . \tag{1.22}$$

In this context it has been customary to express all quantities in terms of the radius $r_{\rm s}$ per electron defined by $4\pi r_{\rm s}^3/3 = 1/n$. We can thus write:

$$k_{\rm F} = (\alpha r_{\rm s})^{-1} ,$$

$$\alpha = \left(\frac{4}{9\pi}\right)^{\frac{1}{3}} = 0.521 . \tag{1.23}$$

Looking first at the HF correction defined by (1.18,1.19) one can show that plane waves are eigenstates of the HF operator, i.e.:

$$\int \Sigma_{\rm x}(\boldsymbol{r},\boldsymbol{r}') u_{\boldsymbol{k}}(\boldsymbol{r}') {\rm d}\boldsymbol{r}' = (\Sigma_{\rm x})_{\boldsymbol{k},\boldsymbol{k}} u_{\boldsymbol{k}}(\boldsymbol{r}) ,$$

$$(\Sigma_{\rm x})_{\boldsymbol{k},\boldsymbol{k}} = -\frac{4\pi e^2}{V} \sum_{\boldsymbol{k}'} n_{\boldsymbol{k}'} \frac{1}{|\boldsymbol{k}-\boldsymbol{k}'|^2} . \tag{1.24}$$

We now determine the average exchange energy per electron $\varepsilon_{\rm x}$ by summing $(\Sigma_{\rm x})_{\boldsymbol{k},\boldsymbol{k}}$ over the filled states, divide by the number N of electrons and by a factor 2 for the interactions counted twice. This gives [6, 7] (in atomic units):

$$\varepsilon_{\rm x} = -\frac{3e^2}{4\pi\alpha r_{\rm s}} = -\frac{0.458}{r_{\rm s}} . \tag{1.25}$$

A lot of studies have been devoted to the calculation of the correlation energy $\varepsilon_{\rm c}$ of the electron gas. These are summarized in many reviews, e.g. [6, 7]. These analytical studies have been confirmed by the more accurate Quantum Monte Carlo calculations [8] which now serve as a basis in local density studies as discussed in the following (Sect. 1.1.5).

It is of interest to try to understand these effects in terms of the exchange–correlation hole. We start by analyzing the HF case in the manner of Slater, i.e. rewrite the exchange term of (1.18, 1.19) as:

$$\int \Sigma_{\rm x}(\boldsymbol{x},\boldsymbol{x}') u_{\boldsymbol{k}}(\boldsymbol{x}') {\rm d}\boldsymbol{x}'$$
$$= -\left\{ \frac{\int v(\boldsymbol{r},\boldsymbol{r}') \sum_l n_l u_l(\boldsymbol{x}) u_l^*(\boldsymbol{x}') u_{\boldsymbol{k}}(\boldsymbol{x}') {\rm d}\boldsymbol{x}'}{u_{\boldsymbol{k}}(\boldsymbol{x})} \right\} u_{\boldsymbol{k}}(\boldsymbol{x}) . \tag{1.26}$$

The term in brackets on the right hand side of (1.26) is the one-body effective exchange potential $V_{\text{HF}}(\boldsymbol{x})$ acting on the one-particle eigenstate $u_k(\boldsymbol{x})$. It can be considered as the electrostatic potential induced by the density

$$n_{\text{HF}}(\boldsymbol{x},\boldsymbol{x}') = \sum_l \frac{n_l u_l(\boldsymbol{x}) u_l^*(\boldsymbol{x}') u_k(\boldsymbol{x}')}{u_k(\boldsymbol{x})} \qquad (1.27)$$

which, when integrated over $\boldsymbol{x}'$, gives unity. This corresponds to the fact that, in the N electron system, the electron at $\boldsymbol{r}$ interacts with $N-1$ other electrons, i.e. with the N electrons contained in the Hartree term V_{H} plus one hole called the exchange hole. The question now arises of the localization of this hole around the electron in question. The answer comes from the behavior of the quantity $\sum_l n_l u_l(\boldsymbol{x}) u_l^*(\boldsymbol{x}')$. If the n_l were unity for all l then it would amount to $\delta(\boldsymbol{x}-\boldsymbol{x}')$, i.e. strictly localized on the electron. As this is not the case it looks like a broadened delta function. For the free electron gas its width is of order $\lambda_{\text{F}} = 2\pi/k_{\text{F}} = 3r_{\text{s}}$ which means that, for alkali metals, it extends just beyond the nearest neighbors sphere. Such a conclusion holds true quite generally and can be also understood from the fact that the Slater determinant prevents two electrons with the same spin from being at the same position.

Turning now to correlation effects we can examine them in a simple classical way which will be developed in the section on the GW approximation (Sect. 1.2.4). In the Hartree and HF treatments one electron at $\boldsymbol{r}$ experiences the average field due to the other electrons. However, due to the electron–electron repulsion, its presence at $\boldsymbol{r}$ modifies the distribution of the other electrons creating around it a Coulomb hole which results in a screening of the electron–electron interactions and a lowering of the total energy (Fig. 1.2). It is important for the following to notice that, for a finite metallic system, this Coulomb hole completely compensates the electron charge locally, the screening charge being repelled on the surface. Thus the total charge in the Coulomb hole always amounts to zero. We shall see later that this is of primary importance for nanostructures.

To end up the considerations on the exchange–correlation hole we now give some exact results, discussed for instance in [7, 9, 10]. We start by defining

Fig. 1.2. The exchange–correlation hole around an electron leads to a screening of the electron–electron interactions

the electron density $n(\mathbf{r})$ as the probability per unit volume of finding one electron at $\mathbf{r}$ or equivalently

$$n(\mathbf{r}) = \left\langle \psi \left| \sum_{i=1}^{N} \delta(\mathbf{r} - \mathbf{r}_i) \right| \psi \right\rangle , \qquad (1.28)$$

where $\mathbf{r}_i$ are the electron positions and $|\psi\rangle$ is the N-particle wave function of the ground state. In the same spirit we define the pair correlation function $n(\mathbf{r}, \mathbf{r}')$ as the squared probability per unit volume to find one electron at $\mathbf{r}$ and another one at $\mathbf{r}'$

$$n(\mathbf{r}, \mathbf{r}') = \left\langle \psi \left| \sum_{i \neq j} \delta(\mathbf{r} - \mathbf{r}_i)\delta(\mathbf{r}' - \mathbf{r}_j) \right| \psi \right\rangle , \qquad (1.29)$$

which obviously contains information about inter-electronic correlations. In particular the exact Coulomb energy of the system

$$V_{\text{coul}} = \frac{1}{2} \left\langle \psi \left| \sum_{i \neq j} \frac{e^2}{|\mathbf{r}_i - \mathbf{r}_j|} \right| \psi \right\rangle \qquad (1.30)$$

can be rewritten in terms of (1.29) as

$$V_{\text{coul}} = \frac{e^2}{2} \int \frac{n(\mathbf{r}, \mathbf{r}')\mathrm{d}\mathbf{r}\mathrm{d}\mathbf{r}'}{|\mathbf{r} - \mathbf{r}'|} . \qquad (1.31)$$

When electrons are totally uncorrelated as in Hartree and HF one can show that $n(\mathbf{r}, \mathbf{r}') = n(\mathbf{r})n(\mathbf{r}')$. One can then express $n(\mathbf{r}, \mathbf{r}')$ as

$$n(\mathbf{r}, \mathbf{r}') = n(\mathbf{r})n(\mathbf{r}')(1 + g(\mathbf{r}, \mathbf{r}')) , \qquad (1.32)$$

where all the information about correlations is contained in $g(\mathbf{r}, \mathbf{r}')$. If we integrate $n(\mathbf{r}, \mathbf{r}')$ given from (1.29) over $\mathbf{r}'$ we get $N - 1$. This means that from (1.32) we have

$$\int g(\mathbf{r}, \mathbf{r}')n(\mathbf{r}')\mathrm{d}\mathbf{r}' = -1 . \qquad (1.33)$$

This important sum rule simply states that each electron in the system only interacts with the $N-1$ other electrons. Equation (1.33) then expresses the fact that for each electron at $\mathbf{r}$ there is an exchange–correlation hole surrounding it. One can however wonder if this hole is due mainly to exchange or correlation. To get some information on this we must take the spin of the particle into account. Assuming now that there are $N_\uparrow$ electrons with spin $\uparrow$ and $N_\downarrow$ particles with spin $\downarrow$ it is clear that one electron with spin $\uparrow$ at $\mathbf{r}$ will interact with $N_\uparrow - 1$ electrons of the same spin and with $N_\downarrow$ of opposite spin. If we separate the previous expressions with respect to the spin components we get in obvious notations:

$$\int g_{\uparrow\uparrow}(r,r')n_\uparrow(r')\mathrm{d}r' = -1\,,$$

$$\int g_{\uparrow\downarrow}(r,r')n_\uparrow(r')\mathrm{d}r' = 0\,. \tag{1.34}$$

This generalizes that the exchange hole, corresponding to $g_{\uparrow\uparrow}$, is equal to -1 even including correlation effects. On the other hand for electrons with opposite spin there is a local screening hole, equal to -1 for metals, due to screening effects only but with the compensating charge on the surface of the system confirming that $\int g_{\uparrow\downarrow}(r,r')n_\uparrow(r')\mathrm{d}r' = 0$. This will prove to be of primary importance for nanocrystals.

1.1.5 Local Density Approaches

The idea of replacing the complex one-electron Schrödinger equation with the non-local exchange–correlation potential by a local density approximation started early with the work of Thomas–Fermi [7, 11]. This one is valid for high electron density systems, e.g. for heavy atoms. In this limit the potential energy $V(r)$ of an electron in the system can be considered to vary slowly in space and the electron density can be approximately replaced locally by the value it would take for the free electron gas, i.e. by (1.22) in which k_F^2 is given by $2m_0(\varepsilon_F - V(r))/\hbar^2$:

$$n(r) = \frac{1}{3\pi^2}\left(\frac{2m_0}{\hbar^2}\right)^{3/2}[\varepsilon_F - V(r)]^{3/2}\,. \tag{1.35}$$

$V(r)$ is solution of the Poisson's equation, i.e. one arrives at the self-consistency equation:

$$\Delta V(r) = -\frac{8\sqrt{2}e^2}{3\pi}\left(\frac{m_0}{\hbar^2}\right)^{3/2}[\varepsilon_F - V(r)]^{3/2}\,. \tag{1.36}$$

This one was solved numerically for neutral and ionized atoms and shown to reproduce important trends along the periodic table. We shall use it in the following to get a simple description of screening effects in the usual semiconductors and insulators which can be considered as high electron density systems.

Many refinements have been brought to make the Thomas–Fermi approximation more quantitative but an essential step forward is due to Slater [12, 13]. His basic idea was to replace the non-local exchange term taken under the form of (1.26) by a statistical average over the filled states which is then calculated locally as if the system was an electron gas. The corresponding Slater exchange potential v_{xSlater} is thus locally the same function of $n(r)$ as the electron gas and is also equal to $2\varepsilon_x$ the total exchange energy per electron given by (1.25) with $4\pi r_s^3/3 = (n(\mathbf{r}))^{-1}$. This gives in atomic units:

$$v_{\text{xSlater}} = -\frac{0.916}{r_s[n(\mathbf{r})]} = -\frac{3}{2}\left[\frac{3}{\pi}n(\mathbf{r})\right]^{1/3}. \tag{1.37}$$

Slater realized that, to get more accurate results, it was necessary to scale this local form of exchange by a factor α, close to 1, lying in the interval [0.75,1]. With this he was able to get quite reasonable results for atoms, molecules and solids.

The decisive step in these density based theories came from the basic theory established by Hohenberg and Kohn [14] but also from the work of Kohn and Sham [15] which made it a practical computational tool. The general idea is always the same, i.e. one writes one-particle equations going beyond Slater's exchange v_{xSlater} to also include correlations effects, introducing an exchange–correlation potential v_{xc}. We now summarize the general arguments leading to these density functional theories (DFT) and their local density approximation (LDA). We write the full N electron Hamiltonian

$$H = T + V_{\text{ee}} + V_{\text{ext}}(\mathbf{r}), \tag{1.38}$$

where T is the kinetic energy operator, V_{ee} the electron–electron interaction and V_{ext} a one-particle external potential. We follow the arguments of [14] as well as other authors [9, 10] and summarize first the conclusions of Hohenberg and Kohn [14] :

- Two external potentials V_{ext} and V'_{ext} differing by more than a constant cannot give the same ground state electron density $n(\mathbf{r})$. There is thus one to one mapping between V_{ext} and $n(\mathbf{r})$ which means that $n(\mathbf{r})$ completely determines the ground state properties of the system.
- In the spirit of the variational methods we define the following density functional

$$F(n) = \min_{\psi \to n(\mathbf{r})} \langle \psi | T + V_{\text{ee}} | \psi \rangle, \tag{1.39}$$

where the minimization is performed over all ψ giving the same density $n(\mathbf{r})$. Then the total energy functional

$$E(n) = F(n) + \int V_{\text{ext}}(\mathbf{r})n(\mathbf{r})d\mathbf{r} \tag{1.40}$$

is also minimum since the last term on the right hand side does not vary with ψ if $n(\mathbf{r})$ is fixed.
- The above definitions remain valid for fractional electron numbers if one uses mixtures of states ψ with different total numbers of electrons. Minimizing $E(n)$ with the constraint that the number of electrons is given by N gives, with the method of Lagrange multipliers, the functional derivative

$$\frac{\delta E(n)}{\delta n}[n] = \mu_N, \tag{1.41}$$

where the chemical potential μ_N is discontinuous at each integer value of N.

These general arguments do not give a recipe for an actual calculation since the functional $F(n)$ is unknown. The idea of Kohn and Sham [15] was then to relate the interacting N electron system to a fictitious non interacting one leading to the same electron density $n(\boldsymbol{r})$. This one corresponds to the following Kohn–Sham Hamiltonian

$$H_{\text{KS}} = T + V_{\text{KS}} , \qquad (1.42)$$

where V_{KS} is the Kohn–Sham external potential. One can repeat the procedure leading to (1.41) for this new Hamiltonian, requiring that it gives the same μ_N. This leads to

$$\frac{\delta T_0(n)}{\delta n} + V_{\text{KS}} = \mu_N \qquad (1.43)$$

with:

$$T_0(n) = \min_{\psi \to n(\boldsymbol{r})} \langle \psi | T | \psi \rangle . \qquad (1.44)$$

Equating (1.41) and (1.43) one gets, with the help of (1.38) and (1.39), the following expression for V_{KS}:

$$V_{\text{KS}} = \frac{\delta}{\delta n} [F(n) - T_0(n)] + V_{\text{ext}} . \qquad (1.45)$$

Writing $F(n) - T_0(n)$ as the sum of a Hartree energy

$$E_{\text{H}} = \frac{e^2}{2} \int \frac{n(\boldsymbol{r})n(\boldsymbol{r}')}{|\boldsymbol{r} - \boldsymbol{r}'|} \mathrm{d}\boldsymbol{r}\mathrm{d}\boldsymbol{r}' , \qquad (1.46)$$

and an exchange–correlation energy $E_{\text{xc}}[n(\boldsymbol{r})]$, we get

$$V_{\text{KS}} = V_{\text{H}} + V_{\text{xc}}[n(\boldsymbol{r})] + V_{\text{ext}} \qquad (1.47)$$

with:

$$V_{\text{H}} = \frac{\delta E_{\text{H}}}{\delta n} = e^2 \int \frac{n(\boldsymbol{r}')}{|\boldsymbol{r} - \boldsymbol{r}'|} \mathrm{d}\boldsymbol{r}' ,$$

$$V_{\text{xc}}[n(\boldsymbol{r})] = \frac{\delta E_{\text{xc}}[n(\boldsymbol{r})]}{\delta n} . \qquad (1.48)$$

All this means that one can construct the ground state electron density by solving the following one-particle Hamiltonian

$$\{t + V_{\text{ext}}(\boldsymbol{r}) + V_{\text{H}}(\boldsymbol{r}) + V_{\text{xc}}[n(\boldsymbol{r})]\} \phi_k(\boldsymbol{r}) = \varepsilon_k \phi_k(\boldsymbol{r}) , \qquad (1.49)$$

where t is the one-electron kinetic energy operator. One then gets the density $n(\boldsymbol{r})$ as for any Slater determinant as

$$n(\boldsymbol{r}) = \sum_k n_k |\phi_k(\boldsymbol{r})|^2 , \qquad (1.50)$$

and the total energy as $T_0(n) + E_\mathrm{H} + E_\mathrm{xc}$. The one-particle equations are of the Slater type discussed before and thus much simpler than HF equations. They also have to be solved self-consistently. However the major difficulty is that $E_\mathrm{xc}[n]$ and thus $V_\mathrm{xc}[n]$ are unknown. The simplest way to overcome this difficulty has been to replace $E_\mathrm{xc}[n(\mathbf{r})]$ locally by the value it would have for the electron gas. This represents an extension of Slater's ideas to include correlation effects and leads to the well-known local density approximation (LDA).

If one considers the exchange-only term the expression of $V_\mathrm{xc} = v_\mathrm{x}$ turns out to be very simple. We have seen earlier that the exchange energy per electron ε_x of the free electron gas is $-3/4\,(3n(\mathbf{r})/\pi)^{1/3}$ from (1.37). The total exchange energy per unit volume E_x is given by $n\varepsilon_\mathrm{x}$, i.e. proportional to $n^{4/3}$. We can thus write:

$$v_\mathrm{x} = \frac{\delta E_\mathrm{x}}{\delta n} = \frac{4}{3}\varepsilon_\mathrm{x}\,. \tag{1.51}$$

This turns out to be only 2/3 of Slater exchange potential v_xSlater as pointed out in early work [15]. This gives some reasonable explanation of an α value smaller than 1 in Slater's self-consistent $X\alpha$ method. To get the full V_xc of LDA one must add the correlation energy of the electron gas. There have been in the past several analytical expressions which have been proposed (see e.g. [9, 10, 16] for a review). A popular one for numerical calculations is a parametrized form [17] which is detailed in Sect. 3.2.2.

LDA turns out to give satisfactory results regarding the predictions of the structural properties of molecules and solids. For instance, in solids (either with sp bonds or d bonds, as in transition metals) the inter-atomic distances and the elastic properties are predicted with a precision better than 5% in general. This remains true for diatomic molecules, except that the binding energy is overestimated by 0.5 to 1 eV [16]. Attempts to improve on simple LDA have first been based on expansions of V_xc in $\boldsymbol{\nabla} n(\mathbf{r})$ [18, 19] but these violate the exchange–correlation sum rule of (1.33) which leads to serious problems [20]. More recently generalized gradient approximations (GGA) have been proposed [20, 21] where V_xc depends on $n(\mathbf{r})$ and $\boldsymbol{\nabla} n(\mathbf{r})$ but not in a simple $\boldsymbol{\nabla} n$ expansion. The various GGA forms which have been proposed are constrained to fulfill exact sum rules. GGA often leads to better results than LDA for total energies but sometimes worse results for bulk moduli and phonon frequencies. For lattice constants they seem to be of equal quality with an accuracy of order 1%.

1.2 Quasi-particles and Excitons

One of the main problems which arise with the ground state theories described earlier is that the eigenvalues of the one-particle equations provide poor information concerning excitation energies. We shall discuss this on the

HF case first and summarize what happens with LDA. We then introduce the concept of quasi-particles and discuss the most efficient method to calculate their properties which is the GW approximation. Finally we treat the case of excitons which correspond to pairs of quasi-particles, an electron and a hole, bound by their attractive interaction.

1.2.1 One-Particle Eigenvalues

Let us consider the simplest process of ionizing an N electron neutral system. The ionization energy I and electron affinity A are defined as:

$$I = E(N-1) - E(N), \quad A = E(N) - E(N+1). \tag{1.52}$$

In the HF approximation one can even define from which state the electron is ionized. Assuming that it is the state k for which $n_k = 1$ in the N electron system and $n_k = 0$ for the $N-1$ electron system. This defines:

$$I_k = E(N-1, n_k = 0) - E(N). \tag{1.53}$$

Using the HF expression (1.14) one straightforwardly gets:

$$I_k = -\left\{ \langle k|h|k\rangle + \sum_l \left(\langle kl|v|kl\rangle - \langle kl|v|lk\rangle \right) \right\}. \tag{1.54}$$

Writing (1.18) in vector form and projecting on $|k\rangle$ one directly obtains

$$I_k = -\varepsilon_k, \tag{1.55}$$

i.e. the ionization energy is just the opposite of the one-particle eigenvalue, as stated by Koopman's theorem [22]. A similar argument holds true for the electron affinity. In HF one can also obtain the excitation energies in terms of the excitation of an electron in state k' and creation of a hole in state k leading to excitation energies equal to $\varepsilon_{k'} - \varepsilon_k$.

Although these results can serve as a guide the situation is by far not so simple. The above derivation is based on the fact that the u_k remain identical or frozen between the N and the $N \pm 1$ systems. This is true for delocalized states in infinite systems since the influence of the extra electron or hole scales as $1/V$, V being the volume of the system. This is no more correct in atoms, molecules or nanocrystals since the one-particle states will depend on occupation numbers. A change in occupation numbers indeed induces an electronic relaxation in the system corresponding to a screening of the extra electron or hole. However, even for infinite systems, HF excitation energies do not provide correct values of the band gaps of semiconductors. In fact HF consistently overestimates their bandgaps, by more than 50%, the maximum discrepancy being for Si with $(E_g)_{\text{HF}} \approx 5.5$ eV instead of 1.17 eV experimentally.

From the variational derivation of DFT, there is no particular reason why the eigenvalues of the one-particle equations should correspond to excitation

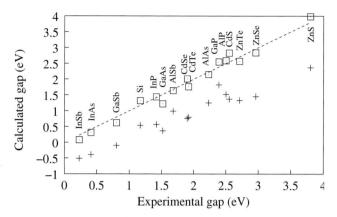

Fig. 1.3. Calculated bandgap in LDA (+) and in GW (□) versus experimental bandgap for several semiconductors. Results for II–VI semiconductors are from [23] and others are from [24]

energies. One exception is the case of the highest occupied eigenvalue which, on the basis of the definition of V_{KS}, should give an accurate value for the ionization energy (it is exact for the true V_{xc}). However, for the excitation energies, LDA, as well as DFT in general, leads to what is called the bandgap problem. Indeed, although the general shape of the bands of semiconductors is correctly predicted by LDA, bandgaps are consistently underestimated. For instance, in silicon, the predicted value is 0.65 eV (1.17 eV experimentally) while in germanium the LDA value is ≈ 0 eV (0.7 eV experimentally). Figure 1.3 shows that this holds true for several semiconductors and insulators.

1.2.2 The Exchange–Correlation Hole and Static Screening

The previous discussions illustrate the fact that one-particle theories cannot yield exact ground state and excitation energies at the same time. In this context a fruitful concept is provided by the notion of quasi-particles. For instance an excess electron injected into the N electron system, due to its interaction with the other electrons, cannot be described exactly by the solution of a one-particle Schrödinger equation. However, in favorable (and fortunately common) cases its spectral representation (see [25–27] for a detailed mathematical discussion) remains strongly peaked around a given energy. If so one can speak of quasi-particles which are approximate solutions of a one-particle Schrödinger equation, but with a broadening due to the interaction with the other particles. The picture in real space is that of an electron surrounded by its exchange–correlation hole.

Before turning to the GW approximation, let us come back to a simple picture of the quasi-particle which we started to discuss in Sect. 1.1.4. If we try to write the Schrödinger equation of the excess electron, we can start from the HF formulation where both the Hartree and exchange terms contain the bare interaction $v(\mathbf{r},\mathbf{r}') = e^2/|\mathbf{r}-\mathbf{r}'|$ between the excess electron at $\mathbf{r}$ and other electrons at $\mathbf{r}'$. Correlation effects correspond to the fact that the electron at $\mathbf{r}$ repels the other electrons to create its Coulomb hole. If one treats this electron as a classical test charge, the resulting effect can be described by linear screening theory. Here the bare potential v due to the test electron charge at $\mathbf{r}$ is at the origin of an induced change in the electron density δn_{ind} which in turn creates an induced potential V_{ind}. The total potential in the system is thus $W = v + V_{\text{ind}}$. The induced electron density depends on W and, if we linearize the dependence, one can write formally

$$\delta n_{\text{ind}} = \chi_0 W ,$$
$$W = v + v_0 \,\delta n_{\text{ind}} , \qquad (1.56)$$

where χ_0 is the polarizability, and v_0 is the potential per unit of induced electron density. As we shall see later (in particular in Sect. 1.2.7), the different one-electron theories give rise to different potentials v_0 and different results for the inverse dielectric function. The simplest case corresponds to the Hartree approximation in which case $v_0 = v$ the bare electron–electron repulsion. The result for $\epsilon^{-1}(\omega)$ corresponds to the so-called random-phase approximation [28–32].

Equations (1.56) lead for W to

$$W = \epsilon^{-1} v ,$$
$$\epsilon^{-1} = (1 - v_0 \chi_0)^{-1} . \qquad (1.57)$$

Of course all these formal relations are integral equations. For instance the screened potential due to the electron at $\mathbf{r}$ is given at $\mathbf{r}'$ by:

$$W(\mathbf{r},\mathbf{r}') = \int \epsilon^{-1}(\mathbf{r},\mathbf{r}'') v(\mathbf{r}'',\mathbf{r}') d\mathbf{r}'' . \qquad (1.58)$$

The simplest idea to take into account correlation effects is thus to replace in the Schrödinger equation of the excess electron the bare potential v by its screened counterpart W. This adds to the Hartree term a correction equal to what corresponds to the Coulomb hole

$$V_{\text{coh}} = \frac{1}{2} V_{\text{ind}}(\mathbf{r},\mathbf{r}) ,$$
$$= \frac{1}{2} \int \left[\epsilon^{-1}(\mathbf{r},\mathbf{r}') - \delta(\mathbf{r}-\mathbf{r}') \right] v(\mathbf{r}',\mathbf{r}) d\mathbf{r}' , \qquad (1.59)$$

1.2.3 Dynamically Screened Interactions

In this section we extend the formulation of Sect. 1.2.2 to the case of time-dependent perturbations. This will play an essential role in the description of quasi-particles and excitons. We first consider the linear response of a system described in a one-particle picture (like Hartree, HF or LDA) to which belongs the so-called random-phase approximation [28–32]. In the second part we give a general expression of the linear response function to be used in our derivation of the GW approximation.

Linear Response Function in One-Electron Pictures. Here we follow the approach of Ehrenreich and Cohen [29] and generalize (1.29) to define the time-dependent density matrix as

$$n(\boldsymbol{x}, \boldsymbol{x}', t) = \sum_k n_k u_k(\boldsymbol{x}, t) u_k^*(\boldsymbol{x}', t) , \qquad (1.60)$$

where again ($\boldsymbol{x} = \boldsymbol{r}, \xi$) and the $u_k(\boldsymbol{x}, t)$ are the solutions of the one-particle equations (in atomic units)

$$\mathrm{i}\frac{\partial}{\partial t} u_k(\boldsymbol{x}, t) = h(\boldsymbol{x}, t) u_k(\boldsymbol{x}, t) , \qquad (1.61)$$

where h can eventually be a non-local operator as in the HF picture. From (1.60) and (1.61), $n(\boldsymbol{x}, \boldsymbol{x}', t)$ obeys the equation of motion:

$$\mathrm{i}\frac{\partial n}{\partial t}(\boldsymbol{x}, \boldsymbol{x}', t) = [h(\boldsymbol{x}, t) - h(\boldsymbol{x}', t)] n(\boldsymbol{x}, \boldsymbol{x}', t) . \qquad (1.62)$$

We now consider the case where the Hamiltonian can be written as the sum of a time independent unperturbed part $h_0(\boldsymbol{x})$ and a small time-dependent perturbation $W(\boldsymbol{x}, t)$ (to be discussed later):

$$h(\boldsymbol{x}, t) = h_0(\boldsymbol{x}) + W(\boldsymbol{x}, t) . \qquad (1.63)$$

It is thus quite natural to expand $n(\boldsymbol{x}, \boldsymbol{x}', t)$ in terms of the eigenstates $u_k(\boldsymbol{x})$ of h_0:

$$n(\boldsymbol{x}, \boldsymbol{x}', t) = \sum_{kk'} n_{kk'}(t) u_k(\boldsymbol{x}) u_{k'}^*(\boldsymbol{x}') . \qquad (1.64)$$

(1.62) is then replaced by

$$\mathrm{i}\frac{\partial}{\partial t} n_{kk'}(t) = [h, n]_{kk'} . \qquad (1.65)$$

We now write this to first order in W and $\delta n(\boldsymbol{x}, \boldsymbol{x}', t)$ the first order induced change in n. We get

$$\mathrm{i}\frac{\partial}{\partial t} \delta n_{kk'}(t) = [h_0, \delta n]_{kk'} + [W, n_0]_{kk'} \qquad (1.66)$$

or, since $(n_0)_{kk'} = n_k \delta_{kk'}$ where n_k is the occupation number:

$$\mathrm{i}\frac{\partial}{\partial t} \delta n_{kk'}(t) = (\varepsilon_k - \varepsilon_{k'}) \delta n_{kk'} + (n_{k'} - n_k) W_{kk'} . \qquad (1.67)$$

We can now Fourier transform this equation by writing

$$W_{kk'}(t) = \frac{1}{2\pi} \int W_{kk'}(\omega) \mathrm{e}^{-\mathrm{i}(\omega + \mathrm{i}\delta)t} d\omega , \qquad (1.68)$$

and a similar expression for $\delta n_{kk'}(t)$ (the infinitesimal δ ensures that all quantities vanish in the remote past). We get:

$$\delta n_{kk'}(\omega) = \frac{n_k - n_{k'}}{\varepsilon_k - \varepsilon_{k'} - \omega - i\delta} W_{kk'}(\omega) . \tag{1.69}$$

This result will be central to our discussion of excitations in the following sections.

We now consider screening by a simple generalization of the static case given by (1.56) writing the induced charge density at frequency ω

$$\delta n_{\mathrm{ind}}(\boldsymbol{r},\omega) = \int \delta n(\boldsymbol{x},\boldsymbol{x},\omega)d\xi = \int \chi_0(\boldsymbol{r},\boldsymbol{r}',\omega)W(\boldsymbol{r}',\omega)d\boldsymbol{r}' \tag{1.70}$$

which, using (1.69) and (1.70) gives for the frequency dependent polarizability

$$\chi_0(\boldsymbol{r},\boldsymbol{r}',\omega) = \sum_{kk'} \frac{n_k - n_{k'}}{\varepsilon_k - \varepsilon_{k'} - \omega - i\delta} f_{kk'}(\boldsymbol{r}) f_{kk'}^*(\boldsymbol{r}') \tag{1.71}$$

where:

$$f_{kk'}(\boldsymbol{r}) = \int u_k(\boldsymbol{x}) u_{k'}^*(\boldsymbol{x}) d\xi . \tag{1.72}$$

With this the set of equations (1.56) to (1.58) remain valid in terms of frequency dependent quantities which we have just defined.

Exact Formulation. One can derive an exact formal expression for $\epsilon^{-1}(\omega)$ which will prove of fundamental interest in the following. We consider an N interacting electron system of Hamiltonian H_0 subject to a small external potential $\phi(\boldsymbol{r},t)$ switched on adiabatically in time. The first order perturbation is thus

$$V = \sum_i^N \phi(\boldsymbol{r}_i, t) = \int \hat{n}(\boldsymbol{r}) \phi(\boldsymbol{r}, t) d\boldsymbol{r} , \tag{1.73}$$

where $\hat{n}(\boldsymbol{r})$ is the density operator defined by (1.73) as $\sum_i^N \delta(\boldsymbol{r} - \boldsymbol{r}_i)$. We start from the time-dependent Schrödinger equation ($\hbar = 1$)

$$i\frac{\partial |\Psi\rangle}{\partial t} = (H_0 + V)|\Psi\rangle , \tag{1.74}$$

which can be simplified by the transformation

$$|\Psi\rangle = e^{-iH_0 t}|u\rangle , \tag{1.75}$$

to give

$$i\frac{\partial |u\rangle}{\partial t} = \tilde{V}(t)|u\rangle , \tag{1.76}$$

where V is now expressed in the interaction representation:

$$\tilde{V}(t) = e^{iH_0 t} V e^{-iH_0 t} . \tag{1.77}$$

Equation (1.76) can now be integrated from a remote time t_0 in the past where $\phi = 0$ and $|u\rangle$ is the ground state $|u_0\rangle$ of H_0. To first order in $\tilde{V}$ this gives:

$$|u(t)\rangle = \left(1 - i \int_{t_0}^{t} \tilde{V}(t')dt'\right) |u_0\rangle . \tag{1.78}$$

Now the electron density is given by

$$n(\boldsymbol{r}, t) = \langle \Psi | \hat{n}(\boldsymbol{r}) | \Psi \rangle = \langle u | \tilde{n}(\boldsymbol{r}, t) | u \rangle , \tag{1.79}$$

where $\tilde{n}(\boldsymbol{r}, t)$ is also defined in the interaction representation. From the previous two equations, we obtain

$$n(\boldsymbol{r}, t) = \left\langle u_0 \left| \left(1 + i \int_{t_0}^{t} \tilde{V}(t')dt'\right) \tilde{n}(\boldsymbol{r}, t) \left(1 - i \int_{t_0}^{t} \tilde{V}(t')dt'\right) \right| u_0 \right\rangle \tag{1.80}$$

which, when subtracting the unperturbed density $\langle u_0 | \tilde{n}(\boldsymbol{r}, t) | u_0 \rangle$, leads to the induced electron density

$$\delta n_{\text{ind}}(\boldsymbol{r}, t) = i \left\langle u_0 \left| \int_{-\infty}^{t} \left[\tilde{V}(t') \tilde{n}(\boldsymbol{r}, t) - \tilde{n}(\boldsymbol{r}, t) \tilde{V}(t') \right] dt' \right| u_0 \right\rangle \tag{1.81}$$

in which we have taken $t_0 \to -\infty$. Using the fact that the net result must be real, we can transform this expression to:

$$\delta n_{\text{ind}}(\boldsymbol{r}, t) = -i \int_{-\infty}^{t} dt' \int d\boldsymbol{r}' \langle u_0 | [\tilde{n}(\boldsymbol{r}, t), \tilde{n}(\boldsymbol{r}', t')] | u_0 \rangle \phi(\boldsymbol{r}', t') . \tag{1.82}$$

The linear response function $\chi(\boldsymbol{r}, \boldsymbol{r}', t - t')$ defined by

$$\delta n_{\text{ind}}(\boldsymbol{r}, t) = \int_{-\infty}^{+\infty} dt' \int d\boldsymbol{r}' \chi(\boldsymbol{r}, \boldsymbol{r}', t - t') \phi(\boldsymbol{r}', t') \tag{1.83}$$

is thus given by the exact result, obeying the Heavyside function

$$\chi(\boldsymbol{r}, \boldsymbol{r}', t - t') = -i \langle N | [\tilde{n}(\boldsymbol{r}, t), \tilde{n}(\boldsymbol{r}', t')] | N \rangle \theta(t - t') , \tag{1.84}$$

where $|N\rangle = |u_0\rangle$ is the exact ground state of the N electron system. This expression can still be transformed by developing the commutator and inserting the closure relation $\sum_s |Ns\rangle\langle Ns| = I$ summed over all excited states s of the N electron system of energy ω_s above the ground state to give, with $\tau = t - t'$,

$$\chi(\boldsymbol{r}, \boldsymbol{r}', \tau) = -i \sum_s \left[n_s(\boldsymbol{r}) n_s^*(\boldsymbol{r}') e^{-i\omega_s \tau} - n_s(\boldsymbol{r}') n_s^*(\boldsymbol{r}) e^{i\omega_s \tau} \right] \theta(\tau) , \tag{1.85}$$

where:

$$n_s(\boldsymbol{r}) = \langle N | \hat{n}(\boldsymbol{r}) | Ns \rangle . \tag{1.86}$$

In the absence of magnetic fields, $n_s(\boldsymbol{r})$ can be taken to be real [32, 33] so that the Fourier transform of (1.85) writes:

$$\chi(\boldsymbol{r}, \boldsymbol{r}', \omega) = \sum_s n_s(\boldsymbol{r}) n_s(\boldsymbol{r}') \left[(\omega - \omega_s + i\delta)^{-1} - (\omega + \omega_s + i\delta)^{-1} \right] . \tag{1.87}$$

These expressions correspond to the retarded response function. One can also define a time-ordered response function to be used with field–theoretic techniques, their properties being closely related [32].

The poles ω_s of the response function are the excitations of the N electron system. Their low-energy part corresponds to electron–hole excitations. However the coupling to an external field, i.e. screening, is dominated by plasmons which are high energy, long wavelength density fluctuations of the system. This point plays a central role in our derivation of the GW approximation (next section) as well as in actual calculations using this method, often based on plasmon–pole techniques.

It is useful to derive the general expression of the induced potential $V_{\text{ind}}(\boldsymbol{r}, \boldsymbol{r}', \omega)$ created at point $\boldsymbol{r}$ by a test charge at $\boldsymbol{r}'$. We can formally write

$$V_{\text{ind}} = v\delta n_{\text{ind}} = v\chi v\,, \tag{1.88}$$

i.e. in full

$$V_{\text{ind}}(\boldsymbol{r}, \boldsymbol{r}', \omega) = \sum_s \frac{2\omega_s}{(\omega + i\delta)^2 - \omega_s^2} V_s(\boldsymbol{r}) V_s(\boldsymbol{r}')\,, \tag{1.89}$$

where V_s is the potential produced by the density fluctuation n_s:

$$V_s(\boldsymbol{r}) = \int v(\boldsymbol{r}, \boldsymbol{r}') n_s(\boldsymbol{r}') \mathrm{d}\boldsymbol{r}'\,. \tag{1.90}$$

In this formulation the inverse dielectric function is given similarly by

$$\begin{aligned}\epsilon^{-1} &= 1 + v\chi\,,\\ &= 1 + \sum_s V_s n_s \frac{2\omega_s}{(\omega + i\delta)^2 - \omega_s^2}\,.\end{aligned} \tag{1.91}$$

Another interesting point is what one obtains when χ is calculated by using for $|Ns\rangle$ pure electron–hole excitations, i.e. excited Slater determinants, neglecting the effects of plasmons. Then χ reduces to χ_0, the independent particle polarizability, and ϵ^{-1} is given by:

$$\epsilon^{-1} = 1 + v\chi_0\,. \tag{1.92}$$

If this is considered as the first term of a series expansion in powers of $v\chi_0$, summation of the series gives

$$\epsilon^{-1} = (1 - v\chi_0)^{-1}\,, \tag{1.93}$$

which corresponds to the expression in the random-phase approximation (see further discussion in Sect. 3.3.1).

1.2.4 The GW Approximation

This was derived long time ago [32] and its recent success is mostly due to the improvement in computational power which now allows its application to realistic systems (small aggregates, perfect crystals and their surfaces...). It is based on an expansion of the exchange–correlation self-energy in terms of the dynamically screened electron–electron interaction. This can be done systematically with field-theoretic techniques, one and two-particle Green's functions and the use of functional derivatives.

Here we present an approach which evidences the role played by plasmons in renormalizing the electron–electron interactions. This will allow us to obtain an expression for the total energy of the ground state of the N electron system and, from this, to get the quasi-particle energies in a form identical to the GW approximation. As discussed in Sect. 1.2.3, the excitation spectrum consists of single particle excitations and also to plasmons which correspond to electron density fluctuations. We start from the decoupled situation which means that we write the combined states of the N electron system as simple products

$$|\psi, s\rangle = |\psi\rangle|s\rangle , \tag{1.94}$$

where $|\psi\rangle$ is a Slater determinant and $|s\rangle$ is a state where the sth plasmon mode of frequency ω_s has been excited. We are mostly interested in the low-lying states $|\psi_k, 0\rangle$. These are coupled to other states $|\psi_l, s\rangle$ by interactions $\langle \psi_k, 0|V|\psi_l, s\rangle$ (Fig. 1.4). To obtain the coupling potential V we consider one electron at $\boldsymbol{r}_i$ which interacts with the plasma-like fluctuations $\delta\hat{n}(\boldsymbol{r})$ via the potential given in (1.73) i.e.

$$V(\boldsymbol{r}_i) = \int v(\boldsymbol{r}_i, \boldsymbol{r})\delta\hat{n}(\boldsymbol{r})\mathrm{d}\boldsymbol{r} . \tag{1.95}$$

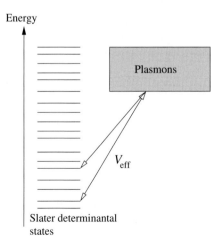

Fig. 1.4. The electron–plasmon coupling induces a renormalization of the electron–electron interactions

The total $V = \sum_i V(\mathbf{r}_i)$ gives rise to the following matrix elements

$$\langle \psi_k, 0 | V | \psi_l, s \rangle = \left\langle \psi_k \left| \left\langle 0 \left| \sum_i V(\mathbf{r}_i) \right| s \right\rangle \right| \psi_l \right\rangle . \tag{1.96}$$

We use the fact that $\delta \hat{n}(\mathbf{r})$ has matrix elements only between the plasmon states, given by (1.86) i.e.

$$\langle 0 | \delta \hat{n}(\mathbf{r}) | s \rangle = \delta n_s(\mathbf{r}) . \tag{1.97}$$

From this we get

$$\langle \psi_k, 0 | V | \psi_l, s \rangle = \left\langle \psi_k \left| \sum_i V_s(\mathbf{r}_i) \right| \psi_l \right\rangle , \tag{1.98}$$

where $V_s(\mathbf{r}_i)$ is the Coulomb potential induced by the fluctuations δn_s defined by (1.90). For the low-lying states $|\psi_k, 0\rangle$, we write

$$(E - E_{k0})|\psi_k, 0\rangle = \sum_{\alpha, s} |\psi_\alpha, s\rangle \langle \psi_\alpha, s | V | \psi_k, 0 \rangle ,$$

$$(E - E_{\alpha s})|\psi_\alpha, s\rangle = \sum_l |\psi_l, 0\rangle \langle \psi_l, 0 | V | \psi_\alpha, s \rangle . \tag{1.99}$$

Injecting the second equation into the first one gives

$$(E - E_{k0})|\psi_k, 0\rangle = \sum_l |\psi_l, 0\rangle \sum_{\alpha, s} \frac{\langle \psi_l, 0 | V | \psi_\alpha, s \rangle \langle \psi_\alpha, s | V | \psi_k, 0 \rangle}{E - E_{\alpha s}} . \tag{1.100}$$

We can now project this set of equations onto $|0\rangle$. Using the fact that $|\psi_k, 0\rangle = |\psi_k\rangle|0\rangle$, using (1.98) and $E_{\alpha s} = E_\alpha + \omega_s$, we get the following set of equations between the Slater determinants

$$(E - E_k)|\psi_k\rangle = \sum_l |\psi_l\rangle \sum_{\alpha, s} \frac{\langle \psi_l | \sum_i V_s(\mathbf{r}_i) | \psi_\alpha \rangle \langle \psi_\alpha | \sum_i V_s(\mathbf{r}_i) | \psi_k \rangle}{E - E_\alpha - \omega_s} . \tag{1.101}$$

This means that the electron–plasmon coupling induces effective interactions between the Slater determinants given by

$$\langle \psi_l | V_{\text{eff}} | \psi_k \rangle = \sum_{\alpha, s} \frac{\langle \psi_l | \sum_i V_s(\mathbf{r}_i) | \psi_\alpha \rangle \langle \psi_\alpha | \sum_i V_s(\mathbf{r}_i) | \psi_k \rangle}{E - E_\alpha - \omega_s} \tag{1.102}$$

which form the renormalizing part of the electron–electron interactions (Fig. 1.4). We now proceed to the calculation of the relevant interactions between the ground state Slater determinant $|\psi_0\rangle$ and those $|\psi_{vc}\rangle$ which differ from it by excitation of one electron from the valence (filled) state u_v to a conduction state u_c. For this we use the rules for the matrix elements of sums of one electron potentials between Slater determinants [12, 13]. This gives us the following results

$$\left\langle \psi_0 \left| \sum_i V_s(\boldsymbol{r}_i) \right| \psi_0 \right\rangle = 0 \, ,$$

$$\left\langle \psi_0 \left| \sum_i V_s(\boldsymbol{r}_i) \right| \psi_{vc} \right\rangle = \langle v|V_s|c\rangle \, . \tag{1.103}$$

In deriving the first contribution to (1.103) we have used the fact that $\langle \psi_0 | \sum_i V_s(\boldsymbol{r}_i)| \psi_0 \rangle = \sum_v \langle v|V_s|v\rangle = 0$. The one-particle matrix elements are

$$\langle \alpha|V_s|\beta\rangle = \int u_\alpha^*(\boldsymbol{r}) V_s(\boldsymbol{r}) u_\beta(\boldsymbol{r}) \mathrm{d}\boldsymbol{r} \, . \tag{1.104}$$

We now directly get the relevant matrix elements $\langle \psi_k|V_{\text{eff}}|\psi_l\rangle$. The basic one is

$$E - E_0 = \langle \psi_0|V_{\text{eff}}|\psi_0\rangle = \sum_{v,c,s} \frac{|\langle v|V_s|c\rangle|^2}{E - E_{vc} - \omega_s} \, , \tag{1.105}$$

where E_0 and E_{vc} as defined in (1.101) are the determinantal energies in the absence of electron–plasmon coupling. To find a simple accurate solution to (1.105) and to the set of equations (1.101) we start from Slater determinants built from single particle equations which match as exactly as possible the excitations energies. We express the denominator of (1.105) as $E - E_0 + E_0 - E_{vc} - \omega_s$. We then write $E_0 - E_{vc} = \varepsilon_v - \varepsilon_c$ the eigenvalues of the single particle equations. If the ground state determinant is correctly chosen $E - E_0$ can be neglected and we get for the ground state energy of the N electron system

$$E_{\text{corr}}(N) = \sum_{v,c,s} \frac{|\langle v|V_s|c\rangle|^2}{\varepsilon_v - \varepsilon_c - \omega_s} \, . \tag{1.106}$$

Note that approximating $E_0 - E_{vc}$ by the difference in eigenvalues neglects the electron–hole attraction which exists even for a determinantal state ψ_{vc}. However this term tends to zero when the size of the system increases and is negligible for large enough systems. This corrective term will be considered in the next section.

To get the quasi-particle energies, we now have to calculate the correlation energy for the $(N+1)$ electron system with an extra electron in state k. The corresponding Slater determinant is now $|k, \psi_0\rangle$. The correlation energy has the same form as (1.106) except that there is one more occupied state k and that one must add the non zero contribution

$$\sum_s \frac{|\langle k, \psi_0 | \sum_i V_s(\boldsymbol{r}_i) | k, \psi_0 \rangle|^2}{-\omega_s} = \sum_s \frac{|\langle k|V_s|k\rangle|^2}{-\omega_s} \, , \tag{1.107}$$

so that one gets

$$E_{\text{corr}}(k, N+1) = \sum_s \frac{|\langle k|V_s|k\rangle|^2}{-\omega_s} + \sideset{}{'}\sum_v \sideset{}{'}\sum_c \sum_s \frac{|\langle v|V_s|c\rangle|^2}{\varepsilon_v - \varepsilon_c - \omega_s} \, , \tag{1.108}$$

where $\sum'$ means that k is included in the sum over v and excluded from that over c. From (1.106) we easily get the correlation contribution $\delta\varepsilon_k$ to the quasi-particle energy ε_k

$$\delta\varepsilon_k = \sum_{c,s} \frac{|\langle k|V_s|c\rangle|^2}{\varepsilon_k - \varepsilon_c - \omega_s} - \sum_{v,s} \frac{|\langle k|V_s|v\rangle|^2}{\varepsilon_v - \varepsilon_k - \omega_s} . \tag{1.109}$$

Adding to this the Hartree–Fock contribution and introducing the occupation numbers we get under vector and operator form

$$\varepsilon_k = \left\langle k \left| h + V_{\mathrm{H}} + \Sigma_{\mathrm{x}} - \sum_{l,s} V_s |l\rangle\langle l| V_s \left\{ \frac{n_l}{\varepsilon_l - \varepsilon_k - \omega_s} - \frac{1 - n_l}{\varepsilon_k - \varepsilon_l - \omega_s} \right\} \right| k \right\rangle . \tag{1.110}$$

In this expression one can consistently use the same ε_k in the correlation contribution since the excitation energies $\varepsilon_k - \varepsilon_l$ of the single-particle Hamiltonian are in principle matched to the exact values. The main difficulty is that the corresponding exact eigenstates $|k\rangle$ are not exactly known. Equation (1.110) is strictly equivalent to the GW approximation [34], as shown in the following. As commonly done, we group together the exchange term and the correlation contribution to get the total self-energy Σ. The quasi-particle Hamiltonian is thus given by the Hartree part plus a total self-energy term given by (1.19) and (1.110) i.e.

$$\Sigma(\boldsymbol{x}, \boldsymbol{x}', \varepsilon_k) = -\sum_l u_l(\boldsymbol{x}) u_l^*(\boldsymbol{x}')$$
$$\times \left[n_l v(\boldsymbol{r}, \boldsymbol{r}') + \sum_s V_s(\boldsymbol{r}) V_s^*(\boldsymbol{r}') \left(\frac{n_l}{\varepsilon_l - \varepsilon_k - \omega_s} - \frac{1 - n_l}{\varepsilon_k - \varepsilon_l - \omega_s} \right) \right] . \tag{1.111}$$

Note that, in this expression, the n_l are the occupation numbers of the N electron system. An identical result is obtained when adding a hole in an occupied state and calculating the corresponding electron energy as $\varepsilon_k = E(N) - E(N-1)$. The relation between the V_s and the frequency dependent screened potential is provided by (1.89). To avoid convergence problems one usually adds a small imaginary part $-i\eta$ to ω_s ($\eta \to 0^+$) which corresponds to the use of a retarded interaction.

To recover the usual expression of the self-energy in terms of the screened Coulomb interaction $W = v + V_{\mathrm{ind}}$, we write from (1.89)

$$-\frac{1}{\pi} \mathrm{Im}\, W(\boldsymbol{r}, \boldsymbol{r}', \omega) = \sum_s V_s(\boldsymbol{r}) V_s(\boldsymbol{r}') [\delta(\omega - \omega_s) - \delta(\omega + \omega_s)] , \tag{1.112}$$

using the fact that:

$$\mathrm{Im}\left(\lim_{\eta \to 0^+} (x + i\eta)^{-1} \right) = -\pi \delta(x) . \tag{1.113}$$

Thus we deduce from (1.111)

$$\Sigma(\boldsymbol{x},\boldsymbol{x}',\omega) = -\sum_l u_l(\boldsymbol{x})u_l^*(\boldsymbol{x}')$$
$$\times \left[n_l v(\boldsymbol{r},\boldsymbol{r}') - \frac{1}{\pi}\int_0^\infty d\omega' \left(\frac{n_l}{\varepsilon_l - \omega - \omega' + i\eta} - \frac{1 - n_l}{\omega - \varepsilon_l - \omega' + i\eta} \right) \right.$$
$$\left. \times \operatorname{Im} W(\boldsymbol{r},\boldsymbol{r}',\omega') \right]. \tag{1.114}$$

A more compact form is obtained by introducing the Green's function

$$G(\boldsymbol{x},\boldsymbol{x}',\omega) = \sum_l \frac{u_l(\boldsymbol{x})u_l^*(\boldsymbol{x}')}{\omega - \varepsilon_l + i\eta \operatorname{Sign}(\varepsilon_l - \varepsilon_F)}, \tag{1.115}$$

where ε_F is the Fermi level. Using an integration along a contour in the complex ω' plane, one can show [34, 35] that the self-energy operator is also given by

$$\Sigma(\boldsymbol{x},\boldsymbol{x}',\omega) = \frac{i}{2\pi}\int_{-\infty}^\infty d\omega' W(\boldsymbol{r},\boldsymbol{r}',\omega') G(\boldsymbol{x},\boldsymbol{x}',\omega+\omega') e^{i\eta\omega'}, \tag{1.116}$$

where the product of G and W is at the origin of the name of the method.

Although quite accurate, the GW expression given by (1.110), (1.111) and (1.114) is only approximate. We shall see in Sect. 1.2.6 how it could be improved along the lines discussed above. Finally one could also obtain (1.110) from resolution of the quasi-particle equation

$$(h_{\mathrm{HF}} + \Sigma_{\mathrm{corr}})|k\rangle = \varepsilon_k|k\rangle, \tag{1.117}$$

where $h_{\mathrm{HF}} = h + V_{\mathrm{H}} + \Sigma_{\mathrm{x}}$ and Σ_{corr} is the correlation part.

Usually GW calculations are performed using perturbation theory from LDA results which provide a quite efficient starting point. Calling ε_k and ε_k^0 the GW and LDA eigenvalues, one gets to first order

$$\varepsilon_k = \varepsilon_k^0 + \langle u_k|\Sigma(\varepsilon_k) - v_{\mathrm{xc}}|u_k\rangle, \tag{1.118}$$

and linearizing this with respect to $\varepsilon_k - \varepsilon_k^0$:

$$\varepsilon_k = \varepsilon_k^0 + \frac{\langle u_k|\Sigma(\varepsilon_k^0) - v_{\mathrm{xc}}|u_k\rangle}{\left(1 - \frac{d}{d\varepsilon_k}\langle u_k|\Sigma(\varepsilon_k)|u_k\rangle\right)_{\varepsilon_k=\varepsilon_k^0}}. \tag{1.119}$$

Successes and failures of GW are discussed in detail in [33, 36]. For our purpose the results of major interest concern the bandgap of semiconductors and insulators. Figure 1.3 shows that the GW predictions are extremely accurate and provide a quantitative tool for analyzing semiconductor nanostructures. We shall come back to these in Chap. 4.

1.2.5 Excitons

In semiconductors and insulators they correspond to electron–hole excitations of the system. To describe their properties the best is to start again from the Hartree–Fock (HF) level where the basic states are Slater determinants. Let us then consider a system for which the ground state is described by a Slater determinant ψ_0 for which all one-particle valence states are filled and all conduction states are empty. ψ_{vc} is the Slater determinant obtained from the previous one by replacing the spin–orbital v by c, i.e. by exciting an electron–hole pair (here we use for v and c full spin–orbitals including the spin part as in (1.6)).

At the HF level one starts from spin orbitals which are solutions of the single particle equations $h_{HF}|k\rangle = 0$ corresponding to (1.18). This makes sure that there is no coupling between ψ_0 and ψ_{vc} since [12, 13]

$$\langle \psi_0 | H | \psi_{vc} \rangle = \langle v | h_{HF} | c \rangle = \varepsilon_c \langle v | c \rangle = 0 \,. \tag{1.120}$$

To obtain the lowest exciton state in the procedure one should then diagonalize H in the manifold span by states ψ_{vc} corresponding to single electron–hole excitations. This obviously neglects coupling with higher excited states via direct electron–electron interactions. In view of their magnitude this is certainly not a good approximation. Nevertheless this will introduce the procedure with use of renormalized interactions. We then calculate $\langle \psi_{vc} | H | \psi_{v'c'} \rangle$ from the general rules given for instance in [12, 13]:

$$\langle \psi_{vc} | H | \psi_{v'c'} \rangle = (\varepsilon_c - \varepsilon_v)\delta_{vv'}\delta_{cc'} + \langle cv' | v | vc' \rangle - \langle cv' | v | c'v \rangle \,. \tag{1.121}$$

In this expression the first term is the difference in HF one-particle energies and the other two correspond to the definition of (1.15). It is of interest to notice that the last term with minus sign corresponds to the direct Coulomb interaction (it involves interaction between charge densities $c^*(\boldsymbol{x})c'(\boldsymbol{x})$ with $v^*(\boldsymbol{x'})v'(\boldsymbol{x'})$ which for the diagonal part give $|c(\boldsymbol{x})|^2$ and $|v(\boldsymbol{x'})|^2$) while the first one represents the exchange term. The natural procedure would then be to diagonalize the matrix given by (1.121) to get the excitation energies.

We now want to parallel this procedure but in terms of renormalized (or dynamically screened) electron–electron interactions. We proceed along the same lines as GW i.e. we now build the Slater determinants from spin–orbitals which are solutions of the quasi-particle equation (1.117). We must then evaluate the matrix elements of the type (1.120) and (1.121) but for the renormalized Hamiltonian $H + V_{\text{eff}}$ where, as before, V_{eff} is due to the renormalizing part induced by the interaction with plasmons. This is worked out in detail in the Appendix A where it is shown that

$$\langle \psi_0 | H + V_{\text{eff}} | \psi_{vc} \rangle \approx \langle v | h_{HF} + \Sigma | c \rangle = \varepsilon_c \langle v | c \rangle = 0 \tag{1.122}$$

and

$$\langle\psi_{vc}|H+V_{\text{eff}}|\psi_{v'c'}\rangle = (\varepsilon_c - \varepsilon_v)\delta_{vv'}\delta_{cc'} + \langle cv'|v|vc'\rangle - \langle cv'|v|c'v\rangle$$
$$+ \sum_s \langle c|V_s|v\rangle\langle v'|V_s|c'\rangle \left\{ \frac{1}{E-E_0-\omega_s} + \frac{1}{E-E_{vc}+\varepsilon_{v'}-\varepsilon_{c'}-\omega_s} \right\}$$
$$- \sum_s \langle c|V_s|c'\rangle\langle v'|V_s|v\rangle \left\{ \frac{1}{E-E_{vc'}-\omega_s} + \frac{1}{E-E_{v'c}-\omega_s} \right\} . \quad (1.123)$$

Note that except for the last two terms in (1.123) these relations only hold approximately if higher order terms can be neglected. We have kept the full energy dependence of these last two terms to compare them with other results but consistency would require that $E - E_{vc}$, $E - E_{v'c}$ and $E - E_{vc'}$ be neglected. The most drastic approximation concerns (1.122) which essentially requires that the bandgap be negligible compared to ω_s.

It is interesting to compare the structure of renormalizing terms in (1.123) to those of the direct terms. We can write:

$$\langle c|V_s|v\rangle\langle v'|V_s|c'\rangle = \int u_c^*(\boldsymbol{x})V_s(\boldsymbol{r})u_v(\boldsymbol{x})u_{v'}^*(\boldsymbol{x}')V_s(\boldsymbol{r}')u_{c'}(\boldsymbol{x}')\mathrm{d}\boldsymbol{x}\mathrm{d}\boldsymbol{x}' ,$$

$$\langle c|V_s|c'\rangle\langle v'|V_s|v\rangle = \int u_c^*(\boldsymbol{x})V_s(\boldsymbol{r})u_{c'}(\boldsymbol{x})u_{v'}^*(\boldsymbol{x}')V_s(\boldsymbol{r}')u_v(\boldsymbol{x}')\mathrm{d}\boldsymbol{x}\mathrm{d}\boldsymbol{x}' . \quad (1.124)$$

The first term has the structure of the exchange interaction and the second that of the Hartree one. One can then rewrite (1.123) as

$$\langle\psi_{vc}|H_{\text{eff}}|\psi_{v'c'}\rangle = (\varepsilon_c - \varepsilon_v)\delta_{vv'}\delta_{cc'} + \langle cv'|v_{\text{SCX}}|vc'\rangle$$
$$- \langle cv'|v_{\text{SCH}}|c'v\rangle , \quad (1.125)$$

where v_{SCX} and v_{SCH} are respectively the renormalized or screened electron–electron interaction for exchange and Hartree terms given by

$$v_{\text{SCX}}(\boldsymbol{r},\boldsymbol{r}')$$
$$= v(\boldsymbol{r},\boldsymbol{r}') + \sum_s V_s(\boldsymbol{r})V_s(\boldsymbol{r}') \left\{ \frac{1}{E-E_0-\omega_s} + \frac{1}{E-E_{vc}+\varepsilon_{v'}-\varepsilon_{c'}-\omega_s} \right\} ,$$
$$v_{\text{SCH}}(\boldsymbol{r},\boldsymbol{r}')$$
$$= v(\boldsymbol{r},\boldsymbol{r}') + \sum_s V_s(\boldsymbol{r})V_s(\boldsymbol{r}') \left\{ \frac{1}{E-E_{vc'}-\omega_s} + \frac{1}{E-E_{v'c}-\omega_s} \right\} . \quad (1.126)$$

The diagonal term $\langle cv|v_{\text{SCX}}|vc\rangle - \langle cv|v_{\text{SCH}}|cv\rangle$ in (1.125) is the correction to $\varepsilon_c - \varepsilon_v$ in single determinant states due to the electron–hole interaction. It tends to zero when the size of the system increases and is thus important for small systems only. The interesting point with (1.125) and (1.126) is that we find that both the Hartree and exchange part are dynamically screened. This is at variance with some work in the literature [36–38] based on approximate resolution of the Bethe–Salpeter equations where only the Hartree term is screened in exactly the same way as in (1.126). We attribute this difference to approximations in these treatments neglecting contributions from excited states $|\psi_\alpha\rangle|s\rangle$ with $|\psi_\alpha\rangle$ outside the manifold of the single excitations $|\psi_{vc}\rangle$.

As shown in Appendix A these excitations are necessary to have a complete renormalization contribution. Their neglect would not allow to get $\varepsilon_c - \varepsilon_v$ in (1.125) as well as the condition $\langle \psi_0 | H + V_{\text{eff}} | \psi_{vc} \rangle \approx 0$ which is necessary for (1.125) to be accurate.

To be coherent with the approximations made on the other terms (equivalent to second order perturbation theory) one should take $E \approx E_{vc} \approx E_{vc'} \approx E_{v'c}$. This is valid when the binding energy of the exciton is weak so that only excitations close to the band gap are important. If $E_g^2 \ll \omega_s^2$, one obtains the simple result that the screening part in both terms of (1.126) is $-2V_s(r)V_s(r')/\omega_s$. From (1.89) this is just the statically induced potential $V_{\text{ind}}(r, r')$ created at r' by a point charge at r, also equal to $(\epsilon^{-1} - 1)v$. Both terms in (1.126) thus involve the statically screened interaction (1.58)

$$W(r, r') = \int \epsilon^{-1}(r, r'')v(r'', r')dr'' . \quad (1.127)$$

Finally let us say a few words about the effect of exchange. From known considerations [2, 7] one can build excitonic states either with total spin $S = 0$ or $S = 1$. In the last case there is no exchange contribution while, for the singlet $S = 0$, the exchange contribution in (1.125) is $2\langle cv'|v_{\text{SCX}}|vc'\rangle$ in which one now only considers the orbital part of the spin–orbitals.

1.2.6 Towards a More Quantitative Theory

One can devise a more quantitative approach along the lines discussed above. The procedure is to build a Slater determinantal basis from an accurate single-particle approach, e.g. the LDA approximation for which excitation energies can be expressed as differences in eigenvalues. One then considers that Slater determinants $|\psi_\alpha\rangle$ and plasmons $|s\rangle$ are eigenstates of separable Hamiltonians. This means that the total eigenstates are products $|\psi_\alpha, s\rangle = |\psi_\alpha\rangle|s\rangle$. As before we want to treat completely the problem for low-lying excitations of the form $|\psi_k\rangle|0\rangle$ where the $|\psi_k\rangle$ are restricted to the ground state $|\psi_0\rangle$ and e.g. to the single particle excitations of the form $|\psi_{vc}\rangle$ (one could eventually include higher excitations but difficulties arise when they overlap the plasmon states so that one should stop at some energy cut-off). We thus write, as before

$$E|\psi_k, 0\rangle = H|\psi_k, 0\rangle + \sum_{\alpha,s} |\psi_\alpha, s\rangle\langle\psi_\alpha, s|V|\psi_k, 0\rangle ,$$

$$(E - E_\alpha - \omega_s)|\psi_\alpha, s\rangle = \sum_l |\psi_l, 0\rangle\langle\psi_l, 0|V|\psi_\alpha, s\rangle . \quad (1.128)$$

Here H is the full Hamiltonian containing the direct electron–electron interactions. The central approximation is that no electron–plasmon interaction is included in the higher excited states, their energy being $E_\alpha - \omega_s$. However such interaction is likely to broaden the spectrum of high energy excitations without modifying substantially the results for low-lying excitations. Injecting the second equation into the first one and projecting onto $|0\rangle$, we get as for (1.101)

$$E|\psi_k\rangle = H|\psi_k\rangle + \sum_l |\psi_l\rangle \sum_{\alpha,s} \frac{\langle\psi_l|\sum_i V_s(\boldsymbol{r}_i)|\psi_\alpha\rangle \langle\psi_\alpha|\sum_j V_s(\boldsymbol{r}_j)|\psi_k\rangle}{E - E_\alpha - \omega_s} . \tag{1.129}$$

This is equivalent to diagonalize the matrix

$$H + \sum_{\alpha,s} \frac{\sum_i V_s(\boldsymbol{r}_i)|\psi_\alpha\rangle\langle\psi_\alpha|\sum_j V_s(\boldsymbol{r}_j)}{E - E_\alpha - \omega_s} \tag{1.130}$$

in the subspace of low-lying determinants $|\psi_0\rangle$, $|\psi_{vc}\rangle$... This is what we have done before but with simplifying assumptions for E equivalent to second order perturbation theory. One could however diagonalize the full energy dependent Hamiltonian (1.130) with the expressions of the matrix elements derived in Appendix A. This would provide to our opinion the most quantitative treatment applied to the exciton problem.

1.2.7 Time-Dependent Density Functional Theory (TDDFT)

This is an extension of the density functional theory (DFT) discussed in Sect. 1.1.5 to time-dependent perturbations. The interest is that, in principle, the frequency dependent response has poles at the excitation energies of the system. One can thus hope to obtain in this way excitation energies which are as accurate as the ground state properties predicted by ordinary DFT even in its local density approximation (LDA).

The extension of DFT to time-dependent problems has been the object of several studies [9, 36, 39–41]. In particular it was shown that there is a one to one correspondence between a time-dependent external potential $V_{\text{ext}}(\boldsymbol{r},t)$ applied to an interacting N electron system and the electron density $n(\boldsymbol{r},t)$ of this system. Furthermore one can work in the Kohn–Sham spirit [15] and obtain the same density for a non-interacting N electron system subject to the so-called Kohn–Sham potential which, as in Sect. 1.1.5 can be written

$$V_{\text{KS}} = V_{\text{ext}} + V_{\text{H}} + V_{\text{xc}} , \tag{1.131}$$

where V_{H} and V_{xc} are now time-dependent Hartree and exchange–correlation potentials. Let us now concentrate on the linear response of the system, i.e. consider a first order external perturbation ϕ (i.e. $V_{\text{ext}} \to V_{\text{ext}} + \phi$) producing a first order change δn_{ind} in electron density (i.e. $n \to n + \delta n_{\text{ind}}$). We can write the formal relation

$$\delta n_{\text{ind}} = \left(\frac{\delta n}{\delta V_{\text{ext}}}\right)_0 \phi = \chi \phi \tag{1.132}$$

which is a short-hand notation for

$$\delta n_{\text{ind}}(\boldsymbol{r}, t) = \int \chi(\boldsymbol{r}, t, \boldsymbol{r}', t')\phi(\boldsymbol{r}', t')\mathrm{d}\boldsymbol{r}'\mathrm{d}t' , \tag{1.133}$$

with

$$\chi(\boldsymbol{r}, t, \boldsymbol{r}', t') = \left(\frac{\delta n(\boldsymbol{r}, t)}{\delta V_{\text{ext}}(\boldsymbol{r}', t')}\right)_0 . \tag{1.134}$$

In all these expressions, the symbol $()_0$ means that functional derivatives are to be taken at the ground state density, χ is the linear density response function or polarizability. Now the same result can be obtained from the non interacting electron system with the Kohn–Sham potential by formally writing

$$\delta n_{\text{ind}} = \left(\frac{\delta n}{\delta V_{\text{KS}}}\frac{\delta V_{\text{KS}}}{\delta V_{\text{ext}}}\right)_0 \phi . \tag{1.135}$$

Realizing that $(\delta n/\delta V_{\text{KS}})_0$ is the Kohn–Sham polarizability χ_0 and using the expression (1.131) for V_{KS} we get

$$\delta n_{\text{ind}} = \chi_0 \left[I + \left(\frac{\delta}{\delta V_{\text{ext}}}(V_{\text{H}} + V_{\text{xc}})\right)_0\right]\phi . \tag{1.136}$$

Writing $\frac{\delta}{\delta V_{\text{ext}}}$ as $\frac{\delta}{\delta n}\frac{\delta n}{\delta V_{\text{ext}}}$ and using (1.136) we obtain

$$\delta n_{\text{ind}} = \chi_0 \phi + \chi_0 \left[\frac{\delta}{\delta n}(V_{\text{H}} + V_{\text{xc}})\right]_0 \delta n_{\text{ind}} , \tag{1.137}$$

which is exactly the condensed notation for equation (9) of [41]. As the polarizabilities depend on time differences, we can Fourier transform this equation over time and write one such equation for each frequency ω. Noticing that $\delta V_{\text{H}}(\boldsymbol{r}, t)/\delta n(\boldsymbol{r}', t') = e^2/|\boldsymbol{r}-\boldsymbol{r}'|\delta(t-t')$ and calling $f_{\text{xc}} = (\delta V_{\text{xc}}/\delta n)_0$, one gets the detailed form of (1.137) as

$$\delta n_{\text{ind}}(\boldsymbol{r}, \omega) = \int \chi_0(\boldsymbol{r}, \boldsymbol{r}', \omega)\left[\frac{e^2}{|\boldsymbol{r}'-\boldsymbol{r}''|} + f_{\text{xc}}(\boldsymbol{r}', \boldsymbol{r}'', \omega)\right]\delta n_{\text{ind}}(\boldsymbol{r}'', \omega)\mathrm{d}\boldsymbol{r}'\mathrm{d}\boldsymbol{r}''$$
$$+ \int \chi_0(\boldsymbol{r}, \boldsymbol{r}', \omega)\phi(\boldsymbol{r}', \omega)\mathrm{d}\boldsymbol{r}' . \tag{1.138}$$

The poles of $\delta n_{\text{ind}}(\boldsymbol{r}, \omega)$ which give the excitation energies thus correspond to the zeros of the left hand side of this integral equation. However to find practical solutions it is convenient to use a matrix formulation. For this one writes δn_{ind} in terms of the Kohn–Sham orbitals

$$\delta n_{\text{ind}}(\boldsymbol{r}, \omega) = \sum_{ij} n_{ij}(\omega)u_i^*(\boldsymbol{r})u_j(\boldsymbol{r}) . \tag{1.139}$$

One can inject this together with the expression (1.71) for χ_0 into the left hand side of (1.138). This leads to the system of equations

$$n_{ij}(\omega) - \frac{f_{ij}}{\omega - \omega_{ji} + i\delta} \sum_{k,l} K_{jk,il} n_{kl}(\omega) = 0 , \qquad (1.140)$$

with $f_{ij} = n_i - n_j$, $\omega_{ji} = \varepsilon_j - \varepsilon_i$ and

$$K_{jk,il} = \int u_j^*(\bm{r}) u_k^*(\bm{r}') \left[\frac{e^2}{|\bm{r} - \bm{r}'|} + f_{\text{xc}}(\bm{r}, \bm{r}', \omega) \right] u_i(\bm{r}) u_l(\bm{r}') \mathrm{d}\bm{r} \mathrm{d}\bm{r}' . \qquad (1.141)$$

Equations (1.140) and (1.141) are the basic equations of TDDFT as given in [41]. The only difference in notations with [41] corresponds to the definition of the $K_{jk,il}$ for which we have used a convention consistent with our notation (1.15) for Coulomb integrals. Practical calculations then require a definition of the kernel K, i.e. of f_{xc}, and different approximations have been used [41]. In the adiabatic LDA approximation [9], the exchange–correlation potential and f_{xc} are expressed in terms of the time-independent exchange–correlation energy

$$V_{\text{xc}}(\bm{r}, t) \cong \frac{\delta E_{\text{xc}}[n]}{\delta n(\bm{r})} ,$$

$$\frac{\delta V_{\text{xc}}(\bm{r}, t)}{\delta n(\bm{r}', t')} \cong \delta(t - t') \frac{\delta^2 E_{\text{xc}}[n]}{\delta n(\bm{r}) \delta n(\bm{r}')} , \qquad (1.142)$$

where, in LDA, $E_{\text{xc}}[n]$ is given locally by its value for the electron gas with the same density (see Sect. 1.1.5).

TDDFT has been applied to the calculation of the polarizability of molecules and clusters (Sect. 3.3.3). It also leads to a substantial improvement of excitation energies with respect to Kohn–Sham eigenvalues for atoms, clusters and molecules [36, 41–47], even using the adiabatic LDA. However, one must note that even the simple random-phase approximation (i.e. neglecting f_{xc}) gives a good agreement compared to experiments for small systems [48]. In contrast, in solids, the wrong Kohn–Sham bandgap remains [36, 49, 50] except if the kernel is deduced from the Bethe–Salpeter equation [48].

Finally, TDDFT has been also used to calculate the electron–vibration coupling in benzene [51].

1.3 Semi-empirical Methods

Up to the advent of the GW approximation, ab initio theories could not accurately predict the band structure of semiconductors. Most of the understanding of these materials was then obtained from less accurate descriptions. Among them, semi-empirical methods have played (and still play) a very important role since they allow us to simulate the true energy bands in terms of a restricted number of adjustable parameters. There are essentially three distinct methods of achieving this goal : the empirical tight binding (ETB) approximation, the empirical pseudopotential method (EPM) and the $\bm{k} \cdot \bm{p}$ approximation which, in its simple form, is equivalent to the effective mass approximation (EMA).

1.3.1 The Empirical Tight Binding Method

The Basis of the Empirical Tight Binding Method. In this approximation, the wave function is written as a combination of localized orbitals centered on each atom,

$$\Psi = \sum_{i,\alpha} c_{i\alpha} \varphi_{i\alpha} , \qquad (1.143)$$

where $\varphi_{i\alpha}$ is the αth free atom orbital of atom i, at position $\boldsymbol{R}_i$. As each complete set of such orbitals belonging to any given atom forms a basis for the Hilbert space, the whole set of $\varphi_{i\alpha}$ is complete, i.e., the $\varphi_{i\alpha}$ are no longer independent and (1.143) can yield the exact wave function of the whole system. In practice one has to truncate the sum of over α in this expansion. In many simplified calculations it has been assumed that the valence states of the system can be described in terms of a minimal basis set which only includes free atom states belonging to the outer shell of the free atom (e.g. $2s$ and $2p$ in diamond). It is that description which provides the most appealing physical picture, allowing us to clearly understand the formation of bands from the atomic limit. The minimal basis set approximation is also used in most semi-empirical calculations.

When the sum over α in (1.143) is limited to a finite number, the energy levels ε of the whole system are given by the secular equation

$$\det |H - \varepsilon S| = 0 , \qquad (1.144)$$

where H is the Hamiltonian matrix in the atomic basis and S the overlap matrix of elements:

$$S_{i\alpha,j\beta} = \langle \varphi_{i\alpha} | \varphi_{j\beta} \rangle . \qquad (1.145)$$

These matrix elements can be readily calculated, especially in the local density theory, and when making use of Gaussian atomic orbitals. The problem, as in the plane wave expansion, is to determine the number of basis states required for good numerical accuracy.

An interesting discussion on the validity of the use of a minimal basis set has been given by Louie [52]. Starting from the minimal basis set $|\varphi_{i\alpha}\rangle$, one can increase the size of the basis set by adding other atomic states $|\chi_{i\mu}\rangle$, called the peripheral states, which must lead to an improvement in the description of the energy levels and wave functions. However, this will rapidly lead to problems related to over-completeness, i.e., the overlap of different atomic states will become more and more important. To overcome this difficulty, Louie proposes three steps to justify the use of a minimum basis set. These are the following :

– Symmetrically orthogonalize the states $|\varphi_{i\alpha}\rangle$ belonging to the minimal basis set between themselves. This leads to an orthogonal set $|\overline{\varphi}_{i\alpha}\rangle$.

- The peripheral states $|\chi_{i\mu}\rangle$ overlap strongly with the $|\overline{\varphi}_{i\alpha}\rangle$. It is thus necessary to orthogonalize them to these $|\overline{\varphi}_{i\alpha}\rangle$, which yield new states $|\overline{\chi}_{i\mu}\rangle$ defined as:

$$|\overline{\chi}_{i\mu}\rangle = |\chi_{i\mu}\rangle - \sum_{j\alpha} |\overline{\varphi}_{j\alpha}\rangle\langle\overline{\varphi}_{j\alpha}|\chi_{i\mu}\rangle \ . \tag{1.146}$$

- The new states $|\overline{\chi}_{i\mu}\rangle$ are then orthogonalized between themselves leading to a final set of states $|\overline{\overline{\chi}}_{i\mu}\rangle$.

Louie has shown that, at least for silicon, the average energies of these atomic states behave in such a way that, after the three steps, the peripheral states $|\overline{\overline{\chi}}_{i\mu}\rangle$ are much higher in energy and their coupling to the minimal set is reduced. They only have a small (although not negligible) influence, justifying the use of the minimal set as the essential step in the calculation.

The quantitative value of LCAO (linear combination of atomic orbitals) techniques for covalent systems such as diamond and silicon was first demonstrated by Chaney et al. [53]. They have shown that the minimal basis set gives good results for the valence bands and slightly poorer (but still meaningful) results for the lower conduction bands. Such conclusions have been confirmed by several groups [52, 54, 55] who worked with pseudopotentials instead of true atomic potentials.

The great interest of the minimal basis set LCAO calculations is that they provide a direct connection between the valence states of the system and the free atom states. This becomes still more apparent with the ETB approximation which we shall later discuss and which allows us to obtain extremely simple, physically sound descriptions of many systems.

ETB can be understood as an approximate version of the LCAO theory. It is generally defined as the use of a minimal atomic basis set neglecting inter-atomic overlaps, i.e., the overlap matrix defined in (1.145) is equal to the unit matrix. The secular equation thus becomes

$$det|H - \varepsilon I| = 0 \ , \tag{1.147}$$

where I is the unit matrix. The resolution of the problem then requires the knowledge of the Hamiltonian matrix elements. In the ETB these are obtained from a fit to the bulk band structure. For this, one always truncates the Hamiltonian matrix in real space, i.e., one only includes inter-atomic terms up to first, second, or, at most, third nearest neighbors. Also, in most cases one makes use of a two-center approximation as discussed by Slater and Koster [56]. In such a case, all Hamiltonian matrix elements $\langle\varphi_{i\alpha}|H|\varphi_{j\beta}\rangle$ can be reduced to a limited number of independent terms which we can call $H_{\alpha\beta}(i,j)$ for the pair of atoms (i,j) and the orbitals (α, β). On an s,p basis, valid for group IV, III–V, and II–VI semiconductors, symmetry considerations applied to the two-center approximation only give the following independent terms [56]

$$H_{\alpha\beta}(i,j) = H_{ss\sigma}(i,j), H_{sp\sigma}(i,j), H_{sp\sigma}(j,i), H_{pp\sigma}(i,j), H_{pp\pi}(i,j) \ , \tag{1.148}$$

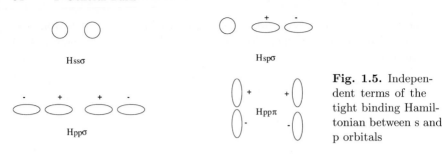

Fig. 1.5. Independent terms of the tight binding Hamiltonian between s and p orbitals

where s stands for the s orbital, σ the p orbital along axis i, j with the positive lobe in the direction of the neighboring atom, and π a p orbital perpendicular to the axis i, j (Fig. 1.5). $H_{\sigma\pi}(i, j)$ is strictly zero in two-center approximation. With these conventions all matrix elements are generally negative.

Similar considerations apply to transition metals with s, p and d orbitals. Simple rules obtained for the $H_{\alpha\beta}(i, j)$ in a nearest neighbor's approximation are given by Harrison [57]. They are based on the use of free atom energies for the diagonal elements of the tight binding Hamiltonian. On the other hand, the nearest neighbor's interactions are taken to scale like d^{-2} (where d is the inter-atomic distance) as determined from the free electron picture of these materials. For s, p systems, this gives numerically (in eV)

$$H_{ss\sigma} = -\frac{10.67}{d^2}, \quad H_{sp\sigma} = -\frac{14.02}{d^2},$$
$$H_{pp\sigma} = -\frac{24.69}{d^2}, \quad H_{pp\pi} = -\frac{6.17}{d^2}, \tag{1.149}$$

where d is expressed in Å.

Viewed as an approximation of LCAO theory, ETB must in principle lead to incorrect energy eigenvalues. For instance the direct neglect of inter-atomic overlaps in (1.144) leads to band structures which are in general narrower than the corresponding LCAO one provided the same Hamiltonian H is used. However one can obtain a simple secular equation like (1.147) in various ways. One of them is to symmetrically orthogonalize the basis set as in Louie's procedure described above. A popular way to achieve this is to use Lowdin's procedure [58] in which the orthogonalized basis is defined as

$$|\overline{\varphi}_{i\alpha}\rangle = (S^{-1/2}|\varphi\rangle)_{i\alpha} . \tag{1.150}$$

This allows to rewrite (1.144) under the following form

$$\det |S^{-1/2} H S^{-1/2} - \varepsilon I| = 0 , \tag{1.151}$$

formally equivalent to the ETB (1.147). As discussed above, in the semi-empirical procedures the parameters are determined from a fit to known band structures which has allowed empirical rules like those of (1.149) to be established. However care must be taken when transferring such parameters from the known bulk situation to other cases like point defect problems for instance. The most obvious case is the vacancy where one simply removes the atom. To describe this correctly one cannot in principle suppress only the matrix elements of $S^{-1/2}HS^{-1/2}$ connected to the missing atom but one should also take into account the change in the matrix S induced by the defect. This point is usually ignored in the ETB calculations of such problems.

Description of Bulk Semiconductors. The parameters discussed above in (1.149) nicely reproduce the valence bands of zinc-blende semiconductors but poorly describe the band gap and the conduction bands. Improvements on this description have been attempted by going to the second nearest neighbors [59] or by keeping the nearest neighbors treatment as it is but adding one s orbital (labeled s*) to the minimal basis set [60]. The role of this latter orbital is to simulate the effect of higher energy d orbitals which have been shown to be essential for a correct simulation of the conduction band. The quality of such a fit can be judged from Figs. 1.6 and 1.7, which show that the lowest conduction bands are reproduced much more correctly than with only first-nearest neighbor sp^3 model.

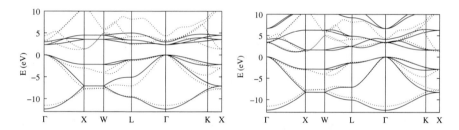

Fig. 1.6. Band structure of silicon. Left: first nearest neighbour ETB [61]; right: sp^{3}s* Vogl's ETB [60]. GW band structure [65] (*dashed line*)

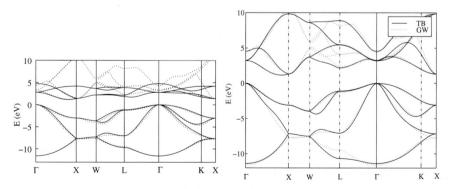

Fig. 1.7. Band structure of silicon. Left: second nearest neighbor ETB [59]; right: third nearest neighbor ETB [64]. GW band structure [65] (*dashed line*)

Although ETB seems in general inferior to the empirical pseudopotential method [62] to obtain accurate valence and conduction bands with a small number of parameters substantial progress has been achieved over the years. For instance, a recent sp^3 ETB model [63] including up to third nearest neighbors interactions and three center integrals has been shown to provide an excellent fit to the silicon band structure both for the valence and the four lower conduction bands. However, as other previous empirical fits, this model is less good in describing the curvature of the bands near their extrema. For instance it provides a Luttinger parameter $\gamma_2 = 1.233$ (see Sect. 1.3.3) instead of 0.320 experimentally and a transverse effective mass $m_t^* = 0.567$ instead of 0.191. As a correct description of these parameters is essential to an application to quantum dots, we have recently [64] used the same method but we have obtained the parameters by minimizing the error on a weighted average of bulk band energies and effective masses taken from an ab initio GW calculation [65]. The corresponding results are given on Fig. 1.7 and Table 1.1. Such a good fit can also be obtained with a first-nearest neighbor $sp^3d^5s^*$ model [66] which is also successful for III–V compound semiconductors with the same degree of accuracy.

Table 1.1. Third nearest neighbors ETB parameters for silicon and first nearest neighbors ETB parameters for Si-H. Neighbour positions are given in units of $a/4$. Δ is the spin–orbit coupling parameter

Si ETB parameters :			
$E_{ss}[000]$	−6.17334 eV	$E_{ss}[111]$	−1.78516 eV
$E_{pp}[000]$	2.39585 eV	$E_{sx}[111]$	0.78088 eV
		$E_{xx}[111]$	0.35657 eV
Δ	0.04500 eV	$E_{xy}[111]$	1.47649 eV
$E_{ss}[220]$	0.23010 eV	$E_{ss}[311]$	−0.06857 eV
$E_{sx}[220]$	−0.21608 eV	$E_{sx}[311]$	0.25209 eV
$E_{sx}[022]$	−0.02496 eV	$E_{sx}[113]$	−0.17098 eV
$E_{xx}[220]$	0.02286 eV	$E_{xx}[311]$	0.13968 eV
$E_{xx}[022]$	−0.24379 eV	$E_{xx}[113]$	−0.04580 eV
$E_{xy}[220]$	−0.05462 eV	$E_{xy}[311]$	−0.03625 eV
$E_{xy}[022]$	−0.12754 eV	$E_{xy}[113]$	0.06921 eV
Si-H ETB parameters :			
E_H	0.17538 eV	$V_{ss\sigma}$	−4.12855 eV
		$V_{sp\sigma}$	3.72296 eV

Total Energies in Tight Binding. Up to now we have discussed how it is possible to get one particle eigenstates, obtained by diagonalization of the tight binding matrix. However it is also possible to derive an empirical technique which allows a determination of total energies. For this let us consider the simpler case of a nearest-neighbor approximation. One then assumes that all independent nearest neighbors $H_{\alpha\beta}(i,j)$ vary in the same way as functions of the inter-atomic distance R_{ij}, for instance as

$$H_{\alpha\beta}(R_{ij}) = H^0_{\alpha\beta} \exp(-qR_{ij}) , \qquad (1.152)$$

or also like some inverse power of R_{ij}, see [57]. Then, by summing over the energies of filled states, one can obtain the band structure energy E_{BS} as a function of the atomic positions. This band structure energy has an attractive character. To determine the crystal stability it is necessary to add a repulsive part corresponding to terms which have been neglected or counted twice in E_{BS}. In a non self-consistent scheme such terms correspond to the repulsion of neutral atoms [67]. They are short-ranged and we can simulate them simply by Born–Mayer pair potentials:

$$V(R_{ij}) = V_0 \exp(-pR_{ij}) . \qquad (1.153)$$

The parameters p, q and V_0 are obtained empirically from a fit to the observed cohesive energy, lattice parameter and compressibility. They can then be used to calculate phonon dispersion curves and relaxation or reconstruction energies near defects or surfaces. This technique has been applied with much success in transition metals [68] as well as in sp bonded semiconductors [67]. When applied to silicon for instance, in its simplest version where the electronic part is treated in the molecular or bond orbital model, it provides a natural justification to the well known Keating Hamiltonian for vibrations [67]. It was also shown that a refinement of the technique leads to a quite good description of the phonon dispersion curves [69, 70].

The empirical laws (1.152) and (1.153) have been generalized to allow more quantitative ETB simulations of total energies versus local environment especially for transition metals and covalent semiconductors. A summary of the present state of the art in this field can be found in [71].

Screening in ETB. One can perform quite efficient calculations of screening in ETB for systems which cannot be handled by ab initio methods, e.g. for clusters with size larger than ≈ 1 nm. The basic assumption is the neglect of differential overlap used long time ago for the description of molecules [72]. It corresponds to the fact that Coulomb integrals of the type (1.15) in a basis of atomic orbitals $\varphi_{i\alpha}$ (i = atom index, α = orbital index) follow the rule

$$\begin{aligned}\langle i\alpha, j\beta | v | k\gamma, l\delta \rangle &= \langle i\alpha, j\beta | v | i\alpha, j\beta \rangle \delta_{i\alpha,k\gamma} \delta_{j\beta,l\delta} , \\ &= \int |\varphi_{i\alpha}(\boldsymbol{r})|^2 \frac{e^2}{|\boldsymbol{r}-\boldsymbol{r'}|} |\varphi_{j\beta}(\boldsymbol{r'})|^2 \mathrm{d}\boldsymbol{r}\mathrm{d}\boldsymbol{r'} \delta_{i\alpha,k\gamma} \delta_{j\beta,l\delta} .\end{aligned}$$
$$(1.154)$$

The remaining Coulomb integrals are usually simplified to

$$\int |\varphi_{i\alpha}(\boldsymbol{r})|^2 \frac{e^2}{|\boldsymbol{r}-\boldsymbol{r'}|} |\varphi_{j\beta}(\boldsymbol{r'})|^2 \mathrm{d}\boldsymbol{r}\mathrm{d}\boldsymbol{r'} = v_{ij} = \frac{e^2}{R_{ij}} \text{ if } i \neq j ,$$
$$= U_i^{\text{at}} \text{ if } i = j , \qquad (1.155)$$

which do not depend on the orbital labels α and β but only on the atomic index. U_i^{at} is an intra-atomic Coulomb energy which can be calculated from atomic wave functions. The expression of the electron density $n(\mathbf{r})$ can be simplified to

$$n(\mathbf{r}) = \sum_{i,\alpha} n_{i\alpha} |\varphi_{i\alpha}(\mathbf{r})|^2 , \qquad (1.156)$$

since we discard overlaps in Coulomb terms, $n_{i\alpha}$ being the electron population of orbital $\varphi_{i\alpha}$. We now look at what happens to the basic relation of static screening described by (1.56)

$$W(\mathbf{r}, \mathbf{r}') = v(\mathbf{r}, \mathbf{r}') + \int v(\mathbf{r}, \mathbf{r}'') \, \delta n_{\text{ind}}(\mathbf{r}'', \mathbf{r}') \mathrm{d}\mathbf{r}'' , \qquad (1.157)$$

corresponding to the potential created at $\mathbf{r}$ by an electronic test charge at $\mathbf{r}'$. In the ETB view, the test charge must be located on atom j, i.e. distributed on the $|\varphi_{j\alpha}(\mathbf{r})|^2$. Multiplying (1.157) by this distribution and integrating over $\mathbf{r}'$ we get:

$$W(\mathbf{r}, \mathbf{R}_j) = v(\mathbf{r}, \mathbf{R}_j) + \int v(\mathbf{r}, \mathbf{r}'') \, \delta n_{\text{ind}}(\mathbf{r}'', \mathbf{R}_j) \mathrm{d}\mathbf{r}'' . \qquad (1.158)$$

We now expand δn_{ind} as in (1.156) to get:

$$W(\mathbf{r}, \mathbf{R}_j) = v(\mathbf{r}, \mathbf{R}_j) + \sum_k v(\mathbf{r}, \mathbf{R}_k) \, \delta n_k(\mathbf{R}_j) . \qquad (1.159)$$

The final step is just to multiply by $|\varphi_{i\alpha}(\mathbf{r})|^2$ and integrate over $\mathbf{r}$ which gives

$$W_{ij} = v_{ij} + \sum_k v_{ik} \, \delta n_{kj} . \qquad (1.160)$$

This means that the continuous equation (1.157) has now been discretized over the atoms replacing integral equations by products of matrices. We can now linearize the δn_{kj} with respect to the W_{lj} in terms of a polarizability matrix χ

$$\delta n_{kj} = \sum_l \chi_{kl} W_{lj} , \qquad (1.161)$$

$$\chi_{ij} = 2 \sum_{k,k'} \frac{n_k - n_{k'}}{\varepsilon_k - \varepsilon_{k'} - \omega - i\delta} \left(\sum_\alpha c_{k,i\alpha} c^*_{k',i\alpha} \right) \left(\sum_\alpha c_{k,j\alpha} c^*_{k',j\alpha} \right) ,$$

which is an obvious transposition. The whole formulation is now in terms of matrices and one can write as usual $W = \epsilon^{-1} v$, $\epsilon = I - v\chi$ now requiring matrix multiplication and inversion. The enormous advantage is that the size of the matrices is $N \times N$, N being the number of atoms. This can be applied to pretty large systems as will be shown by the GW calculations performed in this way for nanocrystals (Sect. 4.4).

1.3.2 The Empirical Pseudopotential Method

The full atomic potentials produce strong divergences at the atomic sites in the solid. These divergences are related to the fact that these potentials must produce the atomic core states as well as the valence states. However, the core states are likely to be quite similar to what they are in the free atom. Thus the use of the full atomic potentials in a band calculation is likely to lead to unnecessary computational complexity since the basis states will have to be chosen in such a way that they describe localized states and extended states at the same time. Therefore, it is of much interest to devise a method which allows us to eliminate the core states, focusing only on the valence states of interest which are easier to describe. This is the basis of the pseudopotential theory.

The pseudopotential concept started with the orthogonalized plane wave theory [73]. Writing the crystal Schrödinger equation for the valence states

$$(T+V)|\Psi\rangle = E|\Psi\rangle , \tag{1.162}$$

one has to recognize that the eigenstate $|\Psi\rangle$ is automatically orthogonal to the core states $|c\rangle$ produced by the same potential V. This means that $|\Psi\rangle$ will be strongly oscillating in the neighborhood of each atomic core, which prevents its expansion in terms of smoothly varying functions, like plane waves, for instance. It is thus interesting to perform the transformation

$$|\Psi\rangle = (1-P)|\varphi\rangle , \tag{1.163}$$

where P is the projector onto the core states

$$P = \sum_c |c\rangle\langle c| . \tag{1.164}$$

$|\Psi\rangle$ is thus automatically orthogonal to the core states and the new unknown $|\varphi\rangle$ does not have to satisfy the orthogonality requirement. The equation for the pseudo-state $|\varphi\rangle$ is:

$$(T+V)(1-P)|\varphi\rangle = E(1-P)|\varphi\rangle . \tag{1.165}$$

Because the core states $|c\rangle$ are eigenstates of the Hamiltonian $T+V$ with energy E_c, one can rewrite (1.165) in the form

$$\left[T + V + \sum_c (E-E_c)|c\rangle\langle c|\right]|\varphi\rangle = E|\varphi\rangle . \tag{1.166}$$

The pseudo-wave function is then solution of a Schrödinger equation with the same energy eigenvalue as $|\Psi\rangle$. This new equation is obtained by replacing the potential V by a pseudopotential

$$V_{\text{ps}} = V + \sum_c (E-E_c)|c\rangle\langle c| . \tag{1.167}$$

This is a complex non-local operator. Furthermore, it is not unique since one can add any linear combination of core states to $|\varphi\rangle$ in (1.166) without changing the eigenvalues. There is a corresponding non-uniqueness in V_{ps} since the modified $|\varphi\rangle$ will obey a new equation with another pseudopotential. This non-uniqueness in V_{ps} is an interesting factor since it can then be optimized to provide the smoothest possible $|\varphi\rangle$, allowing rapid convergence of plane wave expansions for $|\varphi\rangle$. This will be used directly in the empirical pseudopotential method.

First-principle pseudopotentials have been derived for use in quantitative calculations [74]. First of all, they are ion pseudopotentials and not total pseudopotentials as those discussed above. They are deduced from free atom calculations and have the following desirable properties:

- real and pseudo-valence eigenvalues agree for a chosen prototype atomic configuration
- real and pseudo-atomic wave functions agree beyond a chosen core radius r_c
- total integrated charges (norm conservation)
- logarithmic derivatives of the real and pseudo wave functions and their first energy derivatives agree for $r > r_c$.

These properties are crucial to have optimum transferability of the pseudopotential among a variety of chemical environments, allowing self-consistent calculations of a meaningful pseudo-charge density.

Let us now discuss the empirical pseudopotential method (EPM) which consists in a plane wave expansion of the wave function plus the use of a smooth pseudopotential. In EPM one assumes that the self-consistent crystal pseudopotential can be written as a sum of atomic contributions, i.e.,

$$V(\mathbf{r}) = \sum_{j,\alpha} v_\alpha(\mathbf{r} - \mathbf{R}_j - \mathbf{r}_\alpha), \qquad (1.168)$$

where j runs over the unit cells positioned at $\mathbf{R}_j$ and α is the atom index, the atomic position within the unit cell being given by $\mathbf{r}_\alpha$. Let us first assume that the v_α are ordinary functions of $\mathbf{r}_\alpha$ or, in other words, that we are dealing with local pseudopotentials. In that case the matrix elements of V between plane waves become

$$\langle \mathbf{k} + \mathbf{G}|V|\mathbf{k} + \mathbf{G}'\rangle = \frac{1}{\Omega} \sum_\alpha e^{\mathrm{i}(\mathbf{G}'-\mathbf{G})\cdot \mathbf{r}_\alpha} \int v_\alpha(\mathbf{r}) e^{\mathrm{i}(\mathbf{G}'-\mathbf{G})\cdot \mathbf{r}} d\mathbf{r}, \qquad (1.169)$$

where Ω is the volume of the unit cell. Suppose that there can be identical atoms in the unit cell. Then the sum over α can be expressed as a sum over groups β of identical atoms with position specified by a second index γ (i.e. $\mathbf{r}_\alpha = \mathbf{r}_{\gamma\beta}$). Calling n the number of atoms in the unit cell we can write

$$\langle \mathbf{k} + \mathbf{G}|V|\mathbf{k} + \mathbf{G}'\rangle = \sum_\beta S_\beta(\mathbf{G}' - \mathbf{G}) v_\beta(\mathbf{G}' - \mathbf{G}), \qquad (1.170)$$

where S_β and v_β are, respectively, the structure and form factors of the corresponding atomic species, defined by

$$S_\beta(\boldsymbol{G}) = \frac{1}{n}\sum_\gamma e^{i\boldsymbol{G}\cdot\boldsymbol{r}_{\gamma\beta}} \qquad (1.171)$$

and

$$v_\beta(\boldsymbol{G}) = \frac{n}{\Omega}\int v_\beta(\boldsymbol{r})e^{i\boldsymbol{G}\cdot\boldsymbol{r}}\mathrm{d}\boldsymbol{r} \ . \qquad (1.172)$$

In practice, EPM treats the form factors $v_\beta(\boldsymbol{G})$ as disposable parameters. In the case where $v_\beta(\boldsymbol{r})$ are smooth potentials their transforms $v_\beta(\boldsymbol{G})$ will rapidly decay as a function of $|\boldsymbol{G}|$ so that it may be a good approximation to truncate them at a maximum value G_c. For instance, the band structure of tetrahedral covalent semiconductors like Si can be fairly well reproduced using only the three lower Fourier component $v(|\boldsymbol{G}|)$ of the atomic pseudopotential. There are thus two cut-off values for $|\boldsymbol{G}|$ to be used in practice: one, G_M, limits the number of plane waves and thus the size of the Hamiltonian matrix; the other one, G_c, limits the number of Fourier components of the form factors. We shall later give some practical examples.

The use of a local pseudopotential is not fully justified since, from (1.166), it involves, in principle, projection operators. It can be approximately justified for systems with s and p electrons. However, when d states become important, e.g. in the conduction band of semiconductors, it is necessary to use an operator form with a projection operator on the 1-2 angular components.

Let us now discuss the application of EPM to purely covalent materials like silicon or germanium. The basis vectors of the direct zinc–blende lattice are $a/2 \times (110)$, $a/2 \times (011)$ and $a/2 \times (101)$. The corresponding basis vectors of the reciprocal lattice are $2\pi/a \times (11\bar{1})$, $2\pi/a \times (\bar{1}11)$ and $2\pi/a \times (1\bar{1}1)$. The reciprocal lattice vectors $\boldsymbol{G}$ which have the lowest square modulus are given in Table 1.2.

Table 1.2. Reciprocal lattice vectors $\boldsymbol{G}$ with the lowest modulus in the case of a zinc–blende lattice

$\frac{a}{2\pi}(\boldsymbol{G})$	$\left(\frac{a}{2\pi}\right)^2 \boldsymbol{G}^2$
000	0
111	3
200	4
220	8
311	11

For elemental materials like Si and Ge there is only one form factor $v(\boldsymbol{G})$ but we have seen in (1.170) that the matrix element of the potential involves a structure factor which is given here by

$$S(\boldsymbol{G}) = \cos(\boldsymbol{G} \cdot \boldsymbol{\tau}) , \qquad (1.173)$$

where the origin of the unit cell has been taken at the center of a bond in the (111) direction and where $\boldsymbol{\tau}$ is thus the vector $a/8(111)$. For local pseudopotentials this matrix element $\langle \boldsymbol{k} + \boldsymbol{G} | V | \boldsymbol{k} + \boldsymbol{G}' \rangle$ can be written $V(\boldsymbol{G})$ and is thus given by:

$$V(\boldsymbol{G}) = v(\boldsymbol{G}) \cos(\boldsymbol{G} \cdot \boldsymbol{\tau}) . \qquad (1.174)$$

The structure factor part is of importance since, among the lowest values of $|\boldsymbol{G}|$ quoted in Table 1.2, it gives zero for $2\pi/a(200)$. If one indexes $V(\boldsymbol{G})$ by the value taken by the quantity $(a/2\pi)^2 G^2$, then only the values V_3, V_8 and V_{11} are different from zero. It has been shown [75] that the inclusion of these three parameters alone allows to obtain a satisfactory description of the band structure of Si and Ge. This can be understood simply by the consideration of the free electron band structure of these materials which is obtained by neglecting the potential in the matrix elements of the Hamiltonian between plane waves. The eigenvalues are thus the free electron energies $\hbar^2 |\boldsymbol{k} + \boldsymbol{G}|^2 / 2m_0$ which, in the fcc lattice, lead to the energy bands plotted in Fig. 1.8. The similarity is striking, showing that the free electron band structure provides a meaningful starting point.

The formation of gaps in this band structure can be easily understood at least in situations where only two free electron branches cross. To the lowest order in perturbation theory, one will have to solve the 2×2 matrix

$$\begin{vmatrix} \frac{\hbar^2}{2m_0}|\boldsymbol{k} + \boldsymbol{G}|^2 & V(\boldsymbol{G}' - \boldsymbol{G}) \\ V(\boldsymbol{G}' - \boldsymbol{G})^* & \frac{\hbar^2}{2m_0}|\boldsymbol{k} + \boldsymbol{G}'|^2 \end{vmatrix} . \qquad (1.175)$$

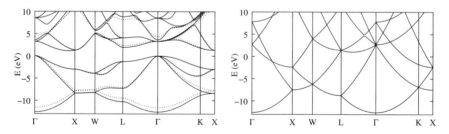

Fig. 1.8. Correspondence between free-electron (*right*) and empirical pseudopotential (*left*) [76] bands, showing how the degeneracies are lifted by the pseudopotential. GW calculation [65] (*dashed line*)

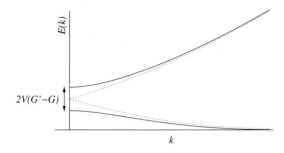

Fig. 1.9. Opening of a gap in the nearly free electron method. The two free electron branches (*dashed line*) are split by the potential Fourier component (*straight line*)

The resulting eigenvalues are

$$E(\mathbf{k}) = \frac{\hbar^2}{2m_0}\left[\frac{|\mathbf{k}+\mathbf{G}|^2 + |\mathbf{k}+\mathbf{G}'|^2}{2}\right]$$
$$\pm\sqrt{\left(\frac{\hbar^2}{2m_0}\frac{|\mathbf{k}+\mathbf{G}'|^2 - |\mathbf{k}+\mathbf{G}|^2}{2}\right)^2 + |V(\mathbf{G}'-\mathbf{G})|^2} \quad (1.176)$$

whose behavior as a function of $\mathbf{k}$ is pictured in Fig. 1.9. The conclusion is that there is formation of a gap at the crossing point equal to $2|V(\mathbf{G}'-\mathbf{G})|$. Note that for this to occur the crossing point at $\mathbf{k} = -(\mathbf{G}'+\mathbf{G})/2$ must lie within the first Brillouin zone or at its boundaries. For points where several branches cross, one will have a higher order matrix to diagonalize but this will generally also result in the formation of gaps. This explains the differences between the free electron band structure and the actual one in Fig. 1.8.

The number of parameters required for fitting the band structures of compounds is different in view of the fact that there are now two different atoms in the unit cell with form factors $v_A(\mathbf{G})$ and $v_B(\mathbf{G})$. The matrix elements $V(\mathbf{G})$ of the total pseudopotential will thus be expressed as

$$V(\mathbf{G}) = V^s(\mathbf{G})\cos(\mathbf{G}\cdot\boldsymbol{\tau}) + iV^a(\mathbf{G})\sin(\mathbf{G}\cdot\boldsymbol{\tau}), \quad (1.177)$$

where V^s and V^a are equal to $(v_A + v_B)/2$ and $(v_A - v_B)/2$, respectively. The number of fitting parameters is then multiplied by 2, the symmetric components V_3^s, V_8^s, and V_{11}^s being close to those of the covalent materials and the antisymmetric components being V_3^a, V_4^a and V_{11}^a since the antisymmetric part of V_8 vanishes.

1.3.3 The $\mathbf{k}\cdot\mathbf{p}$ Description and Effective Masses

The concept of effective masses near a band extremum is very powerful to describe hydrogenic impurities in semiconductors [77]. This will prove still more important for heterostructures and nanostructures that we discuss in the next chapter. In any case it is highly desirable to provide a general framework to analyze this problem. This is obtained directly via the $\mathbf{k}\cdot\mathbf{p}$ method which we present in this section.

The basis of the method is to take advantage of the crystalline structure which allows us to express the eigenfunctions as Bloch functions and to write a Schrödinger-like equation for its periodic part. We start from

$$\left\{\frac{p^2}{2m_0}+V\right\} e^{i\mathbf{k}\cdot\mathbf{r}} u_\mathbf{k}(\mathbf{r}) = E(\mathbf{k}) e^{i\mathbf{k}\cdot\mathbf{r}} u_\mathbf{k}(\mathbf{r}) \, , \tag{1.178}$$

where we have written the wave function in Bloch form. We can rewrite this in the following form

$$\left\{\frac{(\mathbf{p}+\hbar\mathbf{k})^2}{2m_0}+V\right\} u_\mathbf{k}(\mathbf{r}) = E(\mathbf{k}) u_\mathbf{k}(\mathbf{r}) \tag{1.179}$$

which is totally equivalent to the first form. To solve this, we can expand the unknown periodic part $u_\mathbf{k}(\mathbf{r})$ on the basis of the corresponding solutions at a given pointy $\mathbf{k}_0$, which we label $u_{n,\mathbf{k}_0}(\mathbf{r})$:

$$u_\mathbf{k}(\mathbf{r}) = \sum_n c_n(\mathbf{k}) u_{n,\mathbf{k}_0}(\mathbf{r}) \, . \tag{1.180}$$

The corresponding solutions are the eigenstates and eigenvalues of the matrix with the general element

$$A_{n,n'}(\mathbf{k}) = \left\langle u_{n,\mathbf{k}_0} \left| \frac{(\mathbf{p}+\hbar\mathbf{k})^2}{2m_0}+V \right| u_{n',\mathbf{k}_0} \right\rangle \, . \tag{1.181}$$

We now use the fact that $u_{n,\mathbf{k}_0}$ is an eigenfunction of (1.179) for $\mathbf{k}=\mathbf{k}_0$, with energy $E_n(\mathbf{k}_0)$. This allows us to rewrite (1.181) in the simpler form

$$A_{n,n'}(\mathbf{k}) = \left\{E_n(\mathbf{k}_0) + \frac{\hbar^2}{2m_0}(\mathbf{k}-\mathbf{k}_0)^2\right\} \delta_{n,n'}$$
$$+ \frac{\hbar(\mathbf{k}-\mathbf{k}_0)}{m_0} \mathbf{p}_{nn'}(\mathbf{k}_0) \, , \tag{1.182}$$

with

$$\mathbf{p}_{nn'}(\mathbf{k}_0) = \langle u_{n,\mathbf{k}_0} | \mathbf{p} | u_{n',\mathbf{k}_0} \rangle \, . \tag{1.183}$$

Diagonalization of the matrix $A(\mathbf{k})$ given by (1.182) can give the exact band structure (an example of this is given in [78]). However, the power of the method is that it represents the most natural starting point for a perturbation expansion. Let us illustrate this first for the particular case of a single non-degenerate extremum. We thus consider a given non-degenerate energy branch $E_n(\mathbf{k})$ which has an extremum at $\mathbf{k}=\mathbf{k}_0$ and look at its value for $\mathbf{k}$ close to $\mathbf{k}_0$. The last term in (1.182) can then be considered as a small perturbation and we determine the difference $E_n(\mathbf{k})-E_n(\mathbf{k}_0)$ by second order perturbation theory applied to the matrix $A(\mathbf{k})$. This gives

$$E_n(\mathbf{k}) = E_n(\mathbf{k}_0) + \frac{\hbar^2}{2m_0}(\mathbf{k}-\mathbf{k}_0)^2$$
$$+ \frac{\hbar^2}{m_0^2} \sum_{n'\neq n} \frac{[(\mathbf{k}-\mathbf{k}_0)\cdot\mathbf{p}_{nn'}][(\mathbf{k}-\mathbf{k}_0)\cdot\mathbf{p}_{n'n}]}{E_n(\mathbf{k}_0)-E_{n'}(\mathbf{k}_0)} \, , \tag{1.184}$$

which is the second order expansion near $\boldsymbol{k}_0$ leading to the definition of the effective masses. The last term in (1.184) is a tensor. Calling 0α its principal axes, one gets the general expression for the effective masses m_α^*:

$$\frac{m_0}{m_\alpha^*} = 1 + \frac{2}{m_0} \sum_{n' \neq n} \frac{|(p_\alpha)_{nn'}|^2}{E_n(\boldsymbol{k}_0) - E_{n'}(\boldsymbol{k}_0)} \ . \tag{1.185}$$

This shows that when the situation practically reduces to two interacting bands, the upper one has positive effective masses while the opposite is true for the lower one. This is what happens at the Γ point for GaAs, for instance.

Another very important situation is the case of degenerate extremum at the top of the valence band in zinc–blende materials which occurs at $\boldsymbol{k} = 0$. For $\boldsymbol{k} \approx 0$, the last term of (1.182) can still be treated by the second order perturbation theory. By letting i and j be two members of the degenerate set at $\boldsymbol{k} = 0$ and l any other state distant in energy, we now must apply the second order perturbation theory on a degenerate state. As shown in standard textbooks [79], this leads to diagonalization of a matrix

$$A_{ij}^{(2)}(\boldsymbol{k}) = \left[E_i(0) + \frac{\hbar^2}{2m_0}k^2\right]\delta_{ij} + \frac{\hbar^2}{m_0^2}\sum_l \frac{(\boldsymbol{k}\cdot\boldsymbol{p}_{il})(\boldsymbol{k}\cdot\boldsymbol{p}_{lj})}{E_i(0) - E_l(0)} \ . \tag{1.186}$$

The top of the valence band has threefold degeneracy and its basis states behave like atomic p states in cubic symmetry (i.e. like the simple functions x, y, and z). The second order perturbation matrix is thus a 3×3 matrix built from the last term in (1.186) which, from symmetry, can be reduced to [80–83]

$$\begin{vmatrix} Lk_x^2 + M(k_y^2 + k_z^2) & Nk_xk_y & Nk_xk_z \\ Nk_xk_y & Lk_y^2 + M(k_x^2 + k_z^2) & Nk_yk_z \\ Nk_xk_z & Nk_yk_z & Lk_z^2 + M(k_x^2 + k_y^2) \end{vmatrix} , \tag{1.187}$$

where L, M, and N are three real numbers, all of the form:

$$\frac{\hbar^2}{m_0^2}\sum_l \frac{(\boldsymbol{k}\cdot\boldsymbol{p}_{il})(\boldsymbol{k}\cdot\boldsymbol{p}_{lj})}{E_i(0) - E_l(0)} \ . \tag{1.188}$$

It is this matrix plus the term $\hbar^2 k^2/(2m_0)$ on its diagonal which define the matrix to be applied in effective mass theory to a degenerate state. Up to this point we have not included spin effects and in particular spin orbit coupling, which plays an important role in systems with heavier elements. If we add the spin variable, the degeneracy at the top of the valence band is double and the $\boldsymbol{k}\cdot\boldsymbol{p}$ matrix becomes a 6×6 matrix whose detailed form can be found in [84–86] and is given in Sect. 2.2.1. One can slightly simplify its diagonalization when the spin orbit coupling becomes large, from the fact that

$$\boldsymbol{L}\cdot\boldsymbol{S} = (J^2 - L^2 - S^2)/2 \ , \tag{1.189}$$

where $\boldsymbol{J} = \boldsymbol{L} + \boldsymbol{S}$. Because here $L = 1$ and $S = 1/2$, J can take two values $J = 3/2$ and $J = 1/2$. From (1.189), the $J = 3/2$ states will lie at higher energy than the $J = 1/2$ ones and, if the spin orbit coupling constant is large enough, these states can be treated separately. The top of the valence band will then be described by the $J = 3/2$ states leading to a 4×4 matrix whose equivalent Hamiltonian has been shown by Luttinger and Kohn [87] to be

$$H = \frac{\hbar^2}{m_0} \left\{ \left(\gamma_1 + \frac{5}{2} \gamma_2 \right) \frac{k^2}{2} - \gamma_2 \sum_\alpha k_\alpha^2 J_\alpha^2 \right. $$
$$\left. - \gamma_3 \sum_{\alpha \neq \beta} k_\alpha k_\beta \frac{J_\alpha J_\beta + J_\beta J_\alpha}{2} \right\}, \tag{1.190}$$

where $\alpha, \beta = x, y$ or z.

Finally, as shown by Kane [80, 81], it can be interesting to treat the bottom of the conduction band and the top of the valence band at $\boldsymbol{k} = 0$ as a quasi-degenerate system, extending the above described method to a full 8×8 matrix which can be reduced to a 6×6 one if the spin orbit coupling is large enough to neglect the lower valence band.

2 Quantum Confined Systems

The electronic structure of bulk semiconductors is characterized by delocalized electronic states and by a quasi continuous spectrum of energies in the conduction and valence bands. In semiconductor nanostructures, when the electrons are confined in small regions of space in the range of a few tens of nanometers or below, the energy spectrum is profoundly affected by the confinement, with in particular:

- an increase of the width of the bandgap
- the allowed energies become discrete in zero-dimensional (0D) systems and form mini-bands in 1D and 2D systems.

Quantum confinement effects are present in a wide range of systems, e.g. in quantum wells, wires or dots grown by advanced epitaxial techniques, in nanocrystals produced by chemical methods or by ion implantation, or in nanodevices made by state-of-the-art lithographic techniques. This chapter is devoted to the calculation of quantum confinement effects. We begin with a simplified description based on the effective mass approximation (EMA). Then we describe more elaborate methods which allow to get accurate results. We present the results of tight binding calculations on nanostructures of direct and indirect bandgap semiconductors. We compare with the predictions of other methods described in Chap. 1, discussing the advantages and the limits of each one. We also provide analytic expressions for the confinement energies in a large number of semiconductor nanocrystals. Finally, we consider the case of amorphous silicon clusters, to emphasize the interplay between disorder and confinement effects.

2.1 Quantum Confinement and Its Consequences

In this section we describe the effects of the quantum confinement in idealized nanostructures using simplified treatments mainly based on the EMA.

2.1.1 Idealized Quantum Wells

We discuss here the case of square quantum wells, such as those grown by epitaxial techniques. We consider a well made of a semiconductor I

($-L_z/2 \leq z \leq L_z/2$) sandwiched between barrier layers of material II such as the electrons or the holes are confined in the well. Basically, a common way to achieve this is to use for material II a semiconductor or an insulator with a bandgap much larger than in the well. The quantum mechanical description of quantum wells is considerably simplified using the fact that the carriers experience a potential which is almost identical to that of perfect materials I and II in the well and the barriers, respectively. The difference between the potential in the barrier and in the well defines the confining potential $V_{\mathrm{conf}}(\boldsymbol{r})$ which is commonly approximated by a square potential only depending on the coordinate z (Fig. 2.1). Within this approximation, the calculation of the lowest electronic states or of the highest hole states has been extensively done using the envelope function approach [86, 88, 89], i.e. using $\boldsymbol{k} \cdot \boldsymbol{p}$ or EMA (Chap. 1). A basic assumption in these treatments is that the confining potential does not mix the wave functions from different bands, except those which are degenerate. In the simplest case of conduction band states in direct gap semiconductors like GaAs or InP, it can be shown [86, 90] that the electron wave function takes approximately the form

$$\Psi(\boldsymbol{r}) = u_{\boldsymbol{k}_0}^{\mathrm{I,II}}(\boldsymbol{r})\phi(\boldsymbol{r}) , \tag{2.1}$$

where $u_{\boldsymbol{k}_0}^{\mathrm{I,II}}$ is the Bloch wave function at the minimum of the conduction band in the material I or II. $\phi(\boldsymbol{r})$ is an envelope function solution of a Schrödinger-like equation

$$\left(-\frac{\hbar^2}{2m^*}\Delta + V_{\mathrm{conf}}(z)\right)\phi(\boldsymbol{r}) = \varepsilon\phi(\boldsymbol{r}) , \tag{2.2}$$

where m^* is the conduction band effective mass which in principle is material dependent. The origin of the energy corresponds to the bottom of the conduction band of material I. The eigenfunctions are separable in x, y, z directions

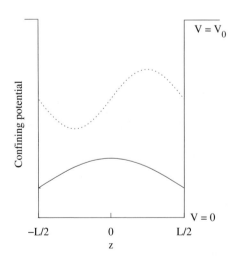

Fig. 2.1. Potential profile and lowest envelope functions $\chi_1(z)$ (*straight line*) and $\chi_2(z)$ (*dashed line*) in an infinite square well ($V_0 \to \infty$)

because V_{conf} only depends on z. Thus the solutions of the Schrödinger equation have the following form

$$\varepsilon_{\boldsymbol{k}n} = \frac{\hbar^2 k^2}{2m^*} + \varepsilon_n^z,$$
$$\phi_{\boldsymbol{k}n}(\boldsymbol{r}) = \frac{1}{\sqrt{S}} e^{i\boldsymbol{k}\cdot\boldsymbol{\rho}} \chi_n(z), \qquad (2.3)$$

where $\boldsymbol{k} = (k_x, k_y)$, $k = |\boldsymbol{k}|$, $\boldsymbol{\rho} = (x, y)$, $S = L_x L_y$ is the sample area and n is an integer. The electrons are free to propagate in the x and y directions. The functions $\chi_n(z)$ are solutions of the 1D equation in the z direction

$$\left(-\frac{\hbar^2}{2m^*}\frac{\partial^2}{\partial z^2} + V_{\text{conf}}(z)\right)\chi_n(z) = \varepsilon_n^z \chi_n(z). \qquad (2.4)$$

The solutions $\chi_n(z)$ are bound states if ε_n^z is smaller than the potential in the barriers and is unbound otherwise [86]. The resolution of (2.4) further requires continuity conditions at the interfaces. One usually imposes that $\chi_n(z)$ and $(1/m^*)(\partial \chi_n(z)/\partial z)$ must be continuous. The condition of continuity of $(1/m^*)(\partial \chi_n(z)/\partial z)$ is required by the conservation of the particle current. If the confining potential is large in the barriers, the problem for the lowest states can be approximated by the simpler one of an infinite square well

$$V_{\text{conf}}(z) = 0 \text{ if } -L_z/2 \le z \le L_z/2,$$
$$V_{\text{conf}}(z) \to \infty \text{ otherwise}, \qquad (2.5)$$

and the solutions are

$$\varepsilon_n^z = \frac{\hbar^2}{2m^*}\left(\frac{n\pi}{L_z}\right)^2,$$
$$\chi_n(z) = \sqrt{\frac{2}{L_z}} \sin\left(\frac{n\pi z}{L_z}\right) \text{ for even } n,$$
$$\chi_n(z) = \sqrt{\frac{2}{L_z}} \cos\left(\frac{n\pi z}{L_z}\right) \text{ for odd } n. \qquad (2.6)$$

The quantity $\hbar^2 \pi^2/(2m^* L_z^2)$ is equal to ≈ 150 meV with $L_z = 100$ Å and $m^* = 0.1 m_0$. Equation (2.3) shows that the 2D confinement leads to the formation of subbands at each energy ε_n^z, the energy $\hbar^2 k^2/(2m^*)$ in a subband corresponding to the kinetic energy for the in-plane motion of the carrier. The effect of the confinement on the electronic structure is evidenced in the density of states $\rho_{\text{2D}}(\varepsilon)$, i.e. the number of allowed states per unit energy around a given energy ε. Using (2.3), we obtain

$$\rho_{\text{2D}}(\varepsilon) = 2 \sum_{k_x, k_y, n} \delta\left[\varepsilon - \varepsilon_n^z - \frac{\hbar^2 k^2}{2m^*}\right], \qquad (2.7)$$

where the factor 2 stands for the spin degeneracy. Applying cyclic boundary conditions in the x and y directions, we have

$$k_x = n_x \frac{2\pi}{L_x}, \quad k_y = n_y \frac{2\pi}{L_y}. \tag{2.8}$$

The summation over k_x, k_y in (2.7) can be converted to an integration and finally

$$\rho_{2D}(\varepsilon) = \frac{m^* S}{\pi \hbar^2} \sum_n \Theta(\varepsilon - \varepsilon_n^z), \tag{2.9}$$

where $\Theta(x)$ is the step function ($= 1$ if $x > 0$, $= 0$ otherwise). Thus the density of states in a 2D system is a staircase function (Fig. 2.2), with a discontinuity at each energy ε_n^z, whereas at 3D it is a continuous function of the energy

$$\rho_{3D}(\varepsilon) = \frac{m^* \Omega}{\pi^2 \hbar^2} \sqrt{\frac{2m^*}{\hbar^2} \varepsilon}, \tag{2.10}$$

where Ω is the volume of the system. On Fig. 2.2, we compare ρ_{2D} calculated for an infinite square well using the energies given in (2.6) with ρ_{3D} calculated for the same volume of material ($\Omega = L_z S$). When going from 3D to 2D, the density of states is reorganized into steps. In particular, $\rho_{2D}(\varepsilon) = 0$ for $\varepsilon < \varepsilon_1^z$ in contrast to $\rho_{3D}(\varepsilon)$. With a similar situation for holes (but the calculation of the eigenstates is more complex in the valence band), we deduce an important consequence of the confinement: the width of the bandgap is increased compared to the bulk. In EMA, the confinement energy, i.e. the difference between the gap at 2D and 3D is proportional to $1/L_z^2$. A large number of physical properties are altered by the 2D confinement (see for example reviews [86, 91]). Among them, it is worth noticing : the blue-shift of the photoluminescence energy and of the optical absorption threshold in

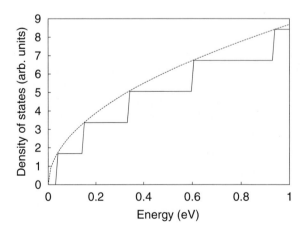

Fig. 2.2. Density of states for an infinite square well (*straight line*) of width L_z and of area S compared to the one for the bulk material (*dashed line*) with the same volume $\Omega = SL_z$ ($L_z = 100$ Å, $m^* = 0.1 \, m_0$)

quantum wells compared to the bulk semiconductor. It is also important to point out that the 2D density of states is finite at the bottom of the lowest subband whereas the 3D one is equal to zero, which has fundamental consequences on the properties of 2D systems, for example for the gain of semiconductor quantum well lasers [92].

2.1.2 Idealized Quantum Wires

In wires, the confinement now takes place in two directions of space (e.g. x, y) and the carrier motion is free in the other direction (z). Using a square potential for $V_{\mathrm{conf}}(x,y)$, i.e. in the form of Fig. 2.1 along x and y, the problem is once again separable into three 1D equations and the solutions become

$$\varepsilon_{kn_x n_y} = \frac{\hbar^2 k^2}{2m^*} + \varepsilon_{n_x}^x + \varepsilon_{n_y}^y ,$$

$$\phi_{kn_x n_y}(\boldsymbol{r}) = \frac{1}{\sqrt{L_z}} e^{ikz} \chi_{n_x}(x) \chi_{n_y}(y) . \qquad (2.11)$$

In the case of an infinite square potential, $\chi_{n_x}(x)$ and $\chi_{n_y}(y)$ are given in (2.6), corresponding to bound states along x and y. The term $\hbar^2 k^2/(2m^*)$ is the kinetic energy for the motion of the carrier along the wire axis. Following the same approach as for wells, we derive the density of states for a wire of length L_z:

$$\rho_{1\mathrm{D}}(\varepsilon) = \sum_{n_x,n_y} \frac{L_z}{\pi} \sqrt{\frac{2m^*}{\hbar^2}} (\varepsilon - \varepsilon_{n_x}^x - \varepsilon_{n_y}^y)^{-1/2} . \qquad (2.12)$$

The density of states is equal to zero when $\varepsilon < \varepsilon_1^x + \varepsilon_1^y$. Thus, the quantum confinement leads to an opening of the band gap like in 2D systems but the 1D density of states is highly peaked, since it presents singularities at each value of $\varepsilon_n^x + \varepsilon_n^y$ (Fig. 2.3). The 1D subbands are often referred to as channels, in particular when discussing transport properties of quantum wires (Chap. 8).

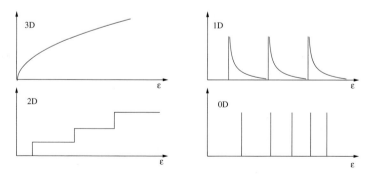

Fig. 2.3. Comparison between the density of states at 3D, 2D, 1D and 0D

This description of the electronic structure of quantum wires is based on a picture of non interacting electrons, which is certainly true for semiconducting wires. However, in metallic wires, or in semiconducting wires with a high level of doping, electron–electron interactions play an important role in contrast to 3D metals where electron–electron interactions are strongly screened and where the Landau's Fermi liquid theory gives a phenomenological description of these systems in terms of the non interacting Fermi gas. In purely 1D electron gas (without lateral dimensions) this theory breaks down because the one-dimensionality restricts the screening of Coulomb interactions. 1D metals are usually described as Luttinger liquids [93, 94] where excitations of the system correspond to collective motions of the electrons. The physics of one-dimensional interacting fermions is presently a very active field of research, in particular due to the fact that good metallic wires are now easily available for experiments (e.g. carbon nanotubes).

2.1.3 Idealized Cubic Quantum Dots

In quantum dots, the confinement takes place in the three directions of space. The main consequence is that the electronic spectrum consists in series of discrete levels, like in isolated atoms. In the simplest case of a square potential like in Fig. 2.1 along x, y and z axes, we easily derive the eigenvalues and eigenstates of the EMA equation:

$$\varepsilon_{n_x n_y n_z} = \varepsilon^x_{n_x} + \varepsilon^y_{n_y} + \varepsilon^z_{n_z},$$
$$\phi_{n_x n_y n_z}(\boldsymbol{r}) = \chi_{n_x}(x)\chi_{n_y}(y)\chi_{n_z}(z). \tag{2.13}$$

The density of states consists of δ functions at the discrete energies:

$$\rho_{0D}(\varepsilon) = 2 \sum_{n_x, n_y, n_z} \delta(\varepsilon - \varepsilon_{n_x n_y n_z}). \tag{2.14}$$

In the case of an infinite square potential, we have:

$$\varepsilon_{n_x n_y n_z} = \frac{\hbar^2 \pi^2}{2m^*} \left[\left(\frac{n_x}{L_x}\right)^2 + \left(\frac{n_y}{L_y}\right)^2 + \left(\frac{n_z}{L_z}\right)^2 \right]. \tag{2.15}$$

In a cubic dot ($L_x = L_y = L_z$), the ground state of the system ε_{111} has a twofold degeneracy (including spin) and the first excited level (ε_{211}, ε_{121}, ε_{112}) is sixfold degenerate. Interestingly, we recover the situation of an atom with twofold degenerate S state and sixfold degenerate P state. Thus quantum dots are often referred to as artificial atoms. Since the electronic structure of these artificial atoms can be tuned by changing their size or their shape, quantum dots are particularly attractive building blocks for the development of nanotechnologies.

2.1.4 Artificial Atoms: Case of Spherical Wells

The similarity between quantum dots and isolated atoms becomes particularly striking in the case of spherical quantum dots, i.e. when the confining potential has a spherical symmetry. For example, nanocrystals in semiconductor doped glasses and colloidal solutions [95–100] often have a spherical shape. When the passivation of the surface is made in such a way that carriers are strongly confined in the nanocrystal, the system is usually correctly described by an infinitely deep spherical well where the confining potential is

$$V_{\text{conf}}(\boldsymbol{r}) = 0 \text{ if } r < R,$$
$$V_{\text{conf}}(\boldsymbol{r}) \to \infty \text{ otherwise}, \quad (2.16)$$

where R is the radius of the nanocrystal. Due to the spherical symmetry of the potential, the orbital momentum operator $\boldsymbol{L}$ commutes with the Hamiltonian and in EMA it is advantageous to rewrite the Schrödinger-like equation for the envelope function ϕ in spherical coordinates:

$$\left[-\frac{\hbar^2}{2m^*} \left(\frac{1}{r^2} \frac{\partial}{\partial r} \left(r^2 \frac{\partial}{\partial r} \right) - \frac{L^2}{r^2} \right) + V_{\text{conf}}(r) \right] \phi(r,\theta,\varphi) = \varepsilon \phi(r,\theta,\varphi). \quad (2.17)$$

The eigenstates are products of the spherical harmonics Y_{lm} and of radial parts. The solutions are

$$\varepsilon_{nl} = \frac{\hbar^2}{2m^*} \left(\frac{X_{nl}}{R} \right)^2, \quad n = 1, 2, 3..., \quad l = 0, 1, 2...,$$

$$\phi_{nlm}(r,\theta,\varphi) = A\, j_l\left(\frac{X_{nl} r}{R} \right) Y_{lm}(\theta,\varphi), \quad (2.18)$$

where j_l is a spherical Bessel function which is related to a normal Bessel function of the first kind with half integer index

$$j_l(x) = \sqrt{\frac{\pi}{2x}} J_{l+1/2}(x), \quad (2.19)$$

with in particular $j_0(x) = \sin(x)/x$. The coefficients X_{nl} are the zeros of the spherical Bessel functions labeled by an integer n in order of increasing energy. Some values of X_{nl} are given in Table 2.1 for the lowest levels defined by n and l. The corresponding levels are shown in Fig. 2.4. The levels can be labeled with the usual atomic notation, e.g. 1S corresponds to $l = 0$ and $n = 1$. Their degeneracy is also the same as in real atoms. However, we must note that there is no restriction on the values of l for a given n like in free atoms where $l < n$, which results from the different nature of the potentials.

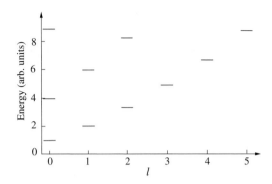

Fig. 2.4. Schematic representation of the lowest levels in an infinite spherical well versus the quantum number l

Table 2.1. Values of X_{nl} for the lowest states in a spherical well

nl	Level	X_{nl}
10	1S	3.142
11	1P	4.493
12	1D	5.763
20	2S	6.283
13	1F	6.988
21	2P	7.725

2.1.5 Electronic Structure from Bulk to Quantum Dots

In this section we present the effects of quantum confinement from a different point of view using the empirical tight binding (ETB) method. The main objective is to show how the electronic structure varies when going from the bulk to quantum dots, i.e. when going from energy bands to discrete levels. In particular, we will see that the distribution of the discrete energy levels in quantum dots is connected to the bulk density of states. For simplicity, we consider a linear chain of atoms, with one atom per unit cell of length a. An atom j at position R_j is described by a single s orbital $|j\rangle$. The Hamiltonian is defined by two parameters, an intra-atomic term $\varepsilon_s = \langle j|H|j\rangle$ and a nearest neighbour interaction term $\beta = \langle j|H|j+1\rangle$ ($\beta < 0$). Interactions with more distant atoms are neglected. The eigenstates of the infinite chain are given by the Bloch theorem:

$$|\Psi_k\rangle = \sum_j e^{ikR_j}|j\rangle \ . \tag{2.20}$$

The electronic structure is characterized by a single band of energy dispersion:

$$\varepsilon_k = \varepsilon_s + 2\beta \cos(ka) \ . \tag{2.21}$$

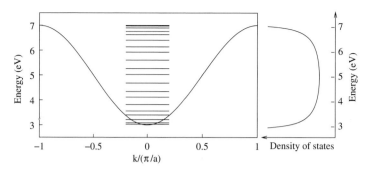

Fig. 2.5. Band structure (*left*) of a linear chain of s orbitals and the electronic levels for a finite chain containing 20 atoms ($\beta = 1$ eV). Density of states for the infinite chain (*right*)

Figure 2.5 presents the band structure and the corresponding density of states which behaves like $1/\sqrt{1 - (\varepsilon - \varepsilon_s)^2/(4\beta^2)}$. Let us consider now the case of a finite chain with N atoms ($1 \leq j \leq N$). The corresponding eigenvalues and eigenstates can be also obtained analytically [101, 102] from the solutions of the infinite chain. For that purpose, we have just to notice that a combination of bulk solutions which vanishes on the atom 0 and on the atom $N + 1$ is also solution of the finite chain, because the inter-atomic couplings are restricted to first nearest neighbors. Combining $|\Psi_k\rangle$ and $|\Psi_{-k}\rangle$, we first form wave functions which always vanish on the atom 0 (setting $R_0 = 0$):

$$|\Psi_k^a\rangle = |\Psi_k\rangle - \mathrm{i}|\Psi_{-k}\rangle \propto \sum_j \sin(kR_j)|j\rangle \ . \tag{2.22}$$

Secondly, $|\Psi_k^a\rangle$ has a zero weight on the atom $N + 1$ ($R_{N+1} = (N + 1)a$) if $k = k_p$ such as

$$k_p R_{N+1} = p\pi \Rightarrow k_p = \frac{p\pi}{(N+1)a} \quad \text{with } p = 1\ldots N \ . \tag{2.23}$$

Using the fact that $|\Psi_{k_p}^a\rangle$ is solution for the energy ε_{k_p} in (2.21), we obtain that the energy levels of the finite chain are just given by the dispersion relation of the infinite chain at discrete values of k. Thus we deduce that the distance between allowed energies in quantum dots strongly depends on two parameters:

– on the size which defines the distribution of allowed k points
– on the density of states in the bulk material.

Figure 2.5 presents the energy levels for a chain containing 20 atoms: in that particular case, the splitting between the levels is smaller near the band edges, in agreement with the fact that the density of states of the infinite chain diverges at the band edges (this is obviously specific to a 1D system).

These conclusions can be generalized to other types of nanostructures, even if the calculation of the energy levels is in general more complex than for the linear chain where the confined states are simply given by a combination of two states of the infinite chain. We will see in the following that in real quantum dots the boundary conditions are quite complex due to the shape of the surface, due to atomic surface reconstructions and due to chemical passivations. For these reasons, the confined states are in general built from a combination of a large number of bulk states, mainly derived from the band under consideration, but also with a non negligible mixing with states from other bands.

It is important to realize that the discretization of the levels occurs in any material. But confinement effects are visible when the energy level spacing typically exceeds kT, where T is the temperature. In a bulk metal, the Fermi level lies in the center of a band where the density of states is usually important. The consequence is that in metal clusters, the level spacing is small, and at temperatures above a few kelvin, the physical properties resemble those of a bulk metal. Thus confinement effects are observed mainly in clusters containing less than hundreds of atoms [103]. In semiconductors, the Fermi level lies between two bands, such that the band edges dominate the low-energy optical and electrical behavior. Near the band edges, the density of states is usually much smaller, and confinement effects remain visible even for large clusters, sometimes containing up to millions of atoms.

If the problem of the linear chain is studied in EMA, the cosine band dispersion is replaced by a parabolic one with the same effective mass at $k = 0$. Thus, in finite chains, it is clear that in EMA the energy of the lowest states would be almost exact in the limit of long chains compared to ETB results because the energy dispersion is well approximated by a parabola at low energy. But for upper levels, or even for the lowest ones in the limit of small chains, EMA overestimates the confinement because the parabolic band acquires too much dispersion compared to the ETB band. Thus we conclude on a general ground that EMA (and $\boldsymbol{k} \cdot \boldsymbol{p}$) is only exact in the limit of large nanostructures, when the confinement energy is small, in an energy range where the band remains parabolic.

2.2 Computational Techniques

This section is concerned with the description of methods for calculating the electronic structure of semiconductor heterostructures and nanostructures. Basic principles of the methods have been presented in Chap. 1. Here we concentrate on more technical aspects of the calculations, specific to confinement effects.

2.2.1 $k \cdot p$ Method and Envelope Function Approximation

There exists extensive literature on how to calculate the electronic structure of quantum confined semiconductor structures on the basis of the $k \cdot p$ method and using the envelope function approximation for the eigenstates. An entire book would not be sufficient to describe all the work done in this field. Many levels of approximation have been used, from the simplest one-band effective mass approximation (EMA) to a multiband approach [85, 86, 92], or by simplifying the Schrödinger-like equation using approximate cylindrical or spherical symmetry of the Hamiltonian. All these approximations are not always completely justified, and in any case are restricted to the treatment of a small number of problems. In this section, we present the numerical method of G.A. Baraff and D. Gershoni [104, 105] for solving the multiband envelope functions in 0D, 1D and 2D systems composed of different types of semiconductors whose properties, e.g. alloy composition and strain, may vary. The method assumes that these material properties only change discontinuously across perpendicular planes. This is not a severe restriction since for problems where a quantity varies continuously one can use a fine mesh of planes. The technique used is Fourier-series expansion of the envelope functions.

We only consider the case of direct gap semiconductors, where the band edges are at $k = 0$. The method presented here is particularly suitable to treat III–V and II–VI semiconductor heterostructures and nanostructures made by epitaxial techniques, such as AlGaAs/GaAs, InGaAs/GaAs or InGaAsP/InP. In the $k \cdot p$ method with the envelope function approximation, the wave function in each compositionally homogeneous region of the structure is assumed to be of the form [87]

$$\Psi(r) = \sum_n u_n(r)\phi_n(r) , \tag{2.24}$$

where the $u_n(r)$ are Bloch waves at $k = 0$ for the material in a particular region. The $\phi_n(r)$ are the envelope functions, and the summation is restricted to bands close to the gap. As shown in Sect. 1.3.3, in general, it is required to include eight Bloch functions, two for the s-like conduction band minimum (including spin) and six for the p-like valence band maximum. We have seen in Sect. 2.1.1 that in one-band EMA the envelope function is solution of a Schrödinger-like equation where the electron mass in the kinetic operator is replaced by the effective mass m^*. Generalizing this in the multiband approximation, the envelope functions are now governed by eight coupled differential equations which can be written in the general form

$$\sum_n H_{mn}\phi_n(r) = \varepsilon \phi_m(r) . \tag{2.25}$$

where H includes operators acting on the envelope functions. In the basis of the eight Bloch waves $|s \uparrow\rangle, |x \uparrow\rangle, |y \uparrow\rangle, |z \uparrow\rangle, |s \downarrow\rangle, |x \downarrow\rangle, |y \downarrow\rangle, |z \downarrow\rangle$ where the arrows indicate the spin, the matrix H can be written as

$$H = \begin{vmatrix} G & \Gamma \\ -\Gamma^* & G^* \end{vmatrix} , \tag{2.26}$$

where G and Γ are both 4×4 matrices. Following Kane [80, 81], we have

$$G = G_1 + G_2 + G_{\text{so}} , \tag{2.27}$$

where

$$G_1 = \begin{vmatrix} \varepsilon_c + \frac{\hbar^2 k^2}{2m_0} & iPk_x & iPk_y & iPk_z \\ -iPk_x & \varepsilon_v + \frac{\hbar^2 k^2}{2m_0} - \Delta/3 & 0 & 0 \\ -iPk_y & 0 & \varepsilon_v + \frac{\hbar^2 k^2}{2m_0} - \Delta/3 & 0 \\ -iPk_z & 0 & 0 & \varepsilon_v + \frac{\hbar^2 k^2}{2m_0} - \Delta/3 \end{vmatrix} , \tag{2.28}$$

$$G_2 = \begin{vmatrix} A'k^2 & Bk_y k_z & Bk_x k_z & Bk_x k_y \\ Bk_y k_z & L'k_x^2 + M(k_y^2 + k_z^2) & N'k_x k_y & N'k_x k_z \\ Bk_x k_z & N'k_x k_y & L'k_y^2 + M(k_x^2 + k_z^2) & N'k_y k_z \\ Bk_x k_y & N'k_x k_z & N'k_y k_z & L'k_z^2 + M(k_x^2 + k_y^2) \end{vmatrix} , \tag{2.29}$$

and

$$G_{\text{so}} = -\frac{\Delta}{3} \begin{vmatrix} 0 & 0 & 0 & 0 \\ 0 & 0 & i & 0 \\ 0 & -i & 0 & 0 \\ 0 & 0 & 0 & 0 \end{vmatrix} . \tag{2.30}$$

The matrix Γ is

$$\Gamma = -\frac{\Delta}{3} \begin{vmatrix} 0 & 0 & 0 & 0 \\ 0 & 0 & 0 & -1 \\ 0 & 0 & 0 & i \\ 0 & 1 & -i & 0 \end{vmatrix} . \tag{2.31}$$

ε_c and ε_v are the band edge energies and Δ is the spin orbit splitting at the top of the valence band. The parameter P is proportional to the momentum matrix element between $|s\rangle$ and $|x\rangle$

$$P = -i\frac{\hbar}{m_0} \langle s | p_x | x \rangle , \tag{2.32}$$

which also defines to the optical matrix elements between the conduction band and the valence band (one must be careful that many definitions of the parameter P can be found in the literature). The parameters A', B, L', M, and N' are defined in [80, 81] (L' and N' differ from L and N of the 6×6 Hamiltonian (1.187) which describes only the valence band). They are all defined in terms of experimental data such as the bandgap $\varepsilon_g = \varepsilon_c - \varepsilon_v$, the conduction band effective mass m_e^*, the heavy ($m_{hh}^*(ijk)$) and light ($m_{lh}^*(ijk)$) hole effective masses in the (ijk) direction and the split-off band effective mass m_{so}^*, or in terms of the Luttinger [87] parameters γ_1, γ_2 and γ_3

$$\frac{m_0}{m_{hh}^*(100)} = \gamma_1 - 2\gamma_2 \quad \frac{m_0}{m_{lh}^*(100)} = \gamma_1 + 2\gamma_2 ,$$

$$\frac{m_0}{m_{hh}^*(111)} = \gamma_1 - 2\gamma_3 \quad \frac{m_0}{m_{lh}^*(111)} = \gamma_1 + 2\gamma_3 ,$$

$$\frac{m_0}{m_e^*} = \frac{2m_0}{\hbar^2}\left(A' + \frac{P^2(\varepsilon_g + 2\Delta/3)}{\varepsilon_g(\varepsilon_g + \Delta)}\right) ,$$

$$\frac{m_0}{m_{so}^*} = \gamma_1 + \frac{2m_0 P^2 \Delta}{3\hbar^2 \varepsilon_g(\varepsilon_g + \Delta)} ,$$

$$L' = -\frac{\hbar^2}{2m_0}(1 + \gamma_1 + 4\gamma_2) + \frac{3P^2}{3\varepsilon_g + \Delta} ,$$

$$M = -\frac{\hbar^2}{2m_0}(1 + \gamma_1 - 2\gamma_2) ,$$

$$N' = -\frac{3\hbar^2}{m_0}\gamma_3 + \frac{3P^2}{3\varepsilon_g + \Delta} . \tag{2.33}$$

In the case of strained crystals, it is possible to add extra terms in the 8×8 Hamiltonian using deformation potentials, i.e. terms which account for the variation of the band edge energies associated with elastic strains [106]. It is also possible to include piezoelectric fields [107].

In the set of differential equations (2.25), the symbol k_j is interpreted as the differential operator

$$k_j \rightarrow \frac{1}{i}\frac{\partial}{\partial x_j} . \tag{2.34}$$

Then one assumes that the system is periodic with periods X, Y, Z along x, y, z directions respectively. The definition of these dimensions is quite natural in the case of superlattices for example. In the case of isolated heterostructures and nanostructures, one must choose X, Y, Z large enough to avoid interactions between the structures. The envelope functions which are solutions of (2.25) are expanded in 1D, 2D or 3D Fourier series according to whether the system has spatial variation in 1D, 2D or 3D [104, 105]. Another possibility is to use a finite difference method to solve the differential equations. The system is divided into different regions which define a mesh of perpendicular planes. The parameters of the Hamiltonian matrix are constant in each region but differ from region to region. A special attention must be given to the problem of matching envelope functions at the boundaries. It was shown [104, 105] that (2.25) leads to an hermitic set of equations if in every term in which a material parameter Q and a derivative both appear, one makes the replacement

Excited state 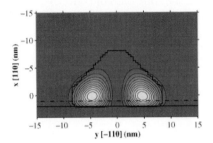   Ground state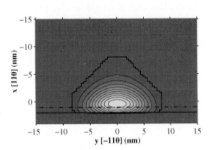

Fig. 2.6. Isosurface plots of the charge densities of the electron ground and first excited states just below the surface of a cleaved InAs box embedded in GaAs [108]. The InAs dot has a truncated pyramid like shape with a 20 nm [100] × [010] square base and {110} faces

$$Q\frac{\partial}{\partial x_j} \to \frac{1}{2}\left[Q(\boldsymbol{r})\frac{\partial}{\partial x_j} + \frac{\partial}{\partial x_j}Q(\boldsymbol{r})\right] ,$$

$$Q\frac{\partial}{\partial x_j}\frac{\partial}{\partial x_k} \to \frac{1}{2}\left[\frac{\partial}{\partial x_j}Q(\boldsymbol{r})\frac{\partial}{\partial x_k} + \frac{\partial}{\partial x_k}Q(\boldsymbol{r})\frac{\partial}{\partial x_j}\right] , \qquad (2.35)$$

where the spatial dependence $Q(\boldsymbol{r})$ is written in terms of a step function Θ

$$Q(x) = Q_{\mathrm{I}} + (Q_{\mathrm{II}} - Q_{\mathrm{I}})\Theta(x - x_0) , \qquad (2.36)$$

for an interface between materials I and II at $x = x_0$. Note that the derivative of a step function produces a delta function which gives rise to a term in the elements of H_{mn} leading to the usual condition that the normal component of the current must be continuous across the interface [104, 105].

After expansion of the envelope functions as Fourier series, (2.25) is reduced to a simple eigenvalue problem, which after projection leads to a set of linear equations whose size is equal to eight times the number of plane waves in the series. The main advantage of the method is that complex problems can be solved, with heterostructures and nanostructures in a wide range of sizes (the limitations of $\boldsymbol{k} \cdot \boldsymbol{p}$ are discussed in Sect. 2.3). For example, we show in Fig. 2.6 the lowest electron states in a cleaved InAs quantum dot embedded in GaAs calculated including the effect of strains on the electronic structure [108].

2.2.2 Tight Binding and Empirical Pseudopotential Methods

The empirical tight binding method (ETB) and the empirical pseudopotential method (EPM) have been basically described in Sects. 1.3.1 and 1.3.2, respectively. These methods are designed to make the best possible approximation to the bulk semiconductor Hamiltonian in the whole Brillouin zone. They involve adjustable parameters that are fitted to experimental data or ab initio band structures. These parameters are then transferred to the nanostructures with appropriate boundary conditions. The better the bulk description and boundary conditions, the better the electronic structure we expect in nanostructures. In the following, we discuss how to achieve this, insisting on the particular case of spherical Si nanocrystals with their surface passivated by hydrogen atoms to saturate the dangling bonds.

Empirical Tight Binding. We have seen in Sect. 2.1.5 that it is important to have a good description of the bands not only over the whole Brillouin zone, but also near the band edges, because ETB and $\bm{k} \cdot \bm{p}$ must be equivalent in large nanostructures. In Sect. 1.3.1, we have presented such a good band structure for Si using a sp^3 ETB model [64] including up to third nearest neighbor interactions and three center integrals (Table 1.1). Other possibilities are to use a $sp^3d^5s^*$ model restricted to first nearest neighbor interactions [66] or a non orthogonal ETB model [109] which give a band structure of similar quality.

To achieve good boundary conditions, it is important to obtain the best possible description of the surface or of the interface. In the case of Si nanocrystals passivated by hydrogen [64], Si-H parameters have been fitted on the SiH_4 experimental gap and charge transfer calculated within LDA. H atoms are described by their 1s orbitals and Si-H parameters are restricted to first nearest neighbor interactions (Table 1.1, Sect. 2.1.5). In the case of interfaces between two semiconductors, e.g. GaAs/AlGaAs, one possible procedure is to fit the ETB parameters in each semiconductor separately, and then to shift the intra-atomic energies of one semiconductor with respect to the other in order to match the experimental band offsets. The inter-atomic terms between the two semiconductors are then estimated using empirical rules [110].

Empirical Pseudopotentials. The application of the EPM to band structure has been reviewed in Sect. 1.3.2. The application of EPM to semiconductor nanostructures has been mainly developed by the group of A. Zunger [76, 111–117]. In the usual EPM, the pseudopotential $V(\bm{G})$ is defined only on the discrete bulk reciprocal lattice vectors. The description of finite quantum dots requires a continuous form $V(\bm{q})$. For Si, Wang and Zunger [76] used a local pseudopotential of the form

$$V_{Si}(q) = a_1(q^2 - a_2)/\left(a_3 \exp(a_4 q^2) - 1\right) \ . \tag{2.37}$$

A fit to the bulk band structure (shown in Fig. 1.8, Sect. 1.3.2), effective masses and the work function gives $a_1 = 0.2685$, $a_2 = 2.19$, $a_3 = 2.06$ and $a_4 = 0.487$ in atomic units. The hydrogen pseudopotential was obtained by fitting the local density of states of H-covered surfaces obtained from experiments and LDA results [76]

$$V_H(q) = -0.1416 + 9.802 \times 10^{-3} q + 6.231 \times 10^{-2} q^2 - 1.895 \times 10^{-2} q^3$$
when $q \leq 2$,

$$V_H(q) = 2.898 \times 10^{-2}/q - 0.3877/q^2 + 0.9692/q^3 - 1.022/q^4$$
when $q > 2$. (2.38)

The atomic positions can be obtained using ab initio calculations or using surface relaxation models. Having determined the pseudopotentials and the atomic coordinates $\boldsymbol{R}_i$, it remains to solve the single particle Schrödinger equation

$$\left(-\frac{\hbar^2}{2m_0}\Delta + \sum_i V_i(|\boldsymbol{r} - \boldsymbol{R}_i|)\right)\phi_j(\boldsymbol{r}) = \varepsilon_j \phi_j(\boldsymbol{r}),$$ (2.39)

where $V_i(|\boldsymbol{r}|)$ is the atomic pseudopotential on the atom i. The nanostructure wave functions are expanded in a basis of plane waves

$$\phi_j(\boldsymbol{r}) = \sum_{\boldsymbol{G}} B_j(\boldsymbol{G}) e^{i\boldsymbol{G}\cdot\boldsymbol{r}}.$$ (2.40)

The cutoff energy used with the pseudopotentials of Si and H is 4.5 Ryd. The transformation between $\phi_j(\boldsymbol{r})$ on a real space grid and $B_j(\boldsymbol{G})$ on a reciprocal space grid is done by numerical Fast Fourier Transform (FFT).

Numerical Methods to Solve Large Eigenvalue Problems. A considerable advantage of empirical methods like ETB or EPM compared to ab initio ones like LDA is that the Schrödinger equation has not to be solved self-consistently. Thus it only remains to calculate the eigenvalues and the eigenstates of the Hamiltonian matrix. However, even with this huge simplification, the size of the matrix directly scales with the number N of atoms in the system and one is rapidly facing computational limits. For example, in EPM, the description of a nanocrystal containing 1315 Si atoms and 460 H atoms requires $100 \times 100 \times 100$ real space FFT grid points [76]. In ETB, with a sp^3 basis for Si and s for H, the size of the matrix is 5720 if spin–orbit coupling is omitted, and is doubled otherwise. Even if the size of the problem in ETB is much smaller than in EPM, in both cases, direct diagonalization or conventional variational minimization methods based on (2.39) are impractical [118], mainly because one is forced to calculate all occupied eigenstates starting from the lowest one (they usually scale as N^3). Thus other approaches are required to solve these large eigenvalue problems.

In ETB, when the shape of the quantum dot is relatively simple (e.g. spheres or cubes), it is often possible to work in the irreducible representations of the point group which characterizes the symmetry of the system. According to Wigner's theorem, this leads to a block-diagonal Hamiltonian. Each block can be diagonalized separately, and the eigenstates of degenerate representations can be deduced using simple symmetry operations. In the best cases, one can gain a factor up to ten on the size of the largest matrix to diagonalize.

Another considerable simplification is that in many cases we are interested only in the highest occupied and in the lowest unoccupied states, around the gap. A few states near the gap can be calculated using a block Lanczos algorithm [119, 120] on $(H - \sigma I)^{-1}$, or a conjugate gradients algorithm [118] on $(H - \sigma I)^2$ (folded spectrum method [76]). The folding energy σ is set in the gap just above the bulk valence band maximum or below the bulk conduction band minimum to directly catch the highest valence or lowest conduction states. In ETB, Jacobi (diagonal) or incomplete Cholesky factorizations [119] ($LL^\dagger$) of $(H - \sigma I)^2$ can be used as pre-conditioners for the conjugate gradients [64]. In the latter case, the incomplete Cholesky factorization is performed on the part of $(H - \sigma I)^2$ having the sparsity pattern of H. Although crude, this pre-conditioner can save up to 75% of the iterations needed to reach convergence depending on the problem. In EPM, preconditioning of the conjugate gradient scheme is described in [121]. Using these methods, EPM can be applied to quantum dots up to a size of 4 nm, and ETB up to 15 nm.

In the case of large semiconductor nanostructures (size $\gtrsim$ 10 nm) embedded in another semiconductor, when the two materials (I, II) have relatively close properties (e.g., InAs quantum dot in GaAs), it is possible to develop the nanostructure wave functions in the basis of bulk states $u_{n,\boldsymbol{k}}^\alpha(\boldsymbol{r})$ where n denotes the band, $\boldsymbol{k}$ the wave vector and α the material (I or II). The number of bulk states required for the convergence may be actually quite small, allowing to treat large systems containing millions of atoms. This method has been mainly designed for EPM calculations [122], but it can be used similarly in ETB.

2.2.3 Density Functional Theory

Only a small number of LDA calculations have been applied to the electronic properties of nanostructures, mainly to Si wires and quantum dots. The main limitation is the computation time which increases rapidly with the number of atoms. Typically, LDA calculations using plane wave basis are limited to a maximum of 50 atoms [123–126], and more recently to 363 atoms using soft pseudopotentials and improved algorithms [127]. Using a basis of atomic orbitals more adapted to finite systems and using clusters with high symmetry, it is possible to treat up to 1000 atoms [128] (more elaborate treatments which include self-energy corrections and excitonic effects will be discussed in Chap. 4).

Another problem in LDA is that the bandgap of bulk semiconductors is underestimated, because LDA – and DFT in general – is not applicable to the calculation of excited states (Sect. 1.2). However, optical spectra of bulk semiconductors are improved in LDA if a rigid shift is applied to the conduction states with respect to the valence ones to get the correct gap. Thus, a common approximation is to apply the same shift to the LDA gap of nanostructures. Justifications of this empirical procedure are discussed in Chap. 4.

2.3 Comparison Between Different Methods

The most widely used techniques to calculate the electronic levels in nanostructures are the EMA and its extension the multiband $\bm{k}\cdot\bm{p}$ method. They have been particularly successful in the case of heterostructrures [86, 92]. Remarkable results have also been obtained in describing the absorption [129], hole burning [130] and photoluminescence excitation spectra [131] in CdSe nanocrystals. However, it is well-kown that $\bm{k}\cdot\bm{p}$ has intrinsic limitations [112–114] which we analyze in this section. We also compare the predictions of the different methods to calculate quantum confinement energies.

We consider the case of Si quantum dots but conclusions can be generalized to other semiconductors. We first compare ETB and $\bm{k}\cdot\bm{p}$ results for spherical and cubic Si dots presented in Fig. 2.7. The energy of the highest occupied state and lowest unoccupied state are plotted versus the effective diameter d of the dot, which for a cube is the diameter of the sphere with the same volume. The ETB sp^3 model [64] applied here is described in Sect. 1.3.1: it includes up to third nearest neighbor interactions and three center integrals (Table 1.1). Spin–orbit coupling is taken into account. Since Si has an indirect bandgap, we assume in $\bm{k}\cdot\bm{p}$ uncoupled valence band maximum and conduction band minima. The valence band is described with a six band

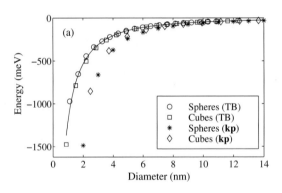

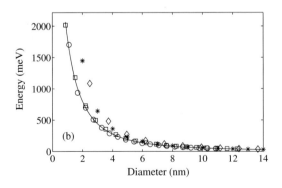

Fig. 2.7. Energy of the highest occupied state (**a**) and of the lowest unoccupied state (**b**) in spherical and cubic ((100)x(010)x(001) faces) Si dots calculated in ETB and in $\bm{k}\cdot\bm{p}$ [64]. The energy is defined with respect to top of the bulk valence band (**a**) or to the bottom of the bulk conduction band (**b**). Fit to ETB results (*straight line*)

$\boldsymbol{k}\cdot\boldsymbol{p}$ model (Dresselhaus–Kip–Kittel Hamiltonian [83]) taking into account the large valence band anisotropy of Si and spin–orbit coupling. The input parameters are the three Luttinger parameters γ_1, γ_2, γ_3, and spin–orbit splitting Δ. The six conduction band minima along ΓX directions are assumed uncoupled and are described in single band EMA. The input parameters are the longitudinal and transverse effective masses m_l^* and m_t^*, and the bulk bandgap energy $\varepsilon_g(\infty)$. Because $\boldsymbol{k}\cdot\boldsymbol{p}$ is not an atomistic description, there is no thorough way to provide a potential consistent with ETB boundary conditions. Thus an infinite barrier is assumed in $\boldsymbol{k}\cdot\boldsymbol{p}$ calculations, and its position is chosen in such a way that the volume of the system is equal to the total volume occupied by the Si atoms.

For consistent comparison, the Luttinger parameters and conduction band effective masses are deduced from the ETB band structure, even if they are very close to experimental ones as discussed in Sect. 1.3.1. The $\boldsymbol{k}\cdot\boldsymbol{p}$ and ETB valence bands of bulk Si are shown in Fig. 2.8. In $\boldsymbol{k}\cdot\boldsymbol{p}$ the valence bands tend to acquire too much dispersion because they miss couplings with other states which are not included in the model [114]. The mean difference between ETB and $\boldsymbol{k}\cdot\boldsymbol{p}$ valence bands is less than 10% in a ~ 250 meV range. In the same way, the conduction band acquires too much dispersion compared to ETB, especially in the transverse directions (not shown). The mean difference between ETB and EMA conduction bands is less than 10% in a ~ 200 meV range.

There is striking evidence for over-confinement of $\boldsymbol{k}\cdot\boldsymbol{p}$ compared to ETB in Fig. 2.7. As shown in [64], $\boldsymbol{k}\cdot\boldsymbol{p}$ predictions obviously get worse from films (2D) to wires (1D) and dots (0D). In spherical Si dots, the error on the confinement energy $\varepsilon_g(d) - \varepsilon_g(\infty)$ is larger than 25% for $d < 8.5$ nm, and 50% for $d < 4.5$ nm. It is still 15% for $d \simeq 12$ nm. The use of a multiband semi-empirical method such as ETB or EPM is therefore recommended in the 5-12 nm range for Si clusters. Indeed, the sp^3 ETB model is not more difficult to solve than the six band $\boldsymbol{k}\cdot\boldsymbol{p}$ model in this range, when valence band anisotropy and spin–orbit coupling are taken into account.

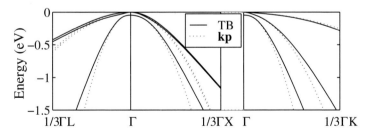

Fig. 2.8. Bulk Si valence band structure within six band $\boldsymbol{k}\cdot\boldsymbol{p}$ (*dashed line*) and ETB models (*straight line*). ETB Luttinger parameters are used for consistent comparison

There are basically three reasons why $\mathbf{k} \cdot \mathbf{p}$ overestimates confinement energies with respect to ETB. The main reason is that in $\mathbf{k} \cdot \mathbf{p}$ valence bands and conduction bands tend to acquire too much dispersion far from the edges (Fig. 2.8). Over-confinement gets worse in small nanostructures that couple Bloch states far from the extrema, and for higher excited states. For example, we present in Fig. 2.9 the projection of the three lowest electron states of A_1 symmetry in a spherical Si dot of diameter $d = 4.89$ nm. The confinement of the wave functions in real space leads to a spread of their projection in momentum space. In Fig. 2.9, the width of the main peak for the lowest state is proportional to $1/d$ and extends over the whole Brillouin zone in the smallest nanostructures. Higher excited states, that have nodal planes in the wavefunction, thus exhibit multiple peaks that extend further in reciprocal space, beyond the range of validity of $\mathbf{k} \cdot \mathbf{p}$ and EMA descriptions of the bulk dispersion curves.

The next reason is the coupling between bulk bands in nanostructures [114]. Indeed, $\mathbf{k} \cdot \mathbf{p}$ assumes that hole states can be decomposed on the six highest bulk valence bands. ETB calculations show that the hole states have non zero projections on other bands, in particular on conduction bands [64]. Interband coupling increases with decreasing nanostructure size.

The last reason is the lack of correct boundary conditions in $\mathbf{k} \cdot \mathbf{p}$. ETB calculations show [64] that electron and hole states may be partly delocalized over hydrogen atoms in small dots. Therefore, hydrogen atoms will contribute to the confinement energy.

We now proceed to the comparison between ETB, EPM and LDA techniques concerning the prediction of the bandgap $\varepsilon_g(d)$ versus size d (Fig. 2.10). ETB [64], EPM [76] and LDA [128] results for Si are in very good agreement. Note that LDA values include a rigid shift of 0.6 eV since it is known that LDA underestimates the bulk bandgap by this amount.

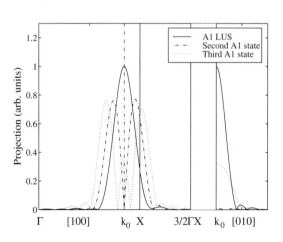

Fig. 2.9. Left part: Bloch decomposition of the three lowest unoccupied states (LUS) of A_1 symmetry in a Si spherical dot of diameter $d = 4.89$ nm. The decomposition is performed along ΓX on the two lowest conduction bands. Results are shown in an extended zone scheme, the first conduction band being on the left of the X point, the second one on the right. Right part: same on the first conduction band in the transverse direction

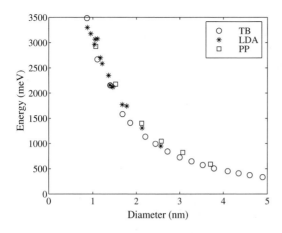

Fig. 2.10. Confinement energy $\varepsilon_g(d) - \varepsilon_g(\infty)$ in spherical Si dots calculated in sp^3 tight binding (TB) [64], using empirical pseudopotentials (PP) [76] and in LDA [128]

Similar agreement is obtained between ETB, EPM and corrected LDA in other semiconductors [132–134]. However, one can wonder why semi-empirical techniques should provide quantitative estimates of $\varepsilon_g(d)$ [135]. This question is of primary importance since there exist other ETB calculations for Si nanocrystals which provide significantly lower values for the gap [136, 137]. The basic point is that semi-empirical calculations are based on the postulate of transferability of the parameters from the known bulk band structure to the unknown nanostructure case. Thus, an essential criterion by which a particular semi-empirical model can be judged is how well it describes the bulk band structure. In this regard ETB and EPM models of [64, 76] both give extremely good fits to the Si band structure. On the contrary, the sp^{3}s* model of [136, 137] gives in comparison a very poor description of the conduction band which is much too flat, and consequently must underestimate the bandgap, as indeed it does.

A second criterion for validity of semi-empirical techniques is that boundary conditions must be correctly simulated in the model. In ETB [64] and EPM [76], the surfaces of the Si nanocrystals are passivated by hydrogen atoms described by appropriate parameters fitted on experimental data or on LDA calculations. This procedure is justified by the agreement with the gaps calculated in LDA since the LDA Hamiltonian plus a rigid shift of 0.6 eV of the conduction states with respect to the valence states gives a quite accurate representation of the bulk Hamiltonian and also of the Si-H terminations.

2.4 Energy Gap of Semiconductor Nanocrystals

The variation with the diameter d of the energy of the lowest unoccupied state and of the highest occupied state in Si spherical nanocrystals shown in Fig. 2.7 can be well fitted by the following expressions:

$$\varepsilon_\mathrm{v}(d) = \varepsilon_\mathrm{v}(\infty) - \frac{1}{a_\mathrm{v}d^2 + b_\mathrm{v}d + c_\mathrm{v}},$$

$$\varepsilon_\mathrm{c}(d) = \varepsilon_\mathrm{c}(\infty) + \frac{1}{a_\mathrm{c}d^2 + b_\mathrm{c}d + c_\mathrm{c}}, \quad (2.41)$$

where $\varepsilon_\mathrm{v}(\infty)$ and $\varepsilon_\mathrm{c}(\infty)$ are the bulk band edges and $a_\mathrm{c}, b_\mathrm{c}, c_\mathrm{c}, a_\mathrm{v}, b_\mathrm{v}, c_\mathrm{v}$ are fitting parameters given in Table 2.2. The mean deviation between the curves and the calculated points is less than 1%. In the limit of large diameters d, the expressions (2.41) behave like $1/d^2$ as predicted in EMA or $\mathbf{k} \cdot \mathbf{p}$ so that they can be considered as valid over the whole range of sizes.

We have performed similar ETB calculations for spherical nanocrystals made in series of III–V and II–VI semiconductors. The variation of the band edges versus size has been also fitted with expressions (2.41) and the corresponding parameters are given in Table 2.2. We have considered that the passivation of the surfaces is such that surface states do not play a role in the confinement energy for nanocrystal with a diameter larger than 2 nm. Thus each dangling bond at the surface is passivated by a pseudo hydrogen atom which repels surface states far from the band edges (for details see [134]).

Comparisons between predicted confinement energies and experimental results are made in [132, 134, 139], and in Sect. 4.5.3 for Si nanocrystals.

Table 2.2. Fits to the energy (eV) of the highest occupied state (HOS) and of the lowest unoccupied state (LUS) in various semiconductor nanocrystals with diameter d (nm) ($sp^3d^5s^*$ ETB model except [†] : sp^3 ETB model). All semiconductors have a zinc–blende structure, except CdSe and ZnO which have a wurtzite structure

Compound	HOS (Valence band)			LUS (Conduction band)			ΔE_g (d = 2 nm)
	a_v	b_v	c_v	a_c	b_c	c_c	
Si[134]	0.15001	0.54779	0.07477	0.20321	0.05673	0.17815	1.463
Si[†][64]	0.16041	0.54395	0.22650	0.17110	0.21798	0.15485	1.295
Ge[134]	0.06996	0.55904	0.07485	0.08368	0.18568	0.22206	1.754
Ge[†][132]	0.06603	0.42691	0.16812	0.08429	0.20154	0.35840	1.689
AlP[134]	0.20639	0.61983	0.08272	0.25820	0.00192	0.23835	1.244
GaP[134]	0.18845	0.64973	0.07095	0.33262	−0.11812	0.36669	1.147
InP[134]	0.16151	0.65387	0.05416	0.04535	0.23340	0.20816	1.674
AlAs[134]	0.17888	0.51807	0.14764	0.24507	−0.01397	0.26441	1.341
GaAs[134]	0.12307	0.55643	0.10362	0.03946	0.22988	0.21366	1.796
InAs[134]	0.10558	0.67644	0.04419	0.01351	0.23309	0.12564	2.101
InAs[†][138]	0.12553	0.65139	0.00829	0.01078	0.24406	0.22099	1.881
AlSb[134]	0.12912	0.89943	−0.06633	0.24907	−0.08260	0.30550	1.297
GaSb[134]	0.08017	0.63268	0.07146	0.02650	0.33745	0.09540	1.747
InSb[134]	0.08177	0.74084	0.01190	0.00754	0.19792	0.13959	2.329
CdSe[†][139]	0.22573	0.63567	−0.13567	0.08292	0.20721	0.33300	1.417
ZnO[†][140]	0.69299	−0.11936	0.30721	0.13745	0.28596	−0.06061	1.294

Results for InAs nanocrystals are discussed in detail in Sect. 4.6.1, showing a good agreement between theory and experiments over a wide range of sizes.

2.5 Confined States in Semiconductor Nanocrystals

In this section, we briefly discuss the nature of the electron and holes states in nanocrystals.

2.5.1 Electron States in Direct Gap Semiconductors

We begin with the simplest case of electron states in semiconductors characterized by a single conduction band minimum at $\boldsymbol{k} = 0$, such as in most III–V and II–VI semiconductors. We plot in Fig. 2.11 the evolution of the lowest levels in spherical ZnO clusters with wurtzite lattice structure. The energies have been calculated in $sp^3d^5s^*$ tight binding including spin–orbit coupling [140]. The lowest levels are grouped into multiplets which correspond to the 1S, 1P, 1D, 2S and 1F states predicted in EMA for the infinite spherical well (Sect. 2.1.4). The multiplicity and the ordering of the levels is the same as in EMA, but the confinement energies are overestimated in EMA for the reasons discussed before. The small energy splittings within each multiplet come from the spin–orbit coupling and from the fact that the symmetry of the system is not spherical since it must be compatible with the symmetry of the lattice of the ZnO crystal.

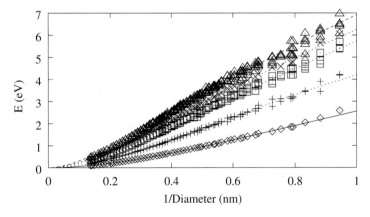

Fig. 2.11. Energy of the lowest states in the conduction band of ZnO spherical clusters. The zero of energy corresponds to the bottom of the bulk ZnO conduction band. The lines are analytical fits of the 1S, 1P, 1D, 2S and 1F levels in order of increasing energy using the expression (2.41) with $a_{1S} = 0.137447$, $b_{1S} = 0.285964$, $c_{1S} = -0.0606154$, $a_{1P} = 0.0685045$, $b_{1P} = 0.145631$, $c_{1P} = 0.00965402$, $a_{1D} = 0.0464297$, $b_{1D} = 0.0741516$, $c_{1D} = 0.0572599$, $a_{2S} = 0.0373006$, $b_{2S} = 0.0722844$, $c_{2S} = 0.05072$, $a_{1F} = 0.030477$, $b_{1F} = 0.06066759$ and $c_{1F} = 0.0582783$

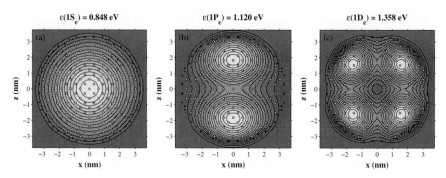

Fig. 2.12. Electron density for s (1S$_e$), p (1P$_e$) and d (1D$_e$) conduction states in an InAs nanocrystal of diameter $d = 6.4$ nm. The white dots represent In or As atoms, black dots pseudo-hydrogen atoms

Similar results are obtained in other semiconductors [138, 141, 142]. The identification of the states is confirmed by plotting the density of probability (i.e. the square of the wave function), as shown in Fig. 2.12 for a 6.4 nm InAs quantum dot. The orbitals are close to their EMA counterparts with angular components given by the spherical harmonics Y_{lm}, in particular for the lowest states. For levels higher in energy, it becomes difficult to identify the nature of the wave functions due to mixing between almost degenerate states.

2.5.2 Electron States in Indirect Gap Semiconductors

The situation for electrons in indirect gap semiconductors like Si is more complex, due to the multiplicity of the conduction band minima. In the following, we consider Si crystallites bounded by (100) equivalent planes of dimensions $L_x L_y L_z$ but similar results are obtained for spherical shapes. We first describe qualitatively the system in EMA and then we substantiate the results using sp^3 tight binding calculations [144].

In a cubic quantum dot made of a direct gap semiconductor with a single conduction band minimum (Fig. 2.13), the lowest confined state is non degenerate (S-like) and the next higher state is threefold degenerate (P-like). In Si, the conduction band is characterized by six equivalent (100) valleys at wave vectors $\bm{k}_{0l}$ ($l \in \{x, \bar{x}, y, \bar{y}, z, \bar{z}\}$ and $|\bm{k}_{0l}| = k_0 \approx 0.85(2\pi/a)$ where a is the bulk lattice parameter, x stands for the (100) valley and $\bar{x}$ for the (−100) valley) and by anisotropic effective masses (transverse mass $m_t = 0.19\ m_0$, longitudinal one $m_l = 0.92\ m_0$). The quantum confinement gives S, P states in each valley [64]. The confined states arising from the valley l are denoted $\langle n_x n_y n_z \rangle_l$ where n_i are integer quantum numbers (≥ 1). The confinement energy for states in valley x or $\bar{x}$ is

2.5 Confined States in Semiconductor Nanocrystals

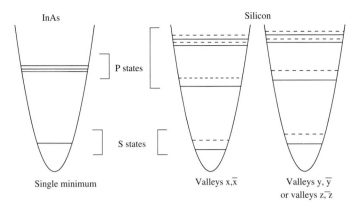

Fig. 2.13. Lowest electron states in a semiconductor nanocrystal with a single conduction band minimum such as InAs or in a Si nanocrystal with six valleys. In Si, the P levels are split due to the anisotropy of the effective masses. The intervalley coupling lifts the degeneracies between x and $\bar{x}$ (resp. y and $\bar{y}$, z and $\bar{z}$) valleys (*dashed lines*). The degeneracy between the different valleys is also lifted when the shape of the nanocrystal is anisotropic

$$\frac{\hbar^2}{2m_l}\left(\frac{n_x\pi}{L_x}\right)^2 + \frac{\hbar^2}{2m_t}\left[\left(\frac{n_y\pi}{L_y}\right)^2 + \left(\frac{n_z\pi}{L_z}\right)^2\right], \tag{2.42}$$

the expressions for the other valleys being obtained by permutation of x, y and z. In each valley l, the ground S state corresponds to $\langle 111 \rangle_l$ and the first excited P states to $\langle 211 \rangle_l$, $\langle 121 \rangle_l$ and $\langle 112 \rangle_l$. Due to the strong anisotropy in the effective masses, the P levels are split in two groups. As shown in Fig. 2.13, the energy of $\langle 211 \rangle_x$ is lower than the one of $\langle 121 \rangle_x$ and $\langle 112 \rangle_x$. If the shape is slightly anisotropic ($L_x \neq L_y \neq L_z$) which is likely in real quantum dots, the degeneracy between x, y and z valleys is also lifted.

We can now compare with the results of tight binding calculations. We plot in Fig. 2.14 the energy of the 25 lowest electron levels in slightly anisotropic crystallites with $L_x = L-a$, $L_y = L$, $L_z = L+a$. The lowest group of six levels contains the S states of the different valleys. The next group of six levels contains P states derived from $\langle 211 \rangle_x$, $\langle 211 \rangle_{\bar{x}}$, $\langle 121 \rangle_y$, $\langle 121 \rangle_{\bar{y}}$, $\langle 112 \rangle_z$ and $\langle 112 \rangle_{\bar{z}}$. The third group contains the other P states as well as other excited states. The gap between the second and the third group comes from the anisotropy in the effective masses. Thus one recovers the simple EMA picture, even if EMA strongly overestimates the confinement energies. However, there is an important difference between EMA and tight binding results. In EMA, there is a complete degeneracy between valleys x and $\bar{x}$ (resp. y and $\bar{y}$, z and $\bar{z}$). In tight binding, and in other microscopic calculations, the degeneracy is lifted due to inter-valley couplings [91, 143]. For example, the coupling between states $\langle 111 \rangle_x$ and $\langle 111 \rangle_{\bar{x}}$ gives two states which, due to the symmetry between $\boldsymbol{k}$ and $-\boldsymbol{k}$ necessarily behave like $(\langle 111 \rangle_x + \langle 111 \rangle_{\bar{x}})/\sqrt{2}$ and

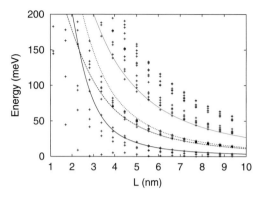

Fig. 2.14. Energy calculated in tight binding of the lowest unoccupied orbitals in the conduction band of Si crystallites bounded by (100) equivalent planes of dimensions $L_x = L - a$, $L_y = L$, $L_z = L + a$ where $a = 5.42$Å. The zero of energy corresponds to the ground state. Analytic fits (*line*) $\varepsilon = b/(L^2 + cL + d)$ of the energy ε of the 6th, 7th, 12th and 13th levels with respectively $(b, c, d) = (0.271, -2.91, 3.18)$, $(0.980, -1.24, 3.85)$, $(1.039, -1.76, 3.23)$ and $(2.593, -0.63, 5.61)$ (ε in eV, L in nm). Results for cubic or spherical nanocrystals are extremely close

$(\langle 111 \rangle_x - \langle 111 \rangle_{\bar{x}})/\sqrt{2}$. All these effects lead to a rich electronic structure, as depicted in Fig. 2.13. In symmetric dots such as cubes or spheres, the six lowest states are degenerate in EMA. In tight binding (not shown), these levels are split by inter-valley couplings into a A_1 level, a twofold degenerate E level and a threefold degenerate T_2 level in clusters with T_d symmetry (A_1, E and T_2 correspond to irreducible representations of the T_d group). More details are given in [64].

2.5.3 Hole States

The calculation of the hole states in quantum dots is more difficult due to the complexity of the bulk electronic structure composed of three bands (heavy hole, light hole and split-off) with anisotropic energy dispersions in $\boldsymbol{k}$-space [145]. For reasons discussed in Sect. 2.2.1, in general it is not possible to decouple the treatment of each band and one must use at least a 6×6 or a 8×8 $\boldsymbol{k} \cdot \boldsymbol{p}$ Hamiltonian [146]. But recent works have shown that the application of $\boldsymbol{k} \cdot \boldsymbol{p}$ to quantum dots may lead to severe discrepancies compared to more elaborate calculations like tight binding or empirical pseudopotentials [112–114, 139]. Therefore tight binding or empirical pseudopotentials must be recommended in the case of 0D systems, whereas in 2D or even 1D systems the performance of $\boldsymbol{k} \cdot \boldsymbol{p}$ methods may be sufficient [86]. In the following we briefly describe the nature of the highest hole states in the particular case of Si nanocrystals and we discuss the problem of the symmetry of the states in the general case.

Hole States in Si Nanocrystals. Tight binding calculations applied to spherical Si nanocrystals show that the nature of the highest hole states depend on the diameter d of the dot [64]. In fact, two situations must be considered depending on the importance of the energy splittings between the discrete levels compared to the spin–orbit splitting Δ at the top of the Si valence band (45 meV).

Without spin–orbit coupling, the states can be labeled with the irreducible representations of the T_d group. The highest hole states have a T_2 symmetry and are approximately s-like (no nodes) with protrusions along {111}-like directions. The next states below have a T_1 symmetry and are approximately p-like, with a nodal plane. Both T_2 and T_1 states are sixfold degenerate, including spin. The ordering of the levels does not change in the whole range of sizes.

When the spin–orbit coupling is included in the calculation, the T_2 and T_1 levels are each split into one fourfold (often labeled by a total momentum $J = 3/2$) and one twofold ($J = 1/2$) degenerate states. When Δ is larger than the splitting between T_2 and T_1 states ($d > 5$ nm), the highest level and the next one below remain s and p-like respectively but they are fourfold degenerate (Fig. 2.15). When Δ is smaller than the splitting between T_2 and T_1 states ($d < 5$ nm), the highest level is fourfold degenerate and s-like, followed by the other twofold degenerate s-like state, then by p-like states.

In III–V and II–VI semiconductors, the situation depends on the nature of the material. However, it seems that the highest level is always fourfold degenerate, corresponding to $J = 3/2$ states.

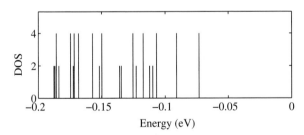

Fig. 2.15. Density of states (DOS) in the valence band of a spherical Si nanocrystal passivated by hydrogen atoms (diameter = 7.6 nm). The zero of energy corresponds to the top of the valence band of bulk Si

Symmetry of the Hole States. In the envelope function approximation, the confined states are written as products of an envelope function with a Bloch function. Recent works have been devoted to the calculation of these states in various types of semiconductor nanocrystals, and in particular to the determination of the symmetry of the envelope function of the highest states

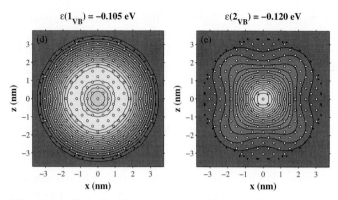

Fig. 2.16. Hole density for the two higher valence states (1_{VB} and 2_{VB}) in an InAs nanocrystal of diameter $d = 6.4$ nm. The two states are fourfold degenerate. The white dots represent In or As atoms, black dots pseudo-hydrogen atoms. The states are calculated in tight binding [138]

in the valence band. In the EMA with a single isotropic band which applies to electron states, we have seen in Sect. 2.1.4 that the angular dependence of the envelope function is given by spherical harmonics Y_{lm}, leading to s, p, d (...) orbitals. In the case of hole states, the $\bm{k} \cdot \bm{p}$ Hamiltonians have a cubic symmetry and l, m are no longer good quantum numbers. Thus the symmetry of the states is more complicated, as shown for example in Fig. 2.16 for the two highest hole states in an InAs quantum dot, even if the highest state is mostly s-like. Recently, $\bm{k} \cdot \bm{p}$ calculations have predicted that the highest hole state may have an envelope function with mostly a p-like symmetry, for example in InP [146, 147], which implies that the optical transition to the s-like conduction state would be dipole-forbidden (see Sect. 5.3.1). In fact, it appears that the ordering of the states in $\bm{k} \cdot \bm{p}$ is quite sensitive to the parameters [146]. In addition, pseudopotential [112–114] and tight binding [139] calculations predict the highest state with mostly a s-like symmetry and p-like states below in energy. The reasons why $\bm{k} \cdot \bm{p}$ calculations fail to predict the correct states are detailed in [114].

2.6 Confinement in Disordered and Amorphous Systems

This section is devoted to the problem of quantum confinement effects in disordered semiconductors such as amorphous silicon. It is well-known that in bulk materials the disorder induces the formation of localized states [148]. Thus disordered systems raise extremely interesting problems:

– Does the confinement induce a blue-shift of the energy gap in clusters of amorphous semiconductors and is it comparable to what is obtained for the crystalline material?
– What is the behavior of the localized states in this regard?

2.6 Confinement in Disordered and Amorphous Systems

To provide answers to these fundamental questions, we summarize the results of tight binding calculations on a-Si and a-Si:H nanoclusters [149]. a-Si is simulated by the well known Wooten–Winer–Weaire (WWW) model [150, 151] which is a periodic model with 4096 atoms per unit cell, representing an optimized distorted continuous random network with essentially fourfold coordination for each atom. The a-Si:H structure is also built from the WWW model by removing Si atoms where there is a strongly localized state and by saturating the dangling bonds with H atoms [149, 152]. The final hydrogen concentration is 8.3 %, which is in the range of experimental values. The clusters of a-Si and a-Si:H are obtained by selecting the atoms belonging to a sphere of a given diameter in the corresponding unit cell. The surface dangling bonds are saturated by hydrogen atoms. The atomic positions are relaxed using a Keating potential [153]. The main results of the calculations are discussed in [149].

A generally accepted picture of the electronic structure of a-Si is that it is still composed of valence and conduction bands separated by an energy gap but with band-tails of defect or disorder induced localized states extending into the gap. Applying this picture to nano-clusters, one expects the boundary conditions to have similar effects on the extended states in both crystalline silicon (c-Si) and a-Si clusters: quantization of the energy and resulting blue-shift. This will not be the case for the localized states belonging to the band-tails. Indeed, the analysis [149] of the variation of the energy levels versus size in a-Si clusters allows to classify the states into three categories:

- delocalized states, experiencing the full confinement effect as for c-Si
- strongly localized states with extension in space much smaller than the cluster diameter and energies deep in the gap, insensitive to the confinement effect and showing no blue-shift
- weakly localized states with extension in space of the order of the cluster diameter and energies near the gap limit, subject to an intermediate blue-shift.

To characterize the effect of the confinement, we have calculated the gap of clusters with 1 to 2.5 nm size. The gap is defined as the distance in energy between the highest occupied level and the lowest unoccupied one. Figure 2.17 shows the average gap versus size for clusters with randomly chosen center in the unit cell of a-Si and a-Si:H compared to the same quantity for c-Si clusters. In both cases, there is an important variation of the gap with size, the blue-shift for a-Si:H being surprisingly close to c-Si clusters. The gap for a-Si clusters is much smaller than for a-Si:H, in particular at large sizes.

To explain this behavior, we plot in Fig. 2.18 the statistical distribution of the gap for two cluster sizes, 2.2 and 1.2 nm. There is a substantial blue-shift in both cases, more important for a-Si than for a-Si:H. Furthermore, the larger a-Si clusters give rise to a two-peak distribution. The lower and upper peaks are, respectively, due to strongly and weakly localized or delocalized states. The relative intensity of the upper peak thus corresponds to the proportion

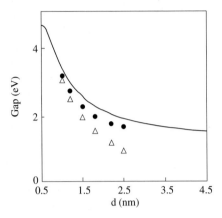

Fig. 2.17. Average gap of amorphous silicon clusters with randomized centres compared to crystallites: a-Si ($\triangle$), a-Si:H ($\bullet$), and c-Si (*straight line*)

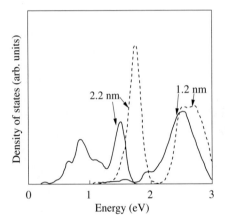

Fig. 2.18. Statistical distribution of gaps $\varepsilon_c - \varepsilon_v$ for 200 clusters with randomized centres, with 1.2 nm and 2.2 nm size: a-Si (*straight line*), a-Si:H (*dashed line*)

of clusters which do not contain strongly localized states. Thus the apparent blue-shift in a-Si clusters has two origins:

– the varying proportion of clusters with strongly localized states [149, 154]
– the normal confinement effect on the other states.

This is confirmed in Fig. 2.18 by the a-Si:H clusters (*dashed lines*), which show only the second type of behavior.

The disorder has also a profound impact on the optical properties: the radiative recombination rates for 2 nm clusters are two orders of magnitude higher in a-Si and a-Si:H than in c-Si [149].

3 Dielectric Properties

In this chapter we deal with the dielectric properties of semiconductor nanostructures. The realization of nanodevices usually requires to combine semiconductors with metals, insulators and molecules in a small region of space. The behavior of these systems strongly depends on the complex repartition of the electric field. Many interesting problems are related to dielectric properties: the current–voltage characteristics of a device, the binding energy of a dopant or an exciton, the energy of a carrier in an ultra-small capacitor, the optical properties and many others. Thus, their simulation at the nanometer scale becomes a critical issue for the development of nanotechnologies.

Simulation at the macroscopic or even the mesoscopic scale usually relies on the macroscopic electrostatic theory of dielectrics where the latter are described by their bulk macroscopic dielectric constant. One can wonder when this macroscopic approach breaks down as the size of the systems diminishes, since it is no longer valid at the molecular scale. Thus, in this chapter, we will briefly review the basic assumptions of the electrostatic theory of macroscopic dielectrics, and we will discuss its validity and its limitations (Sect. 3.1). We will show how the macroscopic approach can be used to describe the quantum mechanics of carriers in dielectrics (Sect. 3.2), for instance in simulators of micro- (and nano-) electronic devices. Then we will move to more accurate treatments (Sect. 3.3) in which the dielectric properties are derived from ab initio and semi-empirical electronic structure calculations. These microscopic methods require to calculate the full dielectric function $\epsilon(r, r')$, which is only possible for small systems. We will present applications of these methods to semiconductor quantum wells and quantum dots (Sect. 3.4). The results will allow us to define two useful quantities for nanostructures:

- a local dielectric constant which depends on the position in the nanostructure and which differs from the bulk one only in the vicinity of the surfaces
- an average dielectric constant, average of the previous quantity over the nanostructure volume, which varies with size of the nanostructure.

We will point out that these physically meaningful quantities must be derived taking into account the effect of the polarization charges at the surfaces or interfaces. Finally, Sect. 3.5 will deal with the important problem of the charging of a semiconductor island with free carriers.

3.1 Macroscopic Approach: The Classical Electrostatic Theory

In this section, we review the bases of the macroscopic electrostatic theory of dielectrics, and we see how it is usually implemented to study the Coulomb interaction between charged particles in small systems. Then we consider specific examples of problems relevant to the field of semiconductor nanostructures.

3.1.1 Bases of the Macroscopic Electrostatic Theory of Dielectrics

The electrostatics of macroscopic conductors and dielectrics is the object of many textbooks [155–158]. We will concentrate on dielectrics (insulators and semiconductors) where the effect of the confinement is expected to be the largest. However the same effects occur in very small metal clusters. The macroscopic theory is based on quantities defined as averages over small volumes, which however are large enough to average the microscopic fluctuations due to the atomic structure of the material. Of course, such a procedure is only valid for systems large compared to the characteristic length of these fluctuations. For bulk crystalline solids, the averaging volume is the unit cell. If these conditions are realized, one can write for instance the macroscopic electric field

$$\boldsymbol{E} = \overline{\boldsymbol{e}}, \tag{3.1}$$

where $\overline{\boldsymbol{e}}$ is the average of the microscopic field $\boldsymbol{e}$. The averaging procedure leaves the Maxwell's equations unchanged (Appendix B). In the absence of macroscopic magnetic fields, one can thus write

$$\boldsymbol{\nabla} \times \boldsymbol{E} = 0, \tag{3.2}$$

$$\boldsymbol{\nabla} \cdot \boldsymbol{E} = \frac{\overline{\rho}}{\epsilon_0}, \tag{3.3}$$

where $\overline{\rho}$ is the mean density of charge in the dielectric. If there is no external charge in the material, then the integral of $\overline{\rho}$ over the whole system remains equal to zero. In that case, $\overline{\rho}$ can be related to the polarization $\boldsymbol{P}$ defined as the dipole moment averaged per unit volume [155]

$$\overline{\rho} = -\boldsymbol{\nabla} \cdot \boldsymbol{P}, \tag{3.4}$$

which leads to

$$\boldsymbol{\nabla} \cdot (\epsilon_0 \boldsymbol{E} + \boldsymbol{P}) = 0. \tag{3.5}$$

When external charges are introduced in the system, their density ρ_{ext} must be added to the second term of (3.3) which leads to

$$\boldsymbol{\nabla} \cdot \boldsymbol{D} = \rho_{\text{ext}}, \tag{3.6}$$

where $\boldsymbol{D} = \epsilon_0 \boldsymbol{E} + \boldsymbol{P}$ is defined as the electric displacement. Equations (3.2) and (3.6) are the well-known macroscopic transpositions of the Maxwell's equations in dielectrics. To solve these equations, a relation between $\boldsymbol{P}$ and $\boldsymbol{E}$ (or $\boldsymbol{D}$ and $\boldsymbol{E}$) is required. In many cases, this relation is linear because the external fields are small compared to the internal molecular fields. In isotropic materials, $\boldsymbol{D}$ and $\boldsymbol{E}$ are simply proportional:

$$\boldsymbol{D} = \epsilon_M \epsilon_0 \boldsymbol{E} \ . \tag{3.7}$$

ϵ_M is the macroscopic dielectric constant of the material. If we introduce the electrostatic potential φ such that $\boldsymbol{E} = -\boldsymbol{\nabla}\varphi$, (3.2) is automatically satisfied. Equation (3.6) gives [155]:

$$\boldsymbol{\nabla} \cdot (\epsilon_M \epsilon_0 \boldsymbol{\nabla}\varphi) = -\rho_{\text{ext}} \ . \tag{3.8}$$

This equation leads to the usual Laplace equation and it remains valid in situations where $\epsilon_M(\boldsymbol{r})$ has a macroscopic space dependence. It can be solved using various numerical approaches, for example using a Green's function formalism [159].

In the following, we will be mainly interested in the Coulomb interaction between point charges in dielectrics. For instance, the interaction energy between two charges q and q' sitting at positions $\boldsymbol{r}$ and $\boldsymbol{r}'$ is $V(\boldsymbol{r},\boldsymbol{r}') = q\varphi(\boldsymbol{r})$ where φ, from (3.8), is given by:

$$\boldsymbol{\nabla}_{\boldsymbol{r}} \cdot (\epsilon_M(\boldsymbol{r})\epsilon_0 \boldsymbol{\nabla}_{\boldsymbol{r}}\varphi(\boldsymbol{r})) = -q'\,\delta(\boldsymbol{r}-\boldsymbol{r}') \ . \tag{3.9}$$

In the case of an homogeneous bulk material where $\epsilon_M(\boldsymbol{r}) = \epsilon_b$ is a constant, we have:

$$V_b(\boldsymbol{r},\boldsymbol{r}') = \frac{qq'}{4\pi\epsilon_b\epsilon_0|\boldsymbol{r}-\boldsymbol{r}'|} \ . \tag{3.10}$$

When the material is inhomogeneous, V differs from V_b. As a result, the Coulomb interactions in heterostructures with large dielectric mismatch can be strongly modified compared to the bulk [160, 161]. Another important consequence is that the electrostatic energy of a charged particle depends on its position $\boldsymbol{r}$. Indeed, the charge polarizes the dielectrics and induces polarization charges at the surfaces and interfaces of the system. Therefore, there is an interaction between the particle and the polarization charges which depends on the position of the particle with respect to the surfaces and interfaces. The interaction energy is called a self-energy because the potential is induced by the own presence of the particle. To calculate this energy $\Sigma(\boldsymbol{r})$, let us consider that one brings successively infinitesimal charges at the position $\boldsymbol{r}$ to build up a charge q, i.e. $q = \int_0^q \mathrm{d}q'$. Since the electrostatic potential due to a charge q' can be written $\alpha q'$, the energy $\mathrm{d}\Sigma$ required to add a charge $\mathrm{d}q'$ ($q' \to q' + \mathrm{d}q'$) is equal to $\alpha q' \mathrm{d}q'$. Thus we obtain after integration

$$\Sigma(\boldsymbol{r}) = \int_0^q \mathrm{d}\Sigma = \alpha\frac{q^2}{2} = \frac{1}{2}q\,\varphi_{\text{ind}}(\boldsymbol{r}) \ , \tag{3.11}$$

where $\varphi_{\text{ind}}(\boldsymbol{r})$ is the macroscopic electrostatic potential due to the polarization charges induced by the charge q. This potential is given by

$$\varphi_{\text{ind}}(\boldsymbol{r}) = \lim_{\boldsymbol{r}'\to\boldsymbol{r}} \varphi(\boldsymbol{r}') - \varphi_{\text{b}}(\boldsymbol{r}') , \qquad (3.12)$$

where φ is the full electrostatic potential, solution of the Poisson's equation for a charge q at the position $\boldsymbol{r}$, and φ_{b} is the potential in the bulk semiconductor (without surfaces and interfaces). Thus, using the notations defined above and taking for ϵ_{b} the dielectric constant at the position $\boldsymbol{r}$, we have

$$\Sigma(\boldsymbol{r}) = \frac{1}{2} \lim_{\boldsymbol{r}'\to\boldsymbol{r}} [V(\boldsymbol{r},\boldsymbol{r}') - V_{\text{b}}(\boldsymbol{r},\boldsymbol{r}')] , \qquad (3.13)$$

where V and V_{b} are calculated with $q' = q$. In some cases where the geometry of the structures is simple, $\Sigma(\boldsymbol{r})$ can be calculated analytically, using for example the image charge method. Specific situations will be analyzed in the next sections. The spatial dependence of $\Sigma(\boldsymbol{r})$ reflects the existence of a macroscopic force induced by the dielectrics on the charge q. We will see that, in some cases, this force is large enough to determine the spatial distribution of the carriers in nanostructures.

3.1.2 Coulomb Interactions in a Dielectric Quantum Well

In this section, we summarize general results concerning Coulomb interactions in dielectric quantum wells, on the basis of macroscopic electrostatics. We describe the image charge method, which leads to analytical expressions for the electrostatic potentials and fields. We start with the simplest case of a single charge close to a dielectric interface, and then we consider quantum wells.

Image Charge Method for a Planar Dielectric Interface. We consider the situation of Fig. 3.1. A charge q is at point O, at a distance z from the planar interface between two dielectrics 1 and 2, with relative dielectric constants ϵ_1 and ϵ_2, respectively. The charge q is in region 1. The discontinuity of the dielectric constant induces polarization charges at the interface.

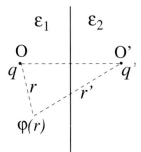

Fig. 3.1. Charge q at a distance z from the interface between two dielectrics

3.1 Macroscopic Approach: The Classical Electrostatic Theory

Following a well-known approach [155, 157], the potential in region 1 can be written as the potential created by two charges in an homogeneous medium of dielectric constant ϵ_1: the charge q, and a fictitious charge q' sitting at the point O', the image of the point O with respect to the interface. Thus, we have

$$\varphi_1(\boldsymbol{r}) = \frac{q}{4\pi\epsilon_1\epsilon_0 r} + \frac{q'}{4\pi\epsilon_1\epsilon_0 r'} , \qquad (3.14)$$

where r and r' are the respective distances from the points O and O'. The potential in region 2 is written as the potential due to a fictitious charge q'' at the point O, in an homogeneous medium of dielectric constant ϵ_2:

$$\varphi_2(\boldsymbol{r}) = \frac{q''}{4\pi\epsilon_2\epsilon_0 r} . \qquad (3.15)$$

Using the boundary conditions on the electric field at the interface, we have [155, 157]:

$$\begin{aligned} q' &= q\frac{\epsilon_1 - \epsilon_2}{\epsilon_1 + \epsilon_2} , \\ q'' &= q\frac{2\epsilon_2}{\epsilon_1 + \epsilon_2} . \end{aligned} \qquad (3.16)$$

From these expressions and (3.13), the electrostatic self-energy of the charge q is:

$$\Sigma(z) = \frac{qq'}{16\pi\epsilon_1\epsilon_0 z} = \frac{q^2}{4z}\frac{\epsilon_1 - \epsilon_2}{4\pi\epsilon_1\epsilon_0(\epsilon_1 + \epsilon_2)} . \qquad (3.17)$$

Figure 3.2 shows the self-energy of an electron as function of the distance z, in the case $\epsilon_1 = 10$ and $\epsilon_2 = 1$. The self-energy is quite substantial at distances in the nanometer range. The electrostatic self-energy of particles cannot be neglected in nanostructures with large dielectric mismatch.

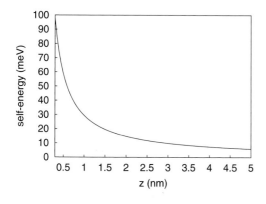

Fig. 3.2. Electrostatic self-energy of an electron or a hole in a material of dielectric constant $\epsilon_1 = 10$ at a distance z from the interface with a material of dielectric constant $\epsilon_2 = 1$

Image Charge Method for a Dielectric Quantum Well. We consider now the problem of a charge q located at the point $(\boldsymbol{r}_{\parallel} = 0, z_0)$, in a semiconductor quantum well. L is the thickness of the well, and the z axis is perpendicular to the interfaces. A schematic structure is shown in Fig. 3.3. The well, with a dielectric constant ϵ_1, is sandwiched by barrier layers having a different dielectric constant ϵ_2. The potential is calculated using the image charge method, following closely [162]. Due to the presence of the two interfaces, there is an infinite series of image charges. The potential in the well is given by regarding the whole structure as having a common dielectric constant ϵ_1, and by placing image charges q_n at the positions:

$$z_n = nL + (-1)^n z_0, \quad n = \pm 1, \pm 2... \, . \tag{3.18}$$

The potential in the left-hand-side barrier layer is given by placing image charges q'_n at z_n, $n = 0, 1, 2...$ and the potential in the right-hand-side barrier layer by placing image charges q''_n at z_n, $n = 0, -1, -2....$ In both cases, the whole structure is seen as having a common dielectric constant ϵ_2. Using the boundary conditions on the electric field at the interfaces, we have:

$$q_n = q\gamma^{|n|}, \quad \gamma = \left(\frac{\epsilon_1 - \epsilon_2}{\epsilon_1 + \epsilon_2}\right),$$

$$q'_n = q''_n = q_n \frac{2\epsilon_2}{\epsilon_1 + \epsilon_2} \, . \tag{3.19}$$

Thus, the potential in the well, at a position $(\boldsymbol{r}_{\parallel}, z)$, is given by:

$$\varphi(\boldsymbol{r}_{\parallel}, z) = \sum_{n=-\infty}^{\infty} \frac{q\gamma^{|n|}}{4\pi\epsilon_0\epsilon_1 \{r_{\parallel}^2 + [z - (-1)^n z_0 - nL]^2\}^{1/2}} \, . \tag{3.20}$$

A physically interesting limit is obtained when $r_{\parallel} \gg (\epsilon_1/\epsilon_2)L$:

$$\varphi(\boldsymbol{r}_{\parallel}, z) \approx \frac{q}{4\pi\epsilon_0\epsilon_2 r_{\parallel}} \, . \tag{3.21}$$

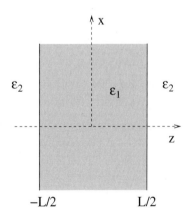

Fig. 3.3. Schematic structure of a dielectric quantum well

Thus, if the semiconductor quantum well is sandwiched between insulators or semiconductors with a small dielectric constant ϵ_2, the long-range Coulomb interactions are strongly enhanced compared to the bulk case. This effect, first pointed out by Keldysh [160], is due to the penetration of the electric field into the barrier with a small dielectric constant. This effect has important consequences, such as the enhancement of the exciton binding energy [161–163], of the excitonic oscillator strength [162], and of the electron–electron interactions [91].

The self-energy of a charge q located at the point $(\boldsymbol{r}_{\|}, z)$ does not depend on $\boldsymbol{r}_{\|}$:

$$\Sigma(z) = \frac{1}{2} \sum_{n=\pm 1, \pm 2 \ldots} \frac{q^2 \gamma^{|n|}}{4\pi\epsilon_0 \epsilon_1 |z - (-1)^n z - nL|} \,. \tag{3.22}$$

This self-energy diverges at the interface. To remedy this divergence, shifted mirror faces are sometimes employed for the lowest order ($n = \pm 1$) image charges [162, 164].

3.1.3 Coulomb Interactions in Dielectric Quantum Dots

We consider Coulomb interactions in a spherical semiconductor quantum dot, where simple analytical results can be obtained. A schematic structure is shown in Fig. 3.4. The dot, of radius R and dielectric constant ϵ_{in}, is surrounded by a medium of dielectric constant ϵ_{out}. This system has been studied in detail in [97, 165–167]. The potential energy $V(\boldsymbol{r}, \boldsymbol{r}')$ of a charge q located at $\boldsymbol{r}$ induced by a charge q' at $\boldsymbol{r}'$ is given by [157]

$$V(\boldsymbol{r}, \boldsymbol{r}') = V_{\text{b}}(\boldsymbol{r}, \boldsymbol{r}') + \delta V(\boldsymbol{r}, \boldsymbol{r}') \,, \tag{3.23}$$

where

$$V_{\text{b}}(\boldsymbol{r}, \boldsymbol{r}') = \frac{qq'}{4\pi\epsilon_{\text{in}}\epsilon_0 |\boldsymbol{r} - \boldsymbol{r}'|} \,,$$

$$\delta V(\boldsymbol{r}, \boldsymbol{r}') = qq' \sum_{n=0}^{\infty} \frac{(\epsilon_{\text{in}} - \epsilon_{\text{out}})(n+1) r^n r'^n P_n(\cos(\theta))}{4\pi\epsilon_0 \epsilon_{\text{in}} [\epsilon_{\text{out}} + n(\epsilon_{\text{in}} + \epsilon_{\text{out}})] R^{2n+1}} \,. \tag{3.24}$$

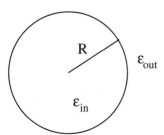

Fig. 3.4. Schematic structure of a dielectric quantum dot

θ is the angle between the two vectors r and r', and P_n the nth Legendre polynomial. In the particular case where a charge q is at the center of the dot, the electrostatic potential in the dot is given by:

$$\varphi(r) = \frac{q}{4\pi\epsilon_0} \left[\frac{1}{\epsilon_{in} r} - \frac{1}{R}\left(\frac{1}{\epsilon_{in}} - \frac{1}{\epsilon_{out}} \right) \right] . \tag{3.25}$$

The second term in the bracket is due to the polarization charge at the surface of the quantum dot. When $\epsilon_{out} \ll \epsilon_{in}$, this constant term becomes the main contribution of the potential in a large part of the dot. We will see in Chap. 6 that it explains the large binding energy of donor and acceptor impurities in quantum dots.

From (3.13), the self-energy of a charge q in the dot is given by

$$\Sigma(r) = \frac{1}{2}\delta V(r,r) , \tag{3.26}$$

with $q' = q$.

At the end of the chapter, we will use this formula to calculate the charging energy of an electron or a hole in a quantum dot.

3.2 Quantum Mechanics of Carriers in Dielectrics: Simplified Treatments

Many problems deal with the energetics and the quantum mechanics of carriers (electrons or holes) in semiconductor quantum structures. Their study, starting from first principles, using for example the density functional theory, is only possible for small systems, typically below 200 atoms. For larger systems, simplified treatments like the effective mass theory, tight binding or empirical pseudopotential methods described in Chap. 1 are required. Our aim in this section is to show how the dielectric properties are handled in these methods, and how they are used to simplify the description of complex systems.

3.2.1 Dielectric Effects in Single-Particle Problems

Dielectric Effects in the Effective Mass Approximation. Let us consider the problem of a single electron or hole in a semiconductor quantum system, like for example a quantum dot embedded in a dielectric matrix. A common approximation consists in writing a single particle Schrödinger equation for the extra charge. In the simplest effective mass approximation, it reduces to

$$\left(-\frac{\hbar^2}{2m^*}\Delta + V_{conf}(r) \right) \phi_\alpha^0(r) = \varepsilon_\alpha^0 \phi_\alpha^0(r) , \tag{3.27}$$

where m^* is the effective mass and V_{conf} is the confining potential. Beyond its well-known limitations discussed in Chap. 1, this effective equation for the extra carrier is only justified when the dielectric constant does not vary too much in the system. This is the case for nanostructures and heterostructures based on GaAs/GaAlAs and GaAs/InGaAs materials, because of the small dielectric mismatch between the constituents [163]. But in systems like Si/SiO$_2$, with a large dielectric mismatch, the dielectric forces on the carrier must be considered. This is usually done [91, 161–163] by adding to the confining potential the self-energy of the carrier, which is given by (3.13):

$$\left(-\frac{\hbar^2}{2m^*}\Delta + V_{\text{conf}}(\mathbf{r}) + \Sigma(\mathbf{r})\right)\phi_\alpha(\mathbf{r}) = \varepsilon_\alpha \phi_\alpha(\mathbf{r}) \ . \tag{3.28}$$

Justifications and limitations of this approach will be discussed in the next chapter.

Dielectric Effects in the Tight Binding Methods. In the same way as in the effective mass approximation, the electrostatic self-energy of an extra electron in a quantum system can be easily included in tight binding or in empirical pseudopotential methods by adding the self-energy $\Sigma(\mathbf{r})$ to the one-particle Hamiltonian. For example, in tight binding, because $\Sigma(\mathbf{r})$ is a slowly variable perturbation, it is reasonable to assume that $\Sigma(\mathbf{r})$ only appears in the diagonal terms of the Hamiltonian matrix written in the basis of the atomic orbitals

$$\begin{aligned}H_{i\alpha,i\alpha} &= H^0_{i\alpha,i\alpha} + \langle i\alpha|\Sigma(\mathbf{r})|i\alpha\rangle \ , \\ H_{i\alpha,j\beta} &= H^0_{i\alpha,j\beta} \text{ if } i\alpha \neq j\beta \ ,\end{aligned} \tag{3.29}$$

where α, β are orbital indices and i, j are atomic indices. $\langle i\alpha|\Sigma(\mathbf{r})|i\alpha\rangle$ can be further approximated by $\Sigma(\mathbf{R}_i)$ where $\mathbf{R}_i$ is the position of the atom i.

Influence of the Self-Energy on One-Particle States. From (3.28) and (3.29), we see that $\Sigma(\mathbf{r})$ contributes to determine the single-particle envelope functions ϕ_α (or, equivalently, the tight binding wave functions). However, in the strong confinement regime, i.e. when the inter-level spacing due to the confinement is large, the one-particle wave-functions are mainly determined by the confining potential, and $\phi_\alpha(\mathbf{r}) \approx \phi^0_\alpha(\mathbf{r})$. Therefore, using first-order perturbation theory, we can write:

$$\varepsilon_\alpha \approx \varepsilon^0_\alpha + \Sigma \ , \ \Sigma = \langle \phi^0_\alpha|\Sigma(\mathbf{r})|\phi^0_\alpha\rangle \ . \tag{3.30}$$

Σ is the average self-energy of the electron in the state ϕ^0_α, i.e. the electrostatic energy of the injected particle. In analogy with the electrostatics of the conductors, Σ is sometimes written in terms of a self-capacitance:

$$\Sigma \equiv \frac{e^2}{2C} \ . \tag{3.31}$$

In the intermediate and weak confinement regimes, the wave functions may be strongly affected by dielectric effects [168]: ϕ_α differs from ϕ^0_α and a perturbative method is no longer valid. To illustrate this effect, we plot in

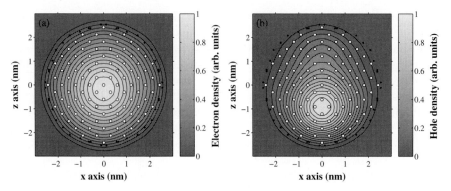

Fig. 3.5. Plot of the electron (a) and hole (b) states in a 4.8 nm InAs spherical nanocrystal situated above a metallic plane ($z = -3$). The white and black dots represent a projection of the atoms in the median plane

Fig. 3.5 the shape of the highest hole state and of the lowest electron state in a 4.8 nm InAs quantum dot deposited on a metallic surface (the metal can be seen here as a dielectric with $\epsilon_b \to +\infty$). Because the effective mass in the conduction band of InAs ($0.023 m_0$) is much smaller than in the valence band ($\approx 0.40 m_0$ in average), the confinement effect is strong in the conduction band and moderate in the valence band. We see in Fig. 3.5(b) that the hole is localized at the bottom of the dot, due to the attractive interaction with the induced polarization charges at the surface of the metal. In contrast, the effect on the electron states is smaller due to stronger confinement (Fig. 3.5a).

3.2.2 Dielectric Effects in Many-Particle Problems

We consider now the addition of several electrons or holes in a quantum system. A considerable simplification consists in writing effective equations for the extra charges, and to replace the effect of the remaining particles (electrons and nuclei) by the corresponding dielectric medium. In the simplest cases, an independent particle approximation is sufficient, and the whole procedure described above remains valid. However, in nanostructures, the addition of charges may be an important perturbation. Thus, Coulomb interactions between extra carriers cannot be ignored, and they must be treated at different levels of approximation which have been described in Chap. 1. Here, we show how dielectric effects can be approximately included in these methods.

Dielectric Effects in the Hartree Approximation. Dielectric effects can be naturally included in the Hartree approximation. The effective one-particle Hamiltonian contains the screened interaction of the particle with the total charge density in the system, leading to a pure problem of electrostatics. Let

3.2 Quantum Mechanics of Carriers in Dielectrics: Simplified Treatments

us consider the case of electrons, the situation for holes being symmetric. In the simplest effective mass approximation, the Hartree equation becomes

$$\left(-\frac{\hbar^2}{2m^*}\Delta + V_{\text{conf}}(\boldsymbol{r}) - e\varphi(\boldsymbol{r})\right)\phi_\alpha(\boldsymbol{r}) = \varepsilon_\alpha \phi_\alpha(\boldsymbol{r}), \tag{3.32}$$

where $\varphi(\boldsymbol{r})$ is the electrostatic potential given by the Poisson's equation

$$\nabla \cdot (\epsilon_M(\boldsymbol{r})\epsilon_0 \nabla \varphi(\boldsymbol{r})) = e\left[\sum_{\alpha \text{ occ.}} |\phi_\alpha(\boldsymbol{r})|^2\right] - \rho_{\text{ext}}, \tag{3.33}$$

where ρ_{ext} describes here the density of fixed charges, like ionized donors or acceptors. In bulk doped semiconductors, it is always replaced by its macroscopic spatial average $\overline{\rho_{\text{ext}}}$. However, in nanostructures, this approximation usually breaks down, and the microscopic distribution of dopants must be considered explicitly.

Equations (3.32) and (3.33) have been extensively used to study doped quantum wells and modulation-doped heterostructures [86, 91, 169] where, in the last case, the charge transfer between the doped barrier and the undoped quantum well determines the confining potential. Also, it has been used to study large quantum dots (> 10 nm), in which a large number (10-100) of carriers can be injected [170].

In tight binding, the application of the Hartree approximation is quite straightforward [77]. Once the eigenvalues and the eigenvectors of the Hamiltonian matrix have been determined, the levels are filled with electrons and, finally, the number of electrons n_i in excess on the atom i can be calculated. Then, the self-consistency is incorporated in a simple manner, where only the diagonal terms of the Hamiltonian matrix are charge dependent (see Sect. 1.3.1)

$$H_{i\alpha,i\alpha} = H^0_{i\alpha,i\alpha} + \sum_j \gamma_{ij} n_j, \tag{3.34}$$

$$H_{i\alpha,j\beta} = H^0_{i\alpha,j\beta} \text{ if } i\alpha \neq j\beta, \tag{3.35}$$

where $H^0_{i\alpha,i\alpha}$ is the matrix element in the absence of net charges. The γ_{ij} are screened Coulomb terms. In the general case of a complex dielectric system, these coefficients must be calculated by solving the Poisson's equation, to obtain the screened Coulomb interaction between two charges on the atoms i and j, respectively. When $i \neq j$, the two charges can be approximated by point charges and γ_{ij} is given by

$$\gamma_{ij} = V(\boldsymbol{R_i}, \boldsymbol{R_j}), \tag{3.36}$$

where V is calculated as in Sect. 3.1.1 with $q = q' = e$. When $i = j$, the above expression is divergent, and the contribution coming from the Coulomb interaction between two densities of charge on the same atom must be considered with some care. To avoid this problem, we can simply write $V = V_b + \delta V$ where V_b is the direct screened electron–electron interaction and δV is the correction due to the polarization charges at the surfaces and interfaces of the system (see Sects. 3.1.1 and 3.1.3). Thus we have

$$\gamma_{ii} = \delta V(\boldsymbol{R}_i, \boldsymbol{R}_i) + U_i^s, \tag{3.37}$$

where U_i^s is given by

$$U_i^s = \int |\varphi_{i\alpha}(\boldsymbol{r})|^2 |\varphi_{i\beta}(\boldsymbol{r'})|^2 V_b(\boldsymbol{r}, \boldsymbol{r'}) d\boldsymbol{r} d\boldsymbol{r'},$$

$$= \int \frac{e^2 |\varphi_{i\alpha}(\boldsymbol{r})|^2 |\varphi_{i\beta}(\boldsymbol{r'})|^2}{4\pi\epsilon_0 \epsilon_M(\boldsymbol{r}-\boldsymbol{r'})|\boldsymbol{r}-\boldsymbol{r'}|} d\boldsymbol{r} d\boldsymbol{r'}, \tag{3.38}$$

where $\varphi_{i\alpha}$ and $\varphi_{i\beta}$ are two orbitals of the atom i. In principle, the result depends on the nature of the orbitals (e.g. $\alpha, \beta = s, p$), but one can use an average value of the Coulomb integrals over several configurations [171]. Because electron–electron interactions within an atom are by nature short-range, one must use the dielectric function $\epsilon_M(\boldsymbol{r}-\boldsymbol{r'})$ given by the inverse Fourier transform of the wave-vector dependent macroscopic dielectric constant $\epsilon_M(\boldsymbol{q})$ of the bulk semiconductor at the position $\boldsymbol{R}_i$. U_i^s is typically between 2 and 5 eV in semiconductors, i.e. about 1/3 of the bare atomic value U_i^{at} (Sect. 1.3.1). The first term in (3.37) is no longer divergent and can be calculated directly. In the case where the dielectric constant is homogeneous in the system, V is simply given by the direct screened Coulomb interaction, which can be approximated by [172]

$$\gamma_{ij} = \frac{e^2}{4\pi\epsilon_0 \epsilon_M R_{ij}} \quad \text{if } i \neq j, \tag{3.39}$$

$$\gamma_{ii} = U_i^s, \tag{3.40}$$

where $\epsilon_M \equiv \epsilon_M(\boldsymbol{q} \to 0)$. This kind of charge-dependent tight binding treatment in the Hartree approximation has been applied to the simulation of the scanning tunneling spectroscopy of semiconductor nanocrystals [138, 173] (Sect. 4.6.1).

Dielectric Effects beyond the Hartree Approximation. The Hartree approximation has important limitations (Chap. 1). A first one concerns the self-interaction term which, in (3.33), comes from the fact that the Hartree potential describes the Coulomb interaction of a particle with the total charge density, including the particle under consideration. This unphysical term can be removed from the total energy of the system at the end of the self-consistent procedure. But only the direct interaction must be removed, not the interaction between the particle and the polarization charge induced by its presence, i.e. a self-energy term, which is physical. However, we must note that the self-energy term obtained in this way is approximated, i.e. is not the same as in the equation (3.28). In the Hartree approximation, the self-energy comes from the screening by the dielectric system of the average charge density of the particle $(-e|\phi_\alpha(\boldsymbol{r})|^2)$ given by its wave function. Actually, the true

3.2 Quantum Mechanics of Carriers in Dielectrics: Simplified Treatments

self-energy $\Sigma(r)$ must correspond to the response of the dielectric system to the point particle at each position r (Sect. 3.1.1).

Another limitation of the Hartree approximation obviously comes from correlation effects. As discussed in Chap. 1, they are often treated with configuration interaction methods, but in a simplified form where only the extra carriers are considered [171, 174–176]. Once again, the effect of the other particles is described by replacing electron–electron interactions by screened interactions. Another possible way to treat correlation effects between extra particles in a nanostructure is to combine the effective mass approximation and the local density approximation [91, 177]. For electrons, this leads to a kind of Kohn–Sham equation (see Sect. 1.1.5)

$$\left(-\frac{\hbar^2}{2m^*}\Delta + V_{\text{conf}}(r) - e\varphi(r) + V_{\text{xc}}(n)\right)\phi_\alpha(r) = \varepsilon_\alpha \phi_\alpha(r), \quad (3.41)$$

where $V_{\text{xc}}(n)$ is the exchange–correlation potential, and n is the density of electrons in excess in the system. Several expressions for $V_{\text{xc}}(n)$ have been used in the literature [91, 177–179]. A conceptual difficulty is to find the effective exchange–correlation potential for the extra particles only. A common procedure is to use expressions of $V_{\text{xc}}(n)$ derived for the homogeneous electron gas, but for electrons with an effective mass m^* and interacting with a screened Coulomb interaction. As an illustration, we describe here the approach of [179]. We have

$$V_{\text{xc}}(n) = \frac{\mathrm{d}}{\mathrm{d}n}[n\varepsilon_{\text{xc}}(n)], \quad (3.42)$$

where $\varepsilon_{\text{xc}}(n)$ is the sum of the exchange $\varepsilon_{\text{ex}}(n)$ and correlation $\varepsilon_{\text{corr}}(n)$ energies per electron, i.e., $\varepsilon_{\text{xc}}(n) = \varepsilon_{\text{ex}}(n) + \varepsilon_{\text{corr}}(n)$:

$$\varepsilon_{\text{ex}}(n) = \frac{-0.4582}{r_s}. \quad (3.43)$$

The radius r_s is expressed in terms of the effective Bohr radius in the semiconductor, $a_0^* = \epsilon_M a_0 / m^*$ and the local electron concentration, $n(r)$ as:

$$r_s = \left[\frac{3}{4\pi n(r)}\right]^{1/3} \frac{1}{a_0^*}. \quad (3.44)$$

The correlation energy is parametrized [17]:

$$\varepsilon_{\text{corr}}(n) = \frac{B}{1 + C\sqrt{r_s} + Dr_s} \text{ if } r_s \geq 1,$$

$$\varepsilon_{\text{corr}}(n) = E + F\ln(r_s) + Gr_s + Hr_s \ln(r_s) \text{ if } r_s < 1,$$

$$B = -0.1423, C = 1.0529, D = 0.3334,$$

$$E = -0.0480, F = 0.0311, G = -0.0116, H = 0.0020. \quad (3.45)$$

The energies and the potentials are expressed in scaled atomic units, i.e. $m^*/(m_0 \epsilon_M^2)$ atomic units. In the case of two-dimensional nanostructures [177], the parametrized form of [180] for the two-dimensional electron gas can be used. In the same way, the spin configuration of charged quantum dots can be studied using spin–density functional theory [177, 179]. In any case, the use of the effective mass and of screened Coulomb interactions in density functional theory is not fully justified. However, recent works seem to show that it gives results in good agreement with configuration interaction methods [177], in spite of a much smaller computational cost.

3.3 Microscopic Calculations of Screening Properties

The macroscopic approach to calculate the dielectric properties of nanostructures is obviously restricted to a small number of problems. For example, it cannot be used for complex systems containing isolated molecules, thin layers of molecules, metals or semiconductors at the atomic scale. In many cases, a microscopic approach is required. Microscopic approach means here that the dielectric response of an external perturbation is computed starting from quantum mechanics. The aim of this section is to review the bases of these calculations.

3.3.1 General Formulation in Linear-Response Theory

As shown in Sect. 1.2.7, an elegant formulation of the linear response of a system to an external perturbation $V_{\text{ext}}(\boldsymbol{r})$ is given by the density functional theory, which is formally exact, and which leads to a one-particle Kohn–Sham equation [9, 40, 41, 181, 182]. One defines the density–response function of non-interacting electrons (χ^0, often written P), which relates the change in density to the total effective potential V_{tot} (we work here in the static limit $\omega \to 0$)

$$\delta n(\boldsymbol{r}) = \int \chi^0(\boldsymbol{r},\boldsymbol{r}') V_{\text{tot}}(\boldsymbol{r}') \mathrm{d}\boldsymbol{r}' , \qquad (3.46)$$

where χ^0 is given by (1.71) and

$$V_{\text{tot}}(\boldsymbol{r}) = V_{\text{ext}}(\boldsymbol{r}) + \int \delta n(\boldsymbol{r}') v(\boldsymbol{r},\boldsymbol{r}') \mathrm{d}\boldsymbol{r}'$$
$$+ \int K_{\text{xc}}(\boldsymbol{r},\boldsymbol{r}') \delta n(\boldsymbol{r}') \mathrm{d}\boldsymbol{r}' . \qquad (3.47)$$

The second term gives the change in the Hartree potential, with $v(\boldsymbol{r},\boldsymbol{r}')$ is the bare Coulomb potential. The kernel $K_{\text{xc}}(\boldsymbol{r},\boldsymbol{r}')$ represents the reduction in the electron–electron interaction due to the existence of exchange–correlation effects. The density functional theory shows that [9] (Sect. 1.2.7)

$$K_{\text{xc}}(\boldsymbol{r},\boldsymbol{r}') = \left[\frac{\delta^2 E_{\text{xc}}[n]}{\delta n(\boldsymbol{r}) \delta n(\boldsymbol{r}')}\right]_{n_0(\boldsymbol{r})} , \qquad (3.48)$$

where $E_{\text{xc}}[n]$ is the exchange–correlation functional and $n_0(\boldsymbol{r})$ is the actual electron density of the system. The full density response function satisfies an integral equation:

$$\chi(\boldsymbol{r},\boldsymbol{r}') = \chi^0(\boldsymbol{r},\boldsymbol{r}') + \int d\boldsymbol{r_1} \int d\boldsymbol{r_2} \chi^0(\boldsymbol{r},\boldsymbol{r_1})[v(\boldsymbol{r_1},\boldsymbol{r_2}) \\ + K_{\text{xc}}(\boldsymbol{r_1},\boldsymbol{r_2})]\chi(\boldsymbol{r_2},\boldsymbol{r}') \ . \qquad (3.49)$$

Finally, one defines the inverse dielectric function, a measure of the screening in the system

$$\epsilon^{-1}(\boldsymbol{r},\boldsymbol{r}') \equiv \frac{\partial V_{\text{scr}}(\boldsymbol{r})}{\partial V_{\text{ext}}(\boldsymbol{r}')} \ , \qquad (3.50)$$

where $V_{\text{scr}}(\boldsymbol{r})$ is the screened potential. If V_{scr} is the potential probed by a test particle (a probe charge), then:

$$V_{\text{scr}}(\boldsymbol{r}) = V_{\text{ext}}(\boldsymbol{r}) + \int \delta n(\boldsymbol{r}') v(\boldsymbol{r},\boldsymbol{r}') d\boldsymbol{r}' \ . \qquad (3.51)$$

Using (1.91), the inverse dielectric function for the probe charge is connected to the polarizability by

$$\epsilon^{-1}(\boldsymbol{r},\boldsymbol{r}') = \delta(\boldsymbol{r}-\boldsymbol{r}') + \int v(\boldsymbol{r},\boldsymbol{r_1}) \chi(\boldsymbol{r_1},\boldsymbol{r}') d\boldsymbol{r_1} \ . \qquad (3.52)$$

Thus the inverse dielectric function can be calculated, provided that the kernel K_{xc} is defined (Sect. 1.2.7).

Equation (3.51) is no longer valid when the external potential is generated by an electron [26]. Details can be found for instance in [183].

3.3.2 Random-Phase Approximation

A widely used approximation is the time-dependent Hartree, or random-phase approximation [3, 6, 28, 32], described in Sect. 1.2.3. It is obtained by setting the exchange–correlation contribution in χ to zero. We simplify the notation by considering the equations as matrices in the continuous labels $\boldsymbol{r}$, and $\boldsymbol{r}'$. From (3.46) and (3.47), we have

$$\delta n = \chi^0 (v \delta n + V_{\text{ext}}) = \chi^{\text{RPA}} V_{\text{ext}} \ , \\ \chi^{\text{RPA}} = (1 - \chi^0 v)^{-1} \chi^0 \ , \\ \epsilon^{-1} = 1 + v \chi^{\text{RPA}} = (1 - v \chi^0)^{-1} \ , \qquad (3.53)$$

and thus, finally:

$$\epsilon(\boldsymbol{r},\boldsymbol{r}') = \delta(\boldsymbol{r}-\boldsymbol{r}') - \int v(\boldsymbol{r},\boldsymbol{r_1}) \chi^0(\boldsymbol{r_1},\boldsymbol{r}') d\boldsymbol{r_1} \ . \qquad (3.54)$$

The random-phase approximation has been applied to a large number of important problems, from atoms, to small molecules and solids. It is at the heart of more sophisticated calculations, like GW described in Sect. 1.2.4.

In spite of its apparent simplicity, its application to nanostructures has been limited to very small systems, in the molecular limit. The calculation of the dielectric response function from first principles remains very demanding, as regards both computer time and memory. As the number N of atoms in the system increases, the size of the basis set usually grows as N, as well as the number of eigenstates. Thus, the double sum on the empty and filled states in (1.71), required to calculate the independent particle polarization χ^0, quickly becomes a bottleneck for systems with a large number of atoms in the unit cell (typically 10), such as surfaces [184, 185], small semiconductor clusters [186, 187] or organic molecules [188].

Random-Phase Approximation in Tight Binding. To study larger systems, it is sometimes possible to work in tight binding (Sect. 1.3.1). This approach has been used to study the dielectric properties of bulk semiconductors [189], and has been applied to semiconductor heterostructures [110] and semiconductor quantum dots [190]. Tight binding leads to a considerable simplification because:

- the equations are considered as matrices in discrete values of r corresponding to the atomic positions $\boldsymbol{R}_n$
- the overlaps between atomic wave functions are neglected.

Thus, ϵ, χ^0, and v are described by matrices with a size given by the number of atoms in the system (or in the unit cell for periodic systems). If, in the calculation of χ^0 in (1.71), the one-electron wave functions $u_i(\boldsymbol{r})$ are defined in an atomic basis $\{\varphi_{n\alpha}\}$, where n denotes the atomic site and α the atomic orbital, we have

$$u_i(\boldsymbol{r}) = \sum_{n,\alpha} c_{i,n\alpha} \varphi_{n\alpha}(\boldsymbol{r}) \,. \tag{3.55}$$

The matrix elements of the independent particle polarization become:

$$\chi^0_{nm} = 2 \sum_{i,j} (n_i - n_j) \frac{\left[\sum_\alpha c_{i,m\alpha} c^*_{j,m\alpha}\right]\left[\sum_\alpha c_{j,n\alpha} c^*_{i,n\alpha}\right]}{\varepsilon_i - \varepsilon_j - i\eta} \,. \tag{3.56}$$

The matrix of the bare Coulomb potential v is given by [171]

$$v_{nm} = \frac{e^2}{4\pi\epsilon_0 |\boldsymbol{R}_n - \boldsymbol{R}_m|} \text{ if } n \neq m \,,$$
$$v_{nn} = U_n^{\text{at}} \,, \tag{3.57}$$

where U_n^{at} is the intra-atomic Coulomb energy on the atom n. Applications of this type of calculation to semiconductor nanostructures will be presented in Sect. 3.4. Recent works show that it is also well adapted to the simulation of molecular systems [191].

3.3.3 Beyond the Random-Phase Approximation

An alternative way to study the dielectric properties of nanostructures is to calculate directly the response of the systems to an external perturbation. The polarizability of molecules has been studied intensively with conventional techniques like Hartree–Fock and density functional theory. The same approach has been applied to GaAs clusters [192]. For time-dependent perturbations, the time-dependent density functional theory allows to compute the evolution of systems with an external field of arbitrary strength [193, 194]. It is applicable to molecules and to semiconductor nanostructures [193], to study excitations of small clusters in intense laser fields [195]. The method, in principle, includes non-linear responses. However, whereas reported errors are typically 10% for dielectric constants, recent results show overestimations of several orders of magnitude for the second hyper-polarizability (γ), due to an incorrect field dependence of the exchange–correlation potential [196].

3.3.4 From Microscopic to Macroscopic Dielectric Function for the Bulk Crystal

As discussed in Sect. 3.1.1, the macroscopic dielectric properties of a material are defined by averages over volumes large enough to remove the fluctuations at the scale of the atoms. When a material is a crystal, the averages can be made in a unit cell. In that case, it is more convenient to work in momentum space and to introduce the Fourier expansion of the inverse dielectric function:

$$\epsilon^{-1}(\boldsymbol{r},\boldsymbol{r}') = \frac{1}{(2\pi)^3} \int_{\mathrm{BZ}} \sum_{\boldsymbol{G},\boldsymbol{G}'} e^{\mathrm{i}(\boldsymbol{q}+\boldsymbol{G})\cdot\boldsymbol{r}} \epsilon^{-1}_{\boldsymbol{G},\boldsymbol{G}'}(\boldsymbol{q}) e^{-\mathrm{i}(\boldsymbol{q}+\boldsymbol{G}')\cdot\boldsymbol{r}'} \, \mathrm{d}\boldsymbol{q} \, . \tag{3.58}$$

The integration runs over the Brillouin zone. Thus, if an external potential $V_{\mathrm{ext}}(\boldsymbol{q})$ is applied, (3.58) shows that the total microscopic potential ($V_{\mathrm{scr}} \equiv \epsilon^{-1} V_{\mathrm{ext}}$) varies with wave vector components $\boldsymbol{q} + \boldsymbol{G}$, where $\boldsymbol{G}$ is a reciprocal lattice vector. This gives rise to microscopic fluctuations, at the origin of the local-field effects. Thus, according to Adler [30] and Wiser [31], the macroscopic dielectric function is obtained by keeping only the $\boldsymbol{G} = \boldsymbol{G}' = 0$ component of V_{scr}:

$$\epsilon_{\mathrm{M}}(\boldsymbol{q}) = \frac{1}{\epsilon^{-1}_{0,0}(\boldsymbol{q})} \, . \tag{3.59}$$

The macroscopic dielectric constant of a bulk material is the value of $\epsilon_{\mathrm{M}}(\boldsymbol{q})$ for $\boldsymbol{q} \to 0$. Its calculation is numerically expensive, because it requires to invert the full dielectric matrix $\epsilon_{\boldsymbol{G},\boldsymbol{G}'}(\boldsymbol{q})$. Thus, a common simplification is to use the independent particle polarization χ^0, and to neglect the local-field effects ($\epsilon_{\mathrm{M}}(\boldsymbol{q}) \approx \epsilon_{0,0}(\boldsymbol{q})$). It gives, after Fourier transform of (1.71)

$$\epsilon_{\mathrm{M}}(\boldsymbol{q}) \approx 1 + \frac{2}{\epsilon_0 \Omega q^2} \sum_{i,j} (n_i - n_j) \frac{\langle j|e^{-\mathrm{i}\boldsymbol{q}\cdot\boldsymbol{r}}|i\rangle \langle i|e^{\mathrm{i}\boldsymbol{q}\cdot\boldsymbol{r}'}|j\rangle}{\varepsilon_i - \varepsilon_j - \mathrm{i}\eta} \, , \tag{3.60}$$

where Ω is the volume of normalization of the one-particle states. This is a standard result, which has been extensively used to study the dielectric function of bulk materials. It gives, for the static dielectric constant of bulk semiconductors, discrepancies of the order of 10%-20% compared to experimental values [197].

3.4 Concept of Dielectric Constant for Nanostructures

We have seen that the dielectric properties of a system are described by $\epsilon^{-1}(\boldsymbol{r},\boldsymbol{r}')$. From this, in a bulk semiconductor, we can deduce directly the macroscopic dielectric constant $\epsilon_M(\boldsymbol{q})$ which contains most of the useful information on the dielectric screening. We can wonder if this macroscopic treatment is possible in semiconductor nanostructures and if it remains meaningful. This important question was addressed in different works [166, 198–200]. In this section, we explain why macroscopic quantities for nanostructures cannot be deduced from $\epsilon^{-1}(\boldsymbol{r},\boldsymbol{r}')$ as simply as in the bulk case. We show that a physically meaningful macroscopic dielectric constant must be derived taking into account explicitly the polarization charges at the surfaces or interfaces (Sect. 3.4.1). Using this prescription, we demonstrate that the macroscopic response is the bulk one a few Fermi wavelengths away from the surface and that the bulk response function $\epsilon_M(\boldsymbol{q})$ provides most of the needed information even for very small nanostructures [200] (Sects. 3.4.2 and 3.4.3). We also show that the average dielectric constant ϵ_{ave} in spherical clusters decreases when going to small radius R and we discuss the origin of this size dependence. At the end of this section (Sect. 3.4.4), we prove that all these conclusions mostly based on microscopic tight binding calculations of the dielectric response in Si quantum wells and dots are in fact completely general for semiconductor nanostructures [200].

3.4.1 The Importance of Surface Polarization Charges

The dielectric function $\epsilon^{-1}(\boldsymbol{r},\boldsymbol{r}')$ provides full information on the screening properties in a system. It relates the screened electrostatic potential to the bare one. From this, in a bulk material, one can directly define the macroscopic dielectric function $\epsilon_M(\boldsymbol{q})$ using (3.59). But, in a nanostructure, such a direct relation does not exit, due to the presence of polarization charges at the surfaces.

To illustrate this point, let us consider for instance the macroscopic limit of a dielectric sphere of dielectric constant ϵ_{in} embedded in vacuum. If we put a charge at the center, the macroscopic potential $\varphi(\boldsymbol{r})$ is given in (3.25) using the correct boundary conditions. From this, the potential screening function $\epsilon_{\text{pot}}(r) = \varphi_b(\boldsymbol{r})/\varphi(\boldsymbol{r})$ is equal to

$$\epsilon_{\text{pot}}(r) = \frac{1}{\epsilon_{\text{in}}^{-1} + \left(1 - \epsilon_{\text{in}}^{-1}\right)(r/R)} \qquad (3.61)$$

inside the cluster of radius R. This is completely different from the macroscopic dielectric constant ϵ_{in} whereas, in this case, the ratio $\boldsymbol{E}_b/\boldsymbol{E}$ of the bare to the screened electric field is equal to ϵ_{in}. $\epsilon_{pot}(r)$ is much smaller than ϵ_{in} on the average when ϵ_{in} is large (e.g. $\epsilon_{pot}(r) \approx 2$ for $r = R/2$). This difference corresponds to the fact that a charge $1 - 1/\epsilon_{in}$ is repelled onto the surface of the finite cluster (Sect. 3.1.3).

Another illustrative example is when an uniform electric field $\boldsymbol{E}_b$ is applied to a dielectric sphere. The screened electric field $\boldsymbol{E}$ inside the sphere is also uniform, and we have [155, 158]

$$\frac{\boldsymbol{E}_b}{\boldsymbol{E}} = \frac{2 + \epsilon_{in}}{3}, \tag{3.62}$$

which also differs from ϵ_{in}. Therefore, a physically meaningful definition of a dielectric constant in a nanostructure must incorporate the effect of the polarization charges at the surfaces. Keeping this in mind, we will present in the following microscopic calculations of the dielectric screening in quantum wells and quantum dots. A dielectric constant $\epsilon(\boldsymbol{r})$ will be calculated at each position $\boldsymbol{r}$ in the nanostructure by imposing that the macroscopic relation between the screened field $\boldsymbol{E}$ and the bare one $\boldsymbol{E}_b$ is verified. Then, ϵ_{ave} will be defined as the average of this local dielectric constant over the volume Ω of the nanostructure ($\epsilon_{ave} = \Omega^{-1} \int \epsilon(\boldsymbol{r}) d\boldsymbol{r}$). Obviously, this definition of ϵ_{ave} is not unique, and slightly different results can be obtained depending on the averaging procedure. It is also important to point out that this approach requires in principle to calculate $\boldsymbol{E}$ as the average of the microscopic field $\boldsymbol{e}$ over volumes not only large enough to cancel the fluctuations due to the atomic structure (Sect. 3.1.1) but also small compared to the volume of the nanostructure. In fact, in the following, we will consider tight binding calculations which naturally provide averages of the quantities over atomic volumes.

3.4.2 Dielectric Screening in Quantum Wells

We consider here the case of Si thin layers with (001) oriented planes submitted to a bare electric field $E_b(z)$ which is uniform inside and is vanishing abruptly between the terminating Si-H planes ($\boldsymbol{E}_b$ is perpendicular to the surfaces). We present the results of tight binding calculations [200]. Figure 3.6 gives the screened electric field along the layer. The most striking feature of these curves is that the local dielectric constant $\epsilon(z)$ defined here as E_b/E keeps its bulk value to a high accuracy except between the last two planes.

It is also interesting to compare these results to the dielectric response of the bulk silicon to the same perturbation. For that purpose, we define E_b as a Heaviside function

$$\begin{aligned} E_b(z) &= +E_0 \text{ if } -d/2 + 2nd \geq z > d/2 + 2nd, \\ &= -E_0 \text{ otherwise}, \end{aligned} \tag{3.63}$$

96   3 Dielectric Properties

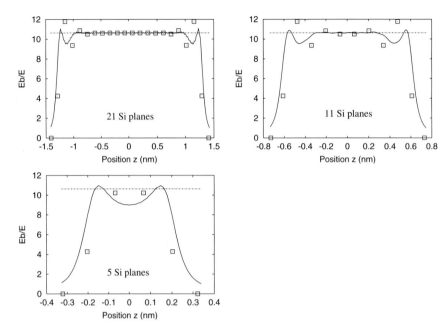

Fig. 3.6. Ratio between the bare electric field E_b and the screened one E in Si layers submitted to a constant electric field perpendicular to the surfaces: tight binding results ($\square$) and continuous model (*straight line*) using the bulk dielectric constant $\epsilon_M(q)$

where d is the width of the layer, E_0 is a constant and n is a positive or negative integer. The electric field and the potential are periodic functions which can be written in Fourier series. Then, screening each Fourier component by the bulk $\epsilon_M(q)$ calculated in tight binding using (3.59), we obtain the results shown in Fig. 3.6. The agreement with the tight binding calculation is striking even near the surfaces where however the oscillatory behavior depends on the nature of the boundary conditions. This is a proof that bulk screening appropriately describes the situation even for very small thicknesses (5 silicon planes). Similar results are obtained for sinusoidal bare electric fields in [200].

3.4.3 Dielectric Screening in Quantum Dots

The situation for Si spherical crystallites is more difficult to analyze than the case of thin layers but the conclusions are similar. We define a bare perturbation which consists in a charge $+q$ on the central atom of the sphere and a neutralizing charge uniformly spread on the outer shell of silicon atoms. Figure 3.7 again compares the local dielectric constants obtained from the direct calculation and from the use of the bulk $\epsilon_M(q)$. Apart from some

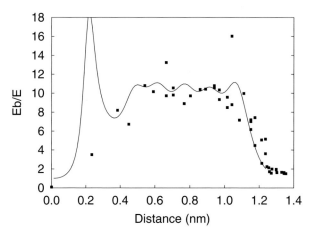

Fig. 3.7. Ratio between the bare electric field E_b and the screened one E versus the distance to the center in a 2.5 nm Si spherical nanocrystal. The bare field is due to a charge $+q$ at the center and a charge $-q$ uniformly spread on the surface of the sphere: tight binding results (■) and continuous model (*straight line*) using the bulk dielectric constant $\epsilon_M(q)$

differences near the boundaries (center and surfaces) and some oscillations, the results are again in close agreement.

Another interesting view on the problem is provided by Fig. 3.8 which gives the average dielectric constant ϵ_{ave} over the volume of the dot. One observes an overall decrease with decreasing radius. This is due to the surface contribution as it can be judged from Figs. 3.6 and 3.7, the major effect being a decrease of the local dielectric constant from the bulk value to 1 over the last two Si layers. Figure 3.8 also gives the average dielectric constant versus size calculated from different situations (e.g. a single donor impurity). All of them give comparable results close to those obtained from the bulk dielectric function $\epsilon_M(q)$. It is interesting to note that quite similar trend was obtained in [198] from a semi-empirical pseudopotential calculation based on the evaluation of (3.60) in the limit $q \to 0$. Equation (3.60) is often used to calculate the polarizability of molecules and thus represents one particular way of calculating ϵ_{ave}.

3.4.4 General Arguments on the Dielectric Response in Nanostructures

An important issue is to know how the previous results can be generalized. We have seen that the dielectric response is the bulk one at typically a few interatomic distance from boundaries. This seems to contradict the general belief that screening becomes less effective in nanocrystals due to the opening of the gap (e.g. [199]). Indeed, the independent particle polarization χ^0 in (1.71) and

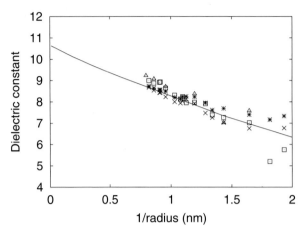

Fig. 3.8. Average dielectric constant ϵ_{ave} of Si spheres defined in different situations: average of E_{b}/E with a constant bare electric field E_{b} ($\times$); average of E_{b}/E with E_{b} due to a charge $+q$ at the center and $-q$ at the surface: tight binding calculations ($\triangle$) and continuous model (*straight line*) using the bulk dielectric constant $\epsilon_{\text{M}}(\boldsymbol{q})$. ϵ_{ave} deduced from the fit of the potential induced by a charge $+q$ at the center, the radius R considered as a parameter ($\square$)

(3.56) is given by a sum of terms which behave like $1/(\varepsilon_i - \varepsilon_j)$ where ε_i and ε_j are the energies of the unoccupied and occupied states, respectively. Thus, we have calculated the matrix element χ_{nm}^0 of the polarization between two first nearest neighbor atoms at the center of Si nanocrystals. We plot in Fig. 3.9 $\chi_{nm}^0/\chi_{nm}^0(\text{bulk})$ and $E_{\text{g}}/E_{\text{g}}(\text{bulk})$ versus size, $E_{\text{g}}(\text{bulk})$ and $\chi_{nm}^0(\text{bulk})$ being the bulk values. The main result is that the polarization is almost independent of the size, and is not at all related to the variation of the gap E_{g}. Similar results are obtained for InAs nanocrystals and for Si quantum wells as shown on the same figure. χ_{nm}^0 is not sensitive to the shift of the band edges induced by the confinement but to the average distance in energy between filled and empty states which remains constant versus size.

Therefore bulk parameters are still pertinent even for very small nanostructures. The decrease of ϵ_{ave} with size is due to a surface contribution, i.e. to the breaking of polarizable bonds at the surface [200]. Thus we can now generalize the results by applying the important theorem due to von Laue [201]. This one states that the electron density recovers its bulk value at distances from boundaries of the order of a few Fermi wave-lengths λ_{F}, i.e. typically the inter-atomic distance. This means that the response function is the bulk one inside a nanocrystal as long as its characteristic size exceeds a few λ_{F}.

The previous results also apply to the electronic part of the dielectric screening in nanostructures of polar materials (e.g. III–V and II–VI semiconductors). For example, the same kind of calculations have been performed

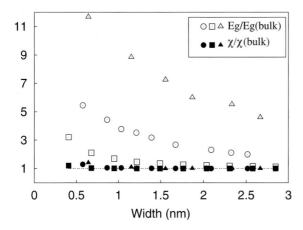

Fig. 3.9. Ratio of the nearest neighbor inter-atomic polarization χ_{nm}^0 and of the bulk value χ_{nm}^0(bulk) at the center of Si layers (*square*), Si spheres (*circle*) and InAs spheres (*triangle*) versus size compared to the ratio of the nanostructure gap and the bulk value

on InAs nanocrystals [138, 173]. Because InAs is a slightly ionic material, the dielectric constant is the sum of two contributions, electronic and ionic:

$$\epsilon_{\text{ave}} = \epsilon_{\text{ave}}^{\text{el}} + \epsilon_{\text{ave}}^{\text{ion}}. \tag{3.64}$$

Since the ionic contribution comes from the displacement of the ions with respect to their equilibrium position under the application of an external field, it is assumed to be weakly dependent on the crystallite size. The size dependence of the average $\epsilon_{\text{ave}}^{\text{el}}$ in InAs nanocrystals is shown in Fig. 3.10.

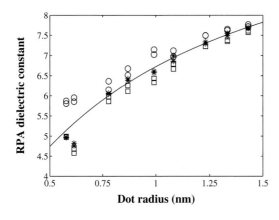

Fig. 3.10. Electronic contribution to the average dielectric constant in InAs quantum dots. Fit (*straight line*) of these values ($\epsilon_{\text{ave}}^{\text{el}} = 1 + \frac{\epsilon_{\text{M}}^{\text{el}} - 1}{1 + (0.79/R)^{1.15}}$) where $\epsilon_{\text{M}}^{\text{el}}$ is the bulk value and R is the radius in nanometer

3.4.5 Conclusions

The results presented in this section allow to establish general rules concerning the screening properties of semiconductor nanostructures:

- a physically meaningful definition of the local macroscopic dielectric constant in a nanostructure must incorporate the effect of the polarization charges at the boundaries
- the macroscopic dielectric response is the bulk one a few Fermi wavelengths away from the boundaries
- the dielectric response of a semiconductor nanostructure can be fairly well described using the macroscopic wave-vector dependent dielectric constant $\epsilon_M(\boldsymbol{q})$
- the local dielectric constant decreases near the boundaries due to the breaking of polarizable bonds
- the average dielectric constant decreases with decreasing size of the nanostructure due to the increasing contribution of the surfaces
- the opening of the bandgap due to the quantum confinement plays no role in these problems.

3.5 Charging of a Nanostructure

3.5.1 Case of a Quantum Dot

One interesting question concerns the self-energy of particles and Coulomb charging effects in nanostructures. In Chap. 2, the level structure of crystallites has been obtained from semi-empirical calculations. Thus, we need to determine the corrections brought by the dielectric effects, due to the finite size of the system. We do this within a macroscopic electrostatic formulation. We concentrate on spherical quantum dots, summarizing the works of [97, 165–167]. In Chap. 4, we describe a more elaborate theory of self-energy corrections which, to a large part, justifies the results presented here. We show in Sect. 4.6.1 that the self-energy of particles and Coulomb charging effects can be measured experimentally using tunneling spectroscopy experiments, corresponding to the so-called Coulomb blockade effects.

We consider a spherical quantum dot of radius R and of macroscopic dielectric constant ϵ_{in}. In Sect. 4.4.2, we will argue that the best value for ϵ_{in} is the bulk macroscopic dielectric constant ϵ_M in that case.

In the case of a strong confinement, we can obtain a fairly good estimation of the self-energy Σ of an electron or a hole ($q = \pm e$) injected in the quantum dot, using (3.30) in a first-order perturbation theory. In the limit of an infinite potential barrier, the one-particle state is well-given by the effective mass solution (Sect. 2.1.4):

$$\phi(\boldsymbol{r}) = \frac{1}{\sqrt{2\pi R}} \frac{\sin(\pi r/R)}{r} \ . \tag{3.65}$$

The self-energy Σ is calculated in Appendix C using (3.30), (3.65) and (3.26). We show that a good approximation of Σ is given by

$$\Sigma \approx \frac{e^2}{8\pi\epsilon_0 R} \frac{\epsilon_{in} - \epsilon_{out}}{\epsilon_{in}[\epsilon_{in} + \epsilon_{out}]} \left(\frac{1}{\eta} + 0.933 - 0.376\eta\right), \quad (3.66)$$

where $\eta = \epsilon_{out}/(\epsilon_{in} + \epsilon_{out})$. When $\eta \ll 1$, which is the usual situation when the quantum dot is embedded in an oxide matrix or in a semiconductor with a large gap, the self-energy becomes

$$\Sigma = \frac{1}{2}\left(\frac{1}{\epsilon_{out}} - \frac{1}{\epsilon_{in}}\right)\frac{e^2}{4\pi\epsilon_0 R} + \delta\Sigma, \quad (3.67)$$

where

$$\delta\Sigma \approx 0.466 \frac{e^2}{4\pi\epsilon_0 \epsilon_{in} R}\left(\frac{\epsilon_{in} - \epsilon_{out}}{\epsilon_{in} + \epsilon_{out}}\right), \quad (3.68)$$

which was already established in [166].

Σ gives the shift in energy of the extra electron (hole) in the lowest conduction (highest valence) state (Fig. 3.11). The injection of a second electron (hole) leads to an additional upwards (downwards) shift U given by the screened repulsion with the other electron (hole) (Fig. 3.11). With the same approximations as for Σ, U is given by

$$U = \int \phi(\mathbf{r})^2 \phi(\mathbf{r}')^2 V(\mathbf{r}, \mathbf{r}') d\mathbf{r} d\mathbf{r}'. \quad (3.69)$$

Using the expression (3.24) of $V(\mathbf{r}, \mathbf{r}')$ with $q = q' = e$, U is the sum of two terms. The first one is given by the average repulsion with the other particle

$$e^2 \int \frac{\phi(\mathbf{r})^2 \phi(\mathbf{r}')^2}{4\pi\epsilon_{in}\epsilon_0 |\mathbf{r} - \mathbf{r}'|} d\mathbf{r} d\mathbf{r}' \approx 1.79 \frac{e^2}{4\pi\epsilon_0 \epsilon_{in} R}, \quad (3.70)$$

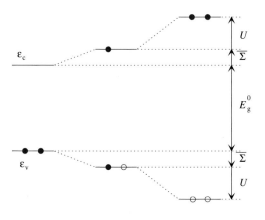

Fig. 3.11. Top: shift of the lowest conduction level due to the injection of one electron (Σ) or two electrons ($\Sigma + U$). Bottom: the situation for holes is symmetrical

and the second one by the average repulsion with the surface polarization charge induced by the other particle:

$$\int \phi(\mathbf{r})^2 \phi(\mathbf{r}')^2 \delta V(\mathbf{r},\mathbf{r}') \mathrm{d}\mathbf{r}\mathrm{d}\mathbf{r}' \,. \tag{3.71}$$

The coefficient 1.79 in (3.70) was obtained numerically [97, 165]. In (3.71), only the term $n = 0$ in the expression of δV given in (3.24) makes a nonzero contribution to the integral. Thus, we obtain:

$$U = \left(\frac{1}{\epsilon_{\mathrm{out}}} + \frac{0.79}{\epsilon_{\mathrm{in}}}\right) \frac{e^2}{4\pi\epsilon_0 R} \,. \tag{3.72}$$

In many situations, the surrounding of the quantum dots is not an homogeneous dielectric medium. Then, Poisson's and Schrödinger equations must be solved self-consistently to calculate the charging energy U. However, U must be necessarily between two bounds, corresponding to $\epsilon_{\mathrm{out}} = 1$ and $\epsilon_{\mathrm{out}} \to \infty$ in (3.72):

$$\frac{0.79}{\epsilon_{\mathrm{in}}} \frac{e^2}{4\pi\epsilon_0 R} < U < \left(1 + \frac{0.79}{\epsilon_{\mathrm{in}}}\right) \frac{e^2}{4\pi\epsilon_0 R} \,. \tag{3.73}$$

These relations are very useful, for example to interpret the $I(V)$ characteristics of devices based on semiconductor quantum dots (see for instance Sect. 4.6.1). Figure 3.12 shows the evolution of these two bounds for U in a Si nanocrystal, as function of its diameter. We see that the values of U can be very large when $\epsilon_{\mathrm{out}} = 1$, such that the injection of more than one carrier becomes difficult.

Each time another electron is injected in the nanocrystal, the conduction states exhibit an energy shift U which can be calculated according to (3.69), using the corresponding wave function. When $\epsilon_{\mathrm{out}} \ll \epsilon_{\mathrm{in}}$, U does not depend too much on the details of the wave function because the dominant term in U is the Coulomb interaction between the electron and its polarization charge at the surface. In that case, (3.72) remains a good approximation for a wide range of charge states.

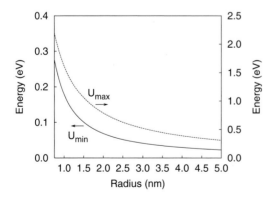

Fig. 3.12. Minimum and maximum value for the charging energy U in Si nanocrystals

In the case of a metallic nanostructure ($\epsilon_{in} \to \infty$), (3.67), (3.68) and (3.72) give $U = e^2/C$ and $\Sigma = U/2$ as it must be for the charging of a metallic sphere of self-capacitance $C = 4\pi\epsilon_0\epsilon_{out}R$. Note that this capacitive model is often extended to the case of semiconductor quantum dots, even if it is not perfectly justified.

3.5.2 Case of a Quantum Well

In the case of quantum wells, the charging energy U is vanishingly small, due to the infinite size of the system. Thus it remains to calculate the self-energy, following the same method as for the dots. For a quantum well of thickness L, the effective mass solution for the one-particle state is:

$$\phi(\mathbf{r}) = \sqrt{\frac{2}{L}} \cos(\pi z/L) \text{ for } -L/2 \leq z \leq L/2 . \tag{3.74}$$

Using (3.22), we have

$$\Sigma \approx \frac{q^2}{4\pi\epsilon_0\epsilon_1 L} \left[\sum_{n=2}^{\infty} \frac{\gamma^n}{n} + \gamma \int_{-1}^{1} \frac{\cos^2(\pi u/2)}{1-u^2} du \right] , \tag{3.75}$$

where γ is defined in (3.19), and we have replaced $|z-(-1)^n z - nL|$ by $|n|L$ when $|n| \geq 3$ in (3.22), which is justified numerically. Calculating the integral numerically, we obtain:

$$\Sigma \approx \frac{q^2}{4\pi\epsilon_0\epsilon_1 L} \left[0.219\, \gamma - \ln(1-\gamma) \right] . \tag{3.76}$$

4 Quasi-particles and Excitons

In this Chapter we discuss the different types of calculations which have been performed for the electronic excitations in semiconductor nanostructures. These range from carrier injection (quasi-particle energies, charging effects) to optical excitation and radiative recombination. We start with basic considerations (Sect. 4.1) where confinement effects and self-energy contributions due to surface polarization are formally separated, which will prove useful when analyzing the results of sophisticated calculations. We then treat excitons (Sect. 4.2) in the effective mass approximation (EMA) which, while simple, remains a powerful tool for the interpretation of many data [202]. This is followed by more refined semi-empirical calculations mainly concentrating on evaluations of the exchange splitting (Sect. 4.3). The two following sections (Sects. 4.4 and 4.5) deal with the application of quantitative methods (GW for quasi-particles, Bethe–Salpeter equations for excitons) to the case of silicon nanostructures. The results allow us to discuss the limits of validity of more approximate methods and to derive useful rules. Finally we describe calculations of charging effects and multi-excitonic transitions which can now be described quite satisfactorily (Sect. 4.6).

4.1 Basic Considerations

When discussing the excitations in bulk systems, it has been common and extremely fruitful to introduce the notion of quasi-particles and subsequently of pairs and even groups of quasi-particles. This notion is useful only if their lifetime is long enough compared to the time characteristic of the experiment to be interpreted. Under these circumstances, a great advantage is that one can write an individual Schrödinger equation for this quasi-particle, as detailed in Chap. 1, leading to the GW approximation. One can also write a separate equation for a quasi-particle pair like an exciton (electron–hole pair). Such an approach which we present here in detail is expected to remain useful for nanostructures down to relatively small sizes. However care should be taken in viewing excitons in few-atom systems (molecules) as well defined electron–hole pairs.

Experimental conditions to which the concept of quasi-particle fully applies occur when there is injection of one electron or one hole (extraction of one electron) into the system. This is the case in tunneling experiments, for instance via a tip of a scanning tunneling microscope, or in photo-emission (direct or inverse). Starting from an N electron neutral system, the lowest energy at which an electron can be injected is

$$\varepsilon_c^{\text{qp}} = E(N+1) - E(N), \tag{4.1}$$

where the index c denotes the lowest conduction state or unoccupied molecular orbital (LUMO) and $E(N)$, $E(N+1)$ are the ground state energies of the N and $N+1$ electron systems. A similar situation occurs for holes, where (in accordance with the GW formulation of Chap. 1) we define the highest electron energy accessible to holes as

$$\varepsilon_v^{\text{qp}} = E(N) - E(N-1), \tag{4.2}$$

where the index v now denotes the highest valence state or occupied molecular orbital (HOMO). It is impossible to split exactly on general general grounds, as was done approximately in Chap. 3, the contributions of the quantum confinement and of the self-energy corrections. However, we shall use in the following the fact that all calculations, even ab initio, start from a single particle calculation performed for the neutral N electron system so that one can partition the quasi-particle energies as

$$\varepsilon_c^{\text{qp}} = \varepsilon_c^0 + \delta\Sigma_c,$$
$$\varepsilon_v^{\text{qp}} = \varepsilon_v^0 + \delta\Sigma_v, \tag{4.3}$$

where, in the spirit of the GW approximation, ε_c^0 and ε_v^0 are the eigenvalues of the single-particle equations appropriate to the neutral N electron system while $\delta\Sigma_c$ and $\delta\Sigma_v$ are the self-energy corrections (Fig. 4.1). Assuming that the particles are confined in the nanostructure, we separate each term into a bulk contribution and a correction due to the boundary conditions defining the nanostructure

$$\varepsilon_c^{\text{qp}} = \varepsilon_{c,\text{bulk}}^0 + \delta\varepsilon_c^0 + \delta\Sigma_{c,\text{bulk}} + \delta\Sigma_{c,\text{surf}},$$
$$\varepsilon_v^{\text{qp}} = \varepsilon_{v,\text{bulk}}^0 + \delta\varepsilon_v^0 + \delta\Sigma_{v,\text{bulk}} + \delta\Sigma_{v,\text{surf}}, \tag{4.4}$$

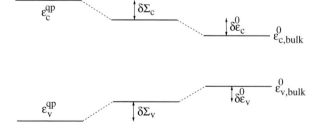

Fig. 4.1. The quasi-particle levels $\varepsilon_c^{\text{qp}}$ and $\varepsilon_v^{\text{qp}}$ with respect to the bulk single-particle conduction and valence band limits $\varepsilon_{c,\text{bulk}}^0$ and $\varepsilon_{v,\text{bulk}}^0$. $\delta\Sigma_c$ and $\delta\Sigma_v$ are the self-energy corrections. $\delta\varepsilon_c^0$ and $\delta\varepsilon_v^0$ correspond to pure confinement effects

where $\varepsilon^0_{c,\text{bulk}}$ and $\varepsilon^0_{v,\text{bulk}}$ stand for the bulk single-particle conduction and valence band limits, $\delta\Sigma_{c,\text{bulk}}$ and $\delta\Sigma_{v,\text{bulk}}$ for the corresponding bulk self-energy corrections. With this partitioning, $\delta\varepsilon^0_c$ and $\delta\varepsilon^0_v$ correspond to pure confinement effects while $\delta\Sigma_{c,\text{surf}}$ and $\delta\Sigma_{v,\text{surf}}$ are self-energies induced by the boundaries, i.e. surface corrections. Depending upon the choice, one can also group terms in (4.4) in different ways, e.g.

$$\varepsilon^{\text{qp}}_c = \varepsilon^{\text{qp}}_{c,\text{bulk}} + \delta\varepsilon^0_c + \delta\Sigma_{c,\text{surf}} ,$$
$$\varepsilon^{\text{qp}}_v = \varepsilon^{\text{qp}}_{v,\text{bulk}} + \delta\varepsilon^0_v + \delta\Sigma_{v,\text{surf}} , \tag{4.5}$$

where $\varepsilon^{\text{qp}}_{c,\text{bulk}}$ and $\varepsilon^{\text{qp}}_{v,\text{bulk}}$ are the exact quasi-particle bulk band limits. For the nanostructure quasi-particle band gap $\varepsilon^{\text{qp}}_g$ (which is the difference $\varepsilon^{\text{qp}}_c - \varepsilon^{\text{qp}}_v$), we can thus write the corresponding expression

$$\varepsilon^{\text{qp}}_g = \varepsilon^{\text{qp}}_{g,\text{bulk}} + \delta\varepsilon^0_g + \delta\Sigma_{g,\text{surf}} , \tag{4.6}$$

where $\delta\varepsilon^0_g = \delta\varepsilon^0_c - \delta\varepsilon^0_v$ and $\delta\Sigma_{g,\text{surf}} = \delta\Sigma_{c,\text{surf}} - \delta\Sigma_{v,\text{surf}}$.

We shall see in the following, both from simplified and ab initio theories, that the essential contribution to the surface self-energy comes from the surface polarization or image charge. From the discussion of Chap. 3, an electron confined in a nanostructure of dielectric constant ϵ_{in} embedded in another dielectric medium of constant ϵ_{out} experiences a self-energy (Σ in Chap. 3) which, when averaged over the probability distribution of the electron takes the form

$$\delta\Sigma_{c,\text{surf}} = \frac{1}{4\pi\epsilon_0} \frac{\alpha_c}{2d}(\epsilon_{\text{in}} - \epsilon_{\text{out}}) , \quad \alpha_c > 0 , \tag{4.7}$$

where d is the characteristic size of the nanostructure. The same is true for the hole. However, $\delta\Sigma_{v,\text{surf}}$ is the self-energy term characterizing the corresponding electron energy which has the opposite sign. Thus

$$\delta\Sigma_{v,\text{surf}} = -\frac{1}{4\pi\epsilon_0} \frac{\alpha_v}{2d}(\epsilon_{\text{in}} - \epsilon_{\text{out}}) , \quad \alpha_v > 0 , \tag{4.8}$$

where the factor $1/2$ explicitly indicates that one deals with a self-energy (Chap. 3). From this, one gets

$$\delta\Sigma_{g,\text{surf}} = \frac{\alpha_1}{4\pi\epsilon_0 d}(\epsilon_{\text{in}} - \epsilon_{\text{out}}) , \quad \alpha_1 = \frac{\alpha_c + \alpha_v}{2} > 0 . \tag{4.9}$$

We shall see later that, for simple EMA models, $\alpha_c \approx \alpha_v$.

We come now to the exciton which can be created by optical excitation across the gap inducing an electron–hole pair. In a qualitative physical picture, the excitation energy will be equal to the quasi-particle gap $\varepsilon^{\text{qp}}_g$ corrected by the average value of the electron–hole pair Hamiltonian

$$\varepsilon^{\text{exc}}_g = \varepsilon^{\text{qp}}_g + \langle H_{\text{eh}} \rangle . \tag{4.10}$$

The most advanced procedure to get $\langle H_{\text{eh}} \rangle$ is discussed in Chap. 1 and in next section but, for the moment, we make use of a more qualitative approach. We write H_{eh} as the sum of two terms:

- a kinetic energy term T due to the fact that the electron–hole wave function mixes excited quasi-particle states
- a potential energy term due to the screened Coulomb interaction between the two quasi-particles.

This last contribution can be split into two parts: the direct electron–hole attraction $-e^2/(4\pi\epsilon_0\epsilon_{in}r_{eh})$ (where r_{eh} is the inter-particle distance) and the interaction between one of the quasi-particles with the image surface charge of the other. This last term is macroscopic and varies slowly along the nanostructure. It can thus be treated by first order perturbation theory and written $-\alpha_{eh}(\epsilon_{in}-\epsilon_{out})/(4\pi\epsilon_0 d)$ which gives

$$\varepsilon_g^{exc} = \varepsilon_{g,bulk}^{qp} + \delta\varepsilon_g^0 + \left\langle T - \frac{e^2}{4\pi\epsilon_0\epsilon_{in}r_{eh}} \right\rangle + \frac{\alpha_1 - \alpha_{eh}}{4\pi\epsilon_0 d}(\epsilon_{in} - \epsilon_{out}) . \quad (4.11)$$

Obviously $\alpha_1 \approx \alpha_{eh}$ (we discuss later their exact values) so that one expects the last term to be small. If it was zero (exact compensation between the surface polarization contributions), one would have the conventional expression

$$\varepsilon_g^{exc*} = \varepsilon_{g,bulk}^{qp} + \delta\varepsilon_g^0 + \left\langle T - \frac{e^2}{4\pi\epsilon_0\epsilon_{in}r_{eh}} \right\rangle , \quad (4.12)$$

which is the basic expression used in many simplified calculations [86]. We shall give evidence that α_{eh} differs slightly from α_1 so that (4.12) might be in error. However, if the dielectric mismatch becomes smaller (i.e. $\epsilon_{in}-\epsilon_{out} \to 0$), then the macroscopic surface contribution vanishes and ε_g^{exc*} becomes exact. This is the case of a lot of conventional heterostructures (e.g. GaAs/GaAlAs) where this correction can be safely neglected.

4.2 Excitons in the Envelope Function Approximation

We now illustrate the notions introduced in the previous section with the simplest description of quasi-particles in a potential well, i.e. the envelope function approximation. Before this, we review the theory of bulk excitons. We then extend the approach to the case of confined systems with no dielectric mismatch. Finally we discuss the influence of the dielectric mismatch in the same approach.

4.2.1 Theory of Bulk Excitons

In the bulk, the description leading to (4.12) fully applies with no confinement effect, i.e. $\delta\varepsilon_g^0 = 0$. The electron–hole interaction thus reduces to $-e^2/(4\pi\epsilon_0\epsilon_M r_{eh})$ where we have taken ϵ_{in} equal to the macroscopic dielectric constant ϵ_M of the bulk semiconductor. This Coulomb potential can give rise to localized gap states like hydrogenic impurities. The justification of

4.2 Excitons in the Envelope Function Approximation

this proceeds via the EMA or envelope function approximation. To find the excitonic wave function, we can write the total wave function of the excited states in the form [158, 203]

$$\Psi_{\text{exc}} = \sum_{\boldsymbol{k}_e, \boldsymbol{k}_h} a(\boldsymbol{k}_e, \boldsymbol{k}_h) \Phi(\boldsymbol{k}_e, \boldsymbol{k}_h) \,, \quad (4.13)$$

where the functions Φ correspond to the excited states obtained from the ground state by exciting a valence band electron of wave vector $\boldsymbol{k}_h$ to a conduction band state $\boldsymbol{k}_e$. In EMA, we introduce a two particle envelope function by the Fourier transform

$$F(\boldsymbol{r}_e, \boldsymbol{r}_h) = \sum_{\boldsymbol{k}_e, \boldsymbol{k}_h} a(\boldsymbol{k}_e, \boldsymbol{k}_h) e^{i(\boldsymbol{k}_e \cdot \boldsymbol{r}_e - \boldsymbol{k}_h \cdot \boldsymbol{r}_h)} \,, \quad (4.14)$$

where $\boldsymbol{r}_e$ and $\boldsymbol{r}_h$ are the electron and hole positions. For simple band extrema with isotropic effective masses, it can be shown that $F(\boldsymbol{r}_e, \boldsymbol{r}_h)$ obeys the effective mass equation [158, 203]

$$\left\{ \varepsilon_g + \frac{p_e^2}{2m_e^*} + \frac{p_h^2}{2m_h^*} - \frac{e^2}{4\pi\epsilon_0\epsilon_M r_{eh}} \right\} F(\boldsymbol{r}_e, \boldsymbol{r}_h) = \varepsilon \, F(\boldsymbol{r}_e, \boldsymbol{r}_h) \,. \quad (4.15)$$

One can separate the center of mass and relative motion in this Hamiltonian in such a way that the total energy becomes

$$\varepsilon = \varepsilon_g + \frac{\hbar^2 k^2}{2M} - \frac{m^* e^4}{2\hbar^2 (4\pi\epsilon_0\epsilon_M)^2} \frac{1}{n^2} \,, \quad (4.16)$$

where M and m^* are the total and reduced masses respectively ($1/m^* = 1/m_e^* + 1/m_h^*$) and $\boldsymbol{k}$ is the wave vector for the center of mass motion. From this, it is clear that the lowest excited states are those for $\boldsymbol{k} = 0$, giving rise to the hydrogenic lines. In this simple model, the lowest exciton wave function is

$$F(\boldsymbol{r}_e, \boldsymbol{r}_h) = \frac{e^{i\boldsymbol{k}\cdot\boldsymbol{R}}}{\sqrt{\Omega}} \frac{e^{-\frac{|r_{eh}|}{a}}}{\sqrt{\pi a^3}} \,, \quad (4.17)$$

where Ω is the volume of the specimen, $\boldsymbol{R}$ is the center of mass position of the two quasi-particles and a is the exciton Bohr radius.

4.2.2 Excitons in Quantum Wells

We now investigate excitons in confined systems with no dielectric mismatch between the well and the barriers. Let us first consider the 1D quantum well of Fig. 4.2 and the effective mass equation (4.15) for the exciton envelope function. The problem is complicated by the fact that there is confinement in the z direction. There can still be free motion of the center of mass in the directions x and y parallel to the layer. The ground state obviously corresponds to no such motion and two variational wave functions for this state have been sought [86]

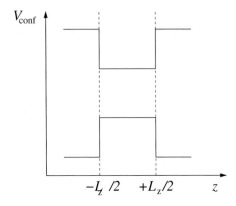

Fig. 4.2. Electron and hole confining potential in a quantum well of width L_z

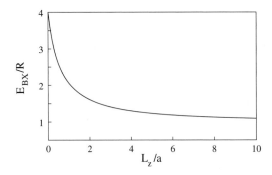

Fig. 4.3. Exciton binding energy in an infinite quantum well in units of the effective Rydberg $R = \frac{m^* e^4}{2\hbar^2 (4\pi\epsilon_0 \epsilon_M)^2}$ with respect to ratio of the well thickness L_z to the bulk exciton Bohr radius a. The energies have been calculated with the variational wave function (4.19)

$$F(\boldsymbol{r}_e, \boldsymbol{r}_h) = \frac{N_1}{\sqrt{S}} \chi_1^{(e)}(z_e) \chi_1^{(h)}(z_h) e^{-\rho/\lambda} , \qquad (4.18)$$

or

$$F(\boldsymbol{r}_e, \boldsymbol{r}_h) = \frac{N_2}{\sqrt{S}} \chi_1^{(e)}(z_e) \chi_1^{(h)}(z_h) \exp\left[-\frac{1}{\lambda}\sqrt{\rho^2 + (z_e - z_h)^2}\right] , \qquad (4.19)$$

where $\rho = \sqrt{(x_e - x_h)^2 + (y_e - y_h)^2}$, N_1 and N_2 are normalization constants, λ is a variational parameter and $\chi_1^{(e)}(z_e)$ and $\chi_1^{(h)}(z_h)$ are the lowest states in each quantum well in the absence of electron hole interaction. As shown in Fig. 4.3, one gets the result that the binding energy in a quantum well with infinite potential barriers increases when the width L_z of the well decreases, to reach a limiting value of four times the bulk one, as for hydrogenic impurities (see Chap. 6).

4.2.3 Exciton Binding Energy in Limiting Situations

It is interesting to say few words about the limiting situations. When the size d of the nanostructure is much larger that the Bohr radius a of the bulk exciton, one obviously recovers the bulk exciton binding energy in the limit $a/d \to$

0. The other limit is more peculiar since the 1D (quantum wires) and 2D (quantum wells) extended systems behave differently from the 0D quantum dot. Indeed, in the first two cases, it is still possible to build hydrogenic bound states in the subspace (1D or 2D) where the system remains infinitely extended. In the last case (0D), the situation becomes totally different. When $d/a \ll 1$, the kinetic energy term which scales like $1/d^2$ dominates over the Coulomb attraction which scales like $1/d$. In this case, we can start from the solution with pure kinetic energy, i.e. the unperturbed electron and hole wave functions that give rise to the confinement energy and we can treat the electron–hole attraction in first order perturbation theory. This means that (4.12) gives

$$\varepsilon_g^{\text{exc}*} = \varepsilon_{g,\text{bulk}}^{\text{qp}} + \delta\varepsilon_g^0 - \left\langle \frac{e^2}{4\pi\epsilon_0\epsilon_{\text{in}}r_{\text{eh}}} \right\rangle, \quad (4.20)$$

where the average is taken over the unperturbed electron–hole wave functions which is just the product of their individual wave functions. It is this situation which we shall treat most often in the following, when discussing more quantitative calculations which can only be performed for relatively small quantum dots.

4.2.4 The Influence of Dielectric Mismatch

Up to now we have only treated the case $\epsilon_{\text{in}} = \epsilon_{\text{out}}$. We now consider the situation where there is a substantial mismatch and we want to determine the extent of the cancellation discussed in Sect. 4.1. We do this for a spherical quantum dot of radius R with zero potential in the central region of dielectric constant ϵ_{in} and infinite potential in the outer region of dielectric constant ϵ_{out} (see also Sect. 2.1.4). As discussed before, in the strong confinement regime, we can treat the electron–hole interaction by perturbation theory. The zero-order wave functions are those of a spherical square well (Sect. 2.1.4), i.e. $\phi_e \propto \sin(\pi r_e/R)/r_e$ and $\phi_h \propto \sin(\pi r_h/R)/r_h$, the combined electron–hole wave function being their product. From this and the discussion of Sect. 4.1, we write

$$\varepsilon_g^{\text{exc}} = \varepsilon_{g,\text{bulk}}^{\text{qp}} + \delta\varepsilon_g^0 - \left\langle \frac{e^2}{4\pi\epsilon_0\epsilon_{\text{in}}r_{\text{eh}}} \right\rangle + \frac{\alpha_1 - \alpha_{\text{eh}}}{4\pi\epsilon_0 R}(\epsilon_{\text{in}} - \epsilon_{\text{out}}). \quad (4.21)$$

From (3.66) of Chap. 3, one gets the term α_1 which corresponds to the surface contribution to the quasi-particle gap

$$\alpha_1 = \frac{e^2}{\epsilon_{\text{in}}\epsilon_{\text{out}}} + \frac{0.933\, e^2}{\epsilon_{\text{in}}(\epsilon_{\text{in}} + \epsilon_{\text{out}})} - \frac{0.376\, e^2\epsilon_{\text{out}}}{\epsilon_{\text{in}}(\epsilon_{\text{in}} + \epsilon_{\text{out}})^2}, \quad (4.22)$$

while α_{eh} only corresponds to the $n = 0$ term of (3.24), i.e. to the first term in α_1:

$$\alpha_{\text{eh}} = \frac{e^2}{\epsilon_{\text{in}}\epsilon_{\text{out}}}. \quad (4.23)$$

Finally the average direct electron–hole interaction exactly corresponds to (3.70)

$$\left\langle \frac{e^2}{4\pi\epsilon_0\epsilon_{in}r_{eh}} \right\rangle = 1.79 \frac{e^2}{4\pi\epsilon_0\epsilon_{in}R} \ . \tag{4.24}$$

The binding energy of the exciton E_{BX} is the opposite of the last two terms in (4.21) and is given by

$$E_{BX} = 1.79 \frac{e^2}{4\pi\epsilon_0\epsilon_{in}R} \left(1 - \frac{0.933}{1.79} \frac{\epsilon_{in} - \epsilon_{out}}{\epsilon_{in} + \epsilon_{out}} + \frac{0.376}{1.79} \frac{(\epsilon_{in} - \epsilon_{out})\epsilon_{out}}{(\epsilon_{in} + \epsilon_{out})^2} \right) . \tag{4.25}$$

In the limit $\epsilon_{in}/\epsilon_{out} \gg 1$, a common situation, we see that the second term in (4.25) can reduce substantially E_{BX}, to about half its value.

4.3 Excitons in More Refined Semi-empirical Approaches

4.3.1 General Discussion

Here we present a natural extension of the previous models, based on the use of the configuration interaction (CI) method. The principle is to start from the eigenstates of a single particle Hamiltonian (like Hartree–Fock) for the N electron neutral system. Its ground state consists of a Slater determinant corresponding to filled valence states and empty conduction states. One builds all Slater determinants obtained from the ground state by excitation of 1,2...N electron–hole pairs and then one diagonalizes the full Hamiltonian matrix in this basis. This way of doing can lead in principle to the exact eigenstates of the system and, in particular, to the ground state and the first excitonic states. However, in practice, CI is impractical except for very small molecules and could not be applied even to relatively small semiconductor nanocrystals due to computational limits.

A way to circumvent this difficulty is to write an expansion in terms of a limited basis set built from all determinants corresponding to one electron–hole excitation as in Sect. 1.2.5. We write

$$|\Psi_{exc}\rangle = \sum_{cv} a_{vc}|\psi_{vc}\rangle , \tag{4.26}$$

where $|\psi_{vc}\rangle$ is the determinant obtained by exciting an electron from the valence spin–orbital $|v\rangle$ to the conduction spin–orbital $|c\rangle$. Standard rules (Sect. 1.2.5) show that the matrix of the Hamiltonian in this basis are given by

$$\langle\psi_{vc}|H|\psi_{v'c'}\rangle = (\varepsilon_c - \varepsilon_v)\delta_{vv'}\delta_{cc'} + \langle cv'|v|vc'\rangle - \langle cv'|v|c'v\rangle , \tag{4.27}$$

where the last term corresponds to the Hartree Coulomb attraction, the second one being the exchange term. This limited expansion can be made exact

4.3 Excitons in More Refined Semi-empirical Approaches

if one uses a folding procedure to include the effect of further excited states. The net effect will be to renormalize the electron–electron interactions as can be shown by the following argument. Let us consider that the Hamiltonian matrix is written as

$$[H] = \begin{bmatrix} H_{00} & H_{0R} \\ H_{R0} & H_{RR} \end{bmatrix}, \tag{4.28}$$

where H_{00} corresponds to the subspace of the $|\psi_{vc}\rangle$. Let us call $|\psi_\alpha\rangle$ and $|\psi_\beta\rangle$ the eigenstates of H_{00} and H_{RR} with energies E_α and E_β, respectively. We write the total eigenstate of H as

$$|\Psi_{\text{exc}}\rangle = \sum_\alpha a_\alpha |\psi_\alpha\rangle + \sum_\beta a_\beta |\psi_\beta\rangle . \tag{4.29}$$

It is solution of the coupled set of equations

$$(E - E_\alpha)a_\alpha = \sum_\beta H_{\alpha\beta} a_\beta ,$$

$$(E - E_\beta)a_\beta = \sum_{\alpha'} H_{\beta\alpha'} a_{\alpha'} , \tag{4.30}$$

which can be rewritten

$$(E - E_\alpha)a_\alpha = \sum_{\beta,\alpha'} \frac{H_{\alpha\beta} H_{\beta\alpha'}}{E - E_\beta} a_{\alpha'} . \tag{4.31}$$

This is the desired set of equations in the restricted subspace and one sees that the folding introduces effective energy dependent interactions between $|\psi_\alpha\rangle$ and $|\psi_{\alpha'}\rangle$ given by

$$\sum_\beta \frac{H_{\alpha\beta} H_{\beta\alpha'}}{E - E_\beta} . \tag{4.32}$$

These terms renormalize the electron–electron interactions in (4.27). This procedure is in principle exact but the evaluation of the terms (4.32) is as complicated as the full CI problem. However in Sects. 1.2.4 and 1.2.5 we have introduced a simplifying procedure in which H_{RR} was approximated by considering only plasmon excitations. We have shown that such a procedure naturally leads to the equations usually derived from field theoretic techniques, i.e. from GW and Bethe–Salpeter-like equations [32, 36]. For single particle states, the GW equations contain an exchange–correlation self-energy involving dynamically screened electron–electron interactions (Sect. 1.2.4).

In Sect. 1.2.5, we have shown that exciton states can be obtained from equations in which both the electron–hole Coulomb and exchange terms are dynamically screened. However the fact of screening or not the electron–hole exchange interaction has remained for long a controversial problem [37, 204, 205]. Our derivation concludes that it should be screened. This conclusion is strengthened by a recent paper on this problem [206].

4.3.2 Excitons in Nanocrystals of Direct Gap Semiconductors

Size-selective spectroscopic techniques such as resonant photoluminescence and photoluminescence excitation allow to get extremely detailed information on the lowest excitonic states of quantum dots or nanocrystals [174, 207–209]. This fine electronic structure is governed by the relative importance of different terms which constitute the Hamiltonian: confinement potential, spin–orbit interaction, Coulomb and exchange interactions. Each of these terms depends on quantum dot size in a different way. Two limiting situations can be considered: the weak confinement regime, when the dot size is larger than a few exciton radii and the strong confinement regime, when the quantum dot size is smaller than the exciton Bohr radius. In the latter regime, the confinement energy is much larger than the Coulomb interaction and both carriers are independently confined. However, even in this case, the Coulomb interaction exists since the electron and the hole are in a finite volume.

Tight binding configuration interaction calculations have been applied to the electronic structure of the lowest excitonic states in CdS nanocrystals with cubic lattice [174]. Here, the nature of the predictions will be analyzed in a simplified effective mass picture in order to identify clearly the evolution from bulk to quantum dots. Figure 4.4 summarizes this simple description of excitons in CdS quantum dots. Apart from the confinement effect, there are two terms in the Hamiltonian of a quantum dot which mainly determine the size dependence of its excitonic structure: the spin–orbit interaction and the electron–hole exchange interaction. Matrix elements of the spin–orbit interaction are constant. However the matrix elements of the electron–hole exchange interaction increase as the nanocrystal radius decreases. To under-

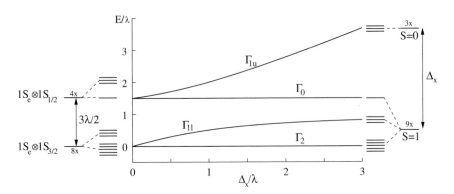

Fig. 4.4. Right and left: energy level diagrams describing the fine structure of the excitonic spectrum; middle: splitting between lowest energy levels of the exciton as a function of the electron–hole exchange interaction Δ_x which itself depends on quantum dot size. All the energies are in units of the spin–orbit coupling parameter λ.

stand the size dependence of the splittings between lowest exciton levels, one first considers two opposite limits: large and small dots.

For large dots, the lowest states converge to the classical situation of the bulk CdS, namely light and heavy-hole excitons and the spin–orbit split exciton. In this situation, the electron–hole exchange interaction is negligible with respect to the spin–orbit interaction (left side of Fig. 4.4). In this limit, the effective mass approximation becomes a powerful tool to calculate the electronic structure of spherical nanocrystals [146, 210, 211]. The first electron level is the $1S_e$ state with twofold spin degeneracy and the first level of holes is the $1S_{3/2}$ state which is fourfold degenerate with respect to the hole angular momentum (Sect. 2.5.3). The next hole level is $1S_{1/2}$ and its distance in energy from the $1S_{3/2}$ level is equal to $3\lambda/2$ where λ is the spin–orbit coupling parameter. Thus the lowest exciton state $1S_e \otimes 1S_{3/2}$ is eightfold degenerate and the next higher state $1S_e \otimes 1S_{1/2}$ is fourfold degenerate. The introduction in a perturbation scheme of the electron–hole exchange interaction splits the $1S_e \otimes 1S_{3/2}$ level into two groups of states. Since the total angular momentum J remains a good quantum number, the lowest exciton state, fivefold degenerate, is characterized by a momentum $J = 2$ and will be denoted hereafter Γ_2. The upper exciton state, threefold degenerate, corresponds to a momentum $J = 1$ and will be denoted Γ_{11}. In the effective mass approximation, the value of the splitting is related to the electron–hole exchange energy in the bulk ($\Delta_x^{bulk} = 0.23$ meV in CdS [212]) and is given by the following expression [213, 215]

$$\Delta_x = A \left(\frac{a}{R}\right)^3 \Delta_x^{bulk} , \qquad (4.33)$$

where a is the Bohr radius (≈ 30 Å in CdS [214]) and A is a constant which depends on the nature of the semiconductor. The $1/R^3$ scaling of Δ_x is a consequence of the effective mass approximation: more elaborate calculations based on empirical pseudopotentials predict exponents between 2 and 3 for InP and CdSe quantum dots [205] (similar trends are obtained in tight binding for Si nanocrystals; see next section).

The higher exciton state $1S_e \otimes 1S_{1/2}$ is also split by the electron–hole exchange interaction into two groups of levels with $J = 0$ and $J = 1$, denoted Γ_0 and Γ_{1u}, respectively.

In the opposite limit of small quantum dots (right side of Fig. 4.4), the electron–hole exchange energy is larger than the spin–orbit interaction. The lowest exciton level, formed by an electron in s-like states and a hole in p-like states, is split by the exchange interaction into a lower triplet state ($S = 1$), ninefold degenerate, and an upper singlet state ($S = 0$), threefold degenerate. The introduction of the spin–orbit interaction in perturbation leads to the splitting of the triplet exciton state into three states with $J = 2$, $J = 1$ and $J = 0$ (Γ_2, Γ_{11} and Γ_0, respectively). The singlet state gives another $J = 1$ state (Γ_{1u}).

The intermediate case, when the electron–hole exchange interaction is comparable with the spin–orbit interaction, can be described by atomistic calculations like tight binding [174, 209]. In the simplified model described above, the Hamiltonian including exchange and spin–orbit interactions can be diagonalized in the basis of exciton states formed by the product of the s states for the electron and of the p states for the hole [209]. It leads to the following expressions for the energy splittings between the exciton states and the lowest state Γ_2

$$E(\Gamma_{11}) - E(\Gamma_2) = \frac{3\lambda}{4} + \frac{\Delta_x}{2} - \sqrt{\left(\frac{\Delta_x}{6} - \frac{3\lambda}{4}\right)^2 + \frac{2\Delta_x^2}{9}},$$

$$E(\Gamma_{1u}) - E(\Gamma_2) = \frac{3\lambda}{4} + \frac{\Delta_x}{2} + \sqrt{\left(\frac{\Delta_x}{6} - \frac{3\lambda}{4}\right)^2 + \frac{2\Delta_x^2}{9}},$$

$$E(\Gamma_0) - E(\Gamma_2) = \frac{3\lambda}{2}, \tag{4.34}$$

where Δ_x, the exchange term, is a function of the nanocrystal radius. Figure 4.4 (middle) shows the energy splittings given by (4.34) with respect to Δ_x/λ, i.e. as a function of size. The model explains qualitatively the evolution of the excitonic levels predicted by tight binding calculations [209] that describe the photoluminescence excitation spectra measured on CdS nanocrystals [174, 209]. However, one must note that the excitonic structure is in fact more complex due the presence of hole states with slightly higher energy which are not included in the model.

This simple model also explains the probabilities of optical transitions. In the limit of small quantum dots where the exchange interaction is larger than the spin–orbit coupling, the optical transitions from the triplet ($S = 1$) derived states, i.e. Γ_2, Γ_{11} and Γ_0, to the ground state are forbidden in the dipole approximation while those from the singlet state Γ_{1u} are allowed (because the spin must be conserved in the transition; see Chap. 5). At increasing size, the exchange interaction decreases and the spin–orbit coupling mixes $S = 0$ and $S = 1$ states, with the important consequence that the transition from Γ_{11} becomes optically allowed [174, 209].

4.3.3 Excitons in Si Nanocrystals

We discuss here experimental data concerning the fine structure of optical spectra obtained for silicon nanocrystals. In particular, several works [216, 217] have shown that the excitation spectrum of the visible luminescence at 2K exhibits a threshold of a few meV and that, at higher energies, one gets phonon assisted transitions. In parallel, several groups have observed that the decay time of the visible luminescence decreases when the temperature increases from 4K to 100–200 K while the photoluminescence intensity also increases [216, 218, 219]. All these experiments have been interpreted on the basis of a two-level model pictured on Fig. 4.5 with a spin triplet as the lowest excitonic state followed by a spin singlet.

To discuss the validity of this model, we follow the tight binding approach of [171]. This was based, as discussed in Sect. 4.3.1, on an expansion in a basis of Slater determinants obtained from the ground state by one electron–hole excitation. The diagonal elements are deduced from the eigenvalues of a single particle tight binding Hamiltonian including spin–orbit coupling. The unscreened Coulomb interactions are given, as discussed in Sects. 1.3.1 and 3.3.2, by a Coulomb term U^{at} on site and $e^2/(4\pi\epsilon_0 R_{ij})$ between two sites i and j distant of R_{ij}. In view of the above discussions (Sect. 4.3.1), we follow [171] by screening the interactions by the bulk inverse dielectric matrix $\epsilon_{bulk}^{-1}(\boldsymbol{r},\boldsymbol{r}')$. The net result is that U^{at} is screened to 0.3 of its bare value while $e^2/(4\pi\epsilon_0 R_{ij})$ is screened by the bulk dielectric constant ϵ_M ([171] describes an essentially equivalent but more refined treatment).

The two-level model proposed in the literature [216, 218–220] is based schematically on the fact that, from the spin 1/2 states of the electron and the hole, one can build one $S = 0$ and one $S = 1$ state. The energy splitting between these two states, due to the electron–hole exchange interaction, is expected to increase with confinement to reach values of order of 10 meV for crystallites of nanometer size [216, 218, 219]. At low temperature (< 20 K), the excitons are mostly in the triplet state (Fig. 4.5) so that the radiative recombination time (from the $S = 1$ triplet exciton to the $S = 0$ ground state) would be infinite (the way to calculate the radiative lifetimes is described in Chap. 5). However spin–orbit coupling slightly mixes the two states so that the recombination time is finite but very long. At higher temperature, the singlet excitonic state becomes populated and the luminescence lifetime decreases. On the other hand, for selectively excited luminescence at 2K [216, 217], the threshold in the excitation spectrum could be due to the fact that the excitons are photo-generated in the $S = 0$ state while the luminescence originates from the $S = 1$ state. This could then explain qualitatively both types of experiments.

Figure 4.6 presents the calculated excitonic spectrum versus energy for a spherical crystallite. The spectrum is complex with many levels within

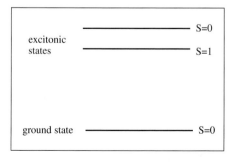

Fig. 4.5. Two-level model for the recombination of excitons in silicon nanocrystals

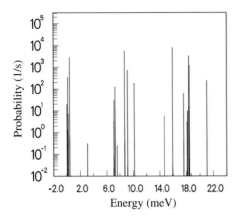

Fig. 4.6. Calculated excitonic spectrum of a spherical Si crystallite (diameter 3.86 nm). The levels are indicated by vertical bars. The zero of energy corresponds to the lowest exciton level. The height of the bars represents the calculated radiative recombination rate

20 meV. This is due to the multiple degeneracies of the conduction and valence bands from which the exciton states are built. The energy intervals between these levels (≈ 1 meV) are induced by the Coulomb and exchange terms [64, 221–223], and by inter-valley coupling (Sect. 2.5.2). The lowest excitonic state is often characterized by a small radiative recombination rate but the states just above can have much more important rates. The relative homogeneity of these rates is due to the spin–orbit coupling. If this one is neglected, the $S = 1$ states would have an infinite lifetime. However, in silicon, the situation is quite different from this limiting case since the spin–orbit coupling constant ($\lambda = 15$ meV) is comparable in magnitude to the other couplings (Coulomb, inter-valley, exchange) so that all triplet and singlet states are mixed. Nevertheless, due to the exchange coupling, a general trend is observed that the lower states have on the average a lower recombination rate. A Boltzmann thermal average of these rates can indeed be roughly simulated by a two-level system with adjusted parameters.

Another interesting effect is the importance of the geometry shown by the studies of spherical, ellipsoidal and undulating ellipsoidal nanocrystals [171]. The ellipsoids are defined by $(x/a)^2 + (y/b)^2 + (z/b)^2 = 1$ with $a > b$. For the undulating ellipsoids, a fluctuation is introduced on the surface by using a random combination of spherical harmonics with arbitrary orientation and amplitude [171]. Such shapes should describe reasonably the undulating wires which have been proposed as luminescent units for porous silicon [216]. The confinement then becomes anisotropic and the degeneracy of the states is lifted. In particular, the states at the top of the valence band, behaving like p_x, p_y and p_z, can be split into three distinct levels and the spin–orbit coupling is quenched when this splitting exceeds 15 meV. Figure 4.7 shows the corresponding exitonic spectrum for $a = b\sqrt{2}$ with the amplitude of the surface fluctuations fixed at 25% of the average radius (35% gives a similar result). The spectra are much simpler than before, the lower excitonic

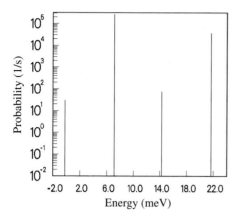

Fig. 4.7. Calculated excitonic spectrum of a spherical Si crystallite with complex shape built from an ellipsoid with a long axis of 2.4 nm, a short axis of 1.8 nm, and 25% of surface undulations. The levels are indicated by vertical bars. The zero of energy corresponds to the lowest exciton level. The height of the bars represents the calculated radiative recombination rate

state having systematically a much lower recombination rate than the first state above. The two-level model then correctly describes the asymmetric nanocrystals.

We plot in Fig. 4.8 the energy interval between the two lowest excitonic states of the same undulating ellipsoids. The results do not depend on the ratio a/b as long as this one is large enough. Calculations have been performed for several orientations of the main axis of the ellipsoids: (100), (110), (111)

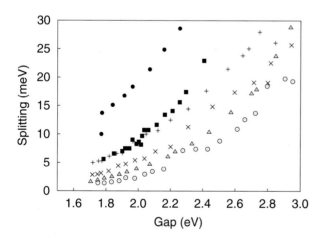

Fig. 4.8. Theory: splitting between the two lowest excitonic levels in Si crystallites with respect to their excitonic bandgap. The crystallites have undulating ellipsoidal shapes with a longer axis in the (100) (∘), (110) (△) and (111) (+) directions (average over the directions: ×). Experiments: onsets measured by selectively excited photoluminescence (■) and energy splittings derived from the fit of the temperature dependence of the luminescence lifetime (•) [216]

together with an average over all directions. The predicted values scale with size approximately like $1/R^{2.8}$ where R is the radius of the crystallite. As shown in [171], the calculations also predict a difference of two to three orders of magnitude between the lifetimes of the two lower levels of the undulating ellipsoids, in agreement with experiment and with the two-level model [216]. However, the calculated lifetimes are one to three orders of magnitude too long which is consistent with the importance of phonon-assisted transitions (see Sect. 5.4).

Figure 4.8 shows that the predicted excitonic splittings are, on the average, half of those measured by selectively excited photoluminescence. However, as shown in [171], there is another contribution with similar magnitude arising from the coupling to acoustic phonons and leading to a Stokes shift between absorption and photoluminescence that scales approximately like $1/R^3$. This contribution is discussed in Sect. 5.2.7.

4.3.4 Screening of the Electron–Hole Interaction and Configuration Interaction

The semi-empirical tight binding approach described in the previous sections parallels that of A. Zunger and collaborators [224] which is based on the empirical pseudopotential method (EPM). As shown in Chap. 2, both methods give identical results for the confinement energies. They also incorporate screening in the exciton problem in a similar way. In the following, we discuss the validity of this common approach and also comment on the inclusion of configuration interaction in this framework.

The calculations described in Sect. 4.3.3 and in [224] in fact make use of an expression like (4.12) for the exciton energy except that $\langle e^2/(4\pi\epsilon_0 \epsilon_{in} r_{eh}) \rangle$ is calculated in a more sophisticated way and that screened exchange terms are incorporated. In practice, both electron–hole Coulomb and exchange interactions are written in terms of a screened interaction $v_{sc}(\boldsymbol{r}_e, \boldsymbol{r}_h)$ which, generally, should be written

$$v_{sc}(\boldsymbol{r}_e, \boldsymbol{r}_h) = \frac{1}{4\pi\epsilon_0} \int \epsilon^{-1}(\boldsymbol{r}_e, \boldsymbol{r}) \frac{e^2}{|\boldsymbol{r} - \boldsymbol{r}_h|} d\boldsymbol{r} , \qquad (4.35)$$

where ϵ^{-1} is the complete inverse dielectric function. However the correct ϵ^{-1} must contain a contribution coming from the induced surface polarization charge (Chap. 3). If one separates this surface contribution, one could define an inner dielectric function ϵ_{in}^{-1} but this cannot be done exactly in a full calculation. Both calculations in Sect. 4.3.3 and in [224] have used this approach, taking different approximations for ϵ_{in}^{-1} and assuming complete cancellation of the surface polarization contributions contained in the quasi-particle energies and the electron–hole interactions. In Sect. 4.3.3, and in [171], the Fourier transform of ϵ_{in}^{-1} has been approximated by its diagonal part taken as $1/\epsilon_M(\boldsymbol{q})$ which is partly justified by the discussion of Sect. 3.4.4. On the other hand [224] approximates $\epsilon^{-1}(\boldsymbol{r}, \boldsymbol{r}')$ by $\delta(\boldsymbol{r} - \boldsymbol{r}')/\epsilon_{TF}(\boldsymbol{r}, \boldsymbol{r}')$

where ϵ_{TF} is a size dependent Thomas–Fermi model dielectric function. Both techniques may have their merits but the neglect of the induced surface polarization is not *a priori* justified, as shown by our simple calculation of (4.22) to (4.25) for the Coulomb terms. Their influence is expected to be weaker for exchange terms but this has not been checked quantitatively.

Another point of concern is the validity of a configuration interaction (CI) calculation as performed in [224] using screened interactions. As discussed in Chap. 1 for the derivation of GW (Sect. 1.2.4) and of the equations for excitons (Sect. 1.2.5), and also from the general considerations of Sect. 4.3.1, screening of the electron–hole interactions occurs as the result of the CI itself. In fact, a full CI interaction treatment would require use of unscreened interactions (e.g., see [206]). It is only when considering low energy excited configurations and folding the effect of higher excited configurations that one gets dynamically screened interactions as shown in Sect. 1.2.5. This means that the concept of screened interactions is valid only if low energy configurations are considered up to a given cut-off in energy lower than the plasmon energy.

In the next section, we apply the GW method for quasi-particle energies and we solve the equations of Sects. 1.2.4 and 1.2.5 for excitons to study nanostructures versus their dimensionality, which allows us to provide some answers to the previously raised problems.

4.4 Quantitative Treatment of Quasi-particles

In this section, the trends of the quasi-particle gap in semiconductor nanostructures versus dimensionality are discussed and compared to the value obtained in the local density approximation (LDA). General arguments are developed based on the GW approach which are then substantiated numerically by a tight binding version of this theory. The gap correction is shown to be dominated by the macroscopic surface self-polarization term and exhibits a non monotonic behavior versus dimensionality.

In the literature, most quantitative calculations deal with the eigenvalue gap ε_g^0 determined from the difference in one-particle eigenvalues $\varepsilon_c^0 - \varepsilon_v^0$ for the neutral system. ε_g^0 is obtained from empirical techniques (tight binding [221], pseudopotentials [76]) or from ab initio calculations in LDA [128] and, as we have seen, differs from $\varepsilon_g^{\mathrm{qp}}$ by large amounts $\delta\Sigma$ corresponding to self-energy corrections. These can be estimated via the GW method [32] derived in Sect. 1.2.4 but the corresponding computations are very time consuming and can be only applied to small systems [186, 187, 225–227]. Therefore simpler methods such as one-particle calculations are highly desirable but, as discussed above, their accuracy is a matter of controversies [228–230]. In principle, the quasi-particle gap $\varepsilon_g^{\mathrm{qp}}$ can be calculated exactly in density functional theory (DFT) as

$$\varepsilon_g^{\mathrm{qp}} = E(n+1) + E(n-1) - 2E(n) , \qquad (4.36)$$

where $E(n)$ is the total energy of the n-electrons neutral system obtained by solving the one-particle Kohn–Sham equations [15] which, as discussed in Sect. 1.1.5, are written in terms of an effective exchange–correlation potential V_{xc}. In LDA, $V_{xc}(\mathbf{r})$ is approximated locally by the corresponding expression of the homogeneous gas of the same electron density $n(\mathbf{r})$. We discuss in the following why the quasi-particle gap $(\varepsilon_g^{qp})_{LDA}$ obtained from (4.36) in LDA differs from the true ε_g^{qp} and, in finite systems, from the LDA eigenvalue gap $(\varepsilon_g^0)_{LDA}$. We write

$$\varepsilon_g^{qp} = (\varepsilon_g^{qp})_{LDA} + \Delta = (\varepsilon_g^0)_{LDA} + \delta\Sigma \ . \tag{4.37}$$

Here, we want to clarify the dependence of $\delta\Sigma$ and Δ upon the dimensionality of the nanostructure. This is important since Δ reflects a discontinuity of the exact V_{xc} of DFT (not contained in LDA) upon addition of one-electron or hole to the neutral system [231, 232]. We shall find that $\delta\Sigma$ exhibits a smooth decreasing behavior with increasing dimensionality. On the contrary Δ presents a peak between 0D and 3D, demonstrating the highly non-local nature of V_{xc} [233]. This behavior can be explained in terms of general arguments based on the GW approximation in which one can isolate a surface long range (macroscopic) contribution to the self-energy. These arguments are then confirmed via a tight binding GW calculation [234, 235], well adapted to quantitatively treat this macroscopic part.

4.4.1 General Arguments

We start from the GW expression (1.114) of the self-energy in terms of the dynamically screened electron–electron interaction W. In bulk metallic systems, W is the potential created by the electron surrounded by a full screening hole with size of the order of the Thomas–Fermi wavelength. In a bulk dielectric system, the long range screened potential is $1/(4\pi\epsilon_0\epsilon_M)|\mathbf{r} - \mathbf{r}'|$ where ϵ_M is the long wavelength dielectric constant. This now corresponds to a screening hole of magnitude $(1 - 1/\epsilon_M)$ around the electron, the corresponding screening charge $-(1 - 1/\epsilon_M)$ being repelled at infinity. However, as shown previously, for finite systems, there is a contribution W_{surf} to W coming from the fact that this screening charge is repelled on the surfaces. This corresponds to the macroscopic surface polarization charge in dielectrics and will give an additional important contribution $\delta\Sigma_{\text{surf}}$ to the self-energy correction. Thus, as discussed in Sect. 4.1, for finite systems the total $\delta\Sigma$ is equal to $\delta\Sigma_{\text{bulk}} + \delta\Sigma_{\text{surf}}$, where $\delta\Sigma_{\text{bulk}}$ is the bulk correction to the energy gap. In simple cases like spherical dots, thin layers or cylindrical wires $\delta\Sigma_{\text{surf}}$ can be readily estimated by the simple classical arguments of Chap. 3.

An interesting feature of W_{surf} is that it varies slowly within the nanostructure and can be treated as a macroscopic potential. From (1.114), the corresponding contribution Σ_{surf} to the self-energy is given by

$$\Sigma_{\text{surf}}(\boldsymbol{r},\boldsymbol{r}',\omega) = \frac{1}{\pi}\sum_k u_k(\boldsymbol{r})u_k^*(\boldsymbol{r}')\int_0^\infty d\omega'\left(\frac{n_k}{\varepsilon_k-\omega-\omega'}\right.$$
$$\left.-\frac{1-n_k}{\omega-\varepsilon_k-\omega'}\right)\text{Im }W_{\text{surf}}(\boldsymbol{r},\boldsymbol{r}',\omega')\ . \qquad (4.38)$$

Let us now consider $\langle u_c|\Sigma_{\text{surf}}(\varepsilon_c)|u_c\rangle$ for the LUMO (lowest unoccupied) state. The macroscopic potential W_{surf} will only mix u_c with states u_k extremely close in energy and local behavior, i.e. the nearby empty states for which $|\varepsilon_c - \varepsilon_k| \ll \omega' \approx \omega_s$ the plasmon energies. From (1.89) applied to the surface induced potential $(V_{\text{ind}})_{\text{surf}} = W_{\text{surf}}$, we see that we can use the static screening limit to obtain

$$\langle u_c|\Sigma_{\text{surf}}(\varepsilon_c)|u_c\rangle = \frac{1}{2}\sum_{c'}\langle u_c u_{c'}|W_{\text{surf}}(\omega=0)|u_{c'}u_c\rangle\ , \qquad (4.39)$$

where the sum is over empty states $u_{c'}$. We now perform a unitary transformation from the delocalized $u_{c'}$ states to localized Wannier functions c_j and get

$$\langle u_c|\Sigma_{\text{surf}}(\varepsilon_c)|u_c\rangle = \frac{1}{2}\sum_j |\langle u_c|c_j\rangle|^2 \langle c_j c_j|W_{\text{surf}}(\omega=0)|c_j c_j\rangle\ . \qquad (4.40)$$

A similar expression holds true with the opposite sign for a hole in the HOMO (highest occupied) state in terms of the Wannier functions of the valence band. $\delta\Sigma_{\text{surf}}$ is then obtained as the difference between these two quantities which should practically be equal to averages of the classical image potential over the quantum state of interest.

One can wonder if this surface contribution is contained in a LDA calculation of nanostructures such as the one performed in [230], i.e. in $(\varepsilon_g^{\text{qp}})_{\text{LDA}}$ obtained from (4.36). The difference $E(n+1) - E(n)$ can be calculated to lowest order (equivalent to linear screening) by using Slater's transition state [236] as detailed in [237], expressing $E(n+1) - E(n) = \varepsilon_c(n+1/2)$, the LUMO calculated self-consistently with 1/2 electron occupation. This corresponds to a bare excess electron density $(1/2)|u_c(\boldsymbol{r})|^2$ which should be screened. For 1D, 2D and 3D systems where the wave function has infinite extension, this is vanishingly small with no net effect on $\varepsilon_c(n+1/2) = \varepsilon_c(n)$. This is not true however for 0D systems where this density is finite, of order $1/\Omega$ (Ω being the quantum dot volume). In that case a total screening charge of $-(1/2)(1-1/\epsilon_{\text{in}})$ is repelled on the surface giving a contribution analogous to (4.40). The same is true for a hole. We thus end up with the conclusion that $\delta\Sigma_{\text{surf}}$ is correctly obtained in LDA for 0D systems but not for 1D and 2D systems.

4.4.2 Tight Binding GW Calculations

We now want to substantiate these general arguments by a more detailed calculation. However ab initio GW calculations for large nanostructures are presently not possible due to computational limits. This is why we discuss here the tight binding formulation of [234, 235] performed for spherical nanocrystals with diameter up to 2.2 nm and for quantum wells with (001) surfaces. In both cases, surface dangling bonds are saturated by hydrogen atoms to avoid spurious states in the bandgap. We have seen in Sects. 1.3.1 and 3.3.2 that tight binding is an efficient way to calculate ϵ^{-1} even for large crystallites [190] and that it can be applied to simplify the GW calculation. In (1.114), the eigenstates u_l are defined in an atomic basis composed of one s and three p orbitals for each silicon atom. As shown in detail in Sect. 1.3.1, due to the neglect of terms involving overlaps of different atomic orbitals, the main advantage of the tight binding method is that all the functions and operators (e.g. ϵ^{-1}, v) are defined by matrices at discrete values of r corresponding to the atomic positions R_i, the size of the matrices being equal to the number of atoms in the system. The matrix of W is equal to the product of the matrix of ϵ^{-1} and v. ϵ itself is equal to $I - v\chi$ where χ is the polarizability matrix [190]. Σ is thus also defined by a matrix which can be calculated from (1.114).

The previous discussion shows that tight binding allows to get information on the self-energy operator Σ. However what is really needed is, starting from a given independent particle Hamiltonian h, to get the corresponding self-energy correction $\delta\Sigma$. Calling v_{xc} the exchange–correlation part of h, $\delta\Sigma$ can be expressed to first order in perturbation as [238]

$$\delta\Sigma = \langle u_c | \Sigma(\varepsilon_c) - v_{xc} | u_c \rangle - \langle u_v | \Sigma(\varepsilon_v) - v_{xc} | u_v \rangle . \tag{4.41}$$

The main problem is then to calculate v_{xc}. The most natural method is to start from the tight binding Hamiltonian $h = h_{TB}$. The corresponding $v_{xc} = (v_{xc})_{TB}$ is simply transferred without change from the bulk to the cluster case. It thus represents the best approximation to Σ_{bulk}, the bulk self-energy. Thus a simple recipe is to replace $(v_{xc})_{TB}$ by $\Sigma_{bulk}(\varepsilon_{c,bulk})$ in the first term of (4.41) and by $\Sigma_{bulk}(\varepsilon_{v,bulk})$ in the second one, where $\varepsilon_{c,bulk}$ and $\varepsilon_{v,bulk}$ are the bulk values. The corresponding results are given on Fig. 4.9 as a function of the cluster radius R.

For obvious reasons one might also want to get $\delta\Sigma$ starting from ab initio LDA calculations, i.e. using in (4.41) Σ deduced from tight binding with the corresponding $(v_{xc})_{LDA} \equiv V_{xc}$. In this case it is more difficult to determine $\Sigma - (v_{xc})_{LDA}$ since a central difficulty in tight binding comes from the use of a minimal basis set, so that the completeness relation $\sum_k u_k(r) u_k^*(r') = \delta(r - r')$ is not verified. The consequence is that the short-range part of Σ (when $r \to r'$) is not correctly described [238] and it is precisely this part which is well approximated by $(v_{xc})_{LDA}$ [239]. We have thus calculated $\Sigma - (v_{xc})_{LDA}$ by two distinct methods:

1. following the arguments of [239], we consider that the short range part of Σ corresponds to $(v_{xc})_{LDA}$ so that the matrix $\Sigma - (v_{xc})_{LDA}$ is simply equal to Σ in which the diagonal terms are removed;

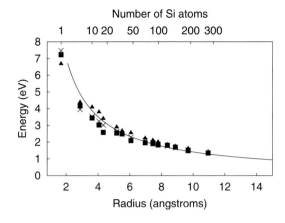

Fig. 4.9. Variation versus size of the self-energy correction $(\delta\Sigma - \delta\Sigma_{\text{bulk}})$ in spherical Si nanocrystals: full tight binding (■), first LDA method (×) and second LDA method (▲). Macroscopic electrostatic energy $\delta\Sigma_{\text{surf}}$ (*straight line*) for the separate addition of an electron and a hole

2. following [238], we replace $(v_{\text{xc}})_{\text{LDA}}$ by the self-energy operator Σ^{hom} of the homogeneous electron gas with the same electron density. The matrix elements of Σ^{hom} are also calculated in tight binding. The homogeneous gas is described using a simple Thomas–Fermi model where the matrix χ of the polarizability is diagonal. Each diagonal term of χ is equal to $N(\varepsilon_F)$, the density of states per atomic volume at the Fermi level of the free electron gas (we use for the atomic volume of silicon the bulk silicon value and for hydrogen a spherical volume of radius 1 Å).

Details on the tight binding calculation are given in [234]. The self-energy corrections to LDA calculated for bulk Si $(\delta\Sigma_{\text{bulk}})$ are respectively 0.41 eV and 0.75 eV with the first and second methods, to compare with an average difference of 0.65 eV between experimental and LDA gaps [225–227] (the full tight binding one is zero by construction). On Fig. 4.9 we plot $\delta\Sigma - \delta\Sigma_{\text{bulk}}$ calculated for nanocrystals containing up to 275 Si atoms. In spite of their differences, the three approaches give very similar results, especially for $R > 0.6$ nm.

Figure 4.9 shows that the main contribution to $\delta\Sigma - \delta\Sigma_{\text{bulk}}$ is actually the classical electrostatic surface polarization effect discussed in Sects. 4.1, 3.5.1 and in [166]: when one puts an extra electron (or hole) into a nanocrystal, the electronic relaxation (screening) induces charges at the surface and the extra particle interacts with this self-image charge distribution leading to a self polarization energy. Averaging this quantity over the cluster with a statistical weight given by the particle wave function, the total result $\delta\Sigma_{\text{surf}}$ is obtained in this way for the separate addition of one electron plus a hole into the cluster. An excellent approximation is obtained by using an effective mass wave function $\sin(\pi r/R)/r$ which, from (3.67) and (3.68) applied to a cluster embedded in vacuum, leads to the analytic expression

$$\delta\Sigma_{\text{surf}} \approx \left(1 - \frac{1}{\epsilon_{\text{in}}}\right) \frac{e^2}{4\pi\epsilon_0 R} + 0.94 \frac{e^2}{4\pi\epsilon_0 \epsilon_{\text{in}} R} \left(\frac{\epsilon_{\text{in}} - 1}{\epsilon_{\text{in}} + 1}\right). \quad (4.42)$$

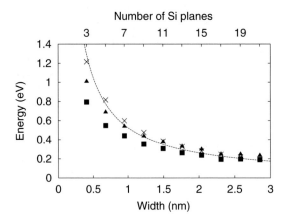

Fig. 4.10. Variation versus size d of the self-energy correction ($\delta\Sigma - \delta\Sigma_{\text{bulk}}$) in Si layers: full tight binding (■), first LDA method (×) and second LDA method (▲). Macroscopic electrostatic energy $\delta\Sigma_{\text{surf}}$ (*straight line*) for the separate addition of an electron and a hole

We have seen in Chap. 3 that, in a nanostructure, one can still define a local macroscopic dielectric constant, equal to its bulk value ϵ_M within the cluster, except in the vicinity of the surface where the probability $[\sin(\pi r/R)/r]^2$ of finding the charge carrier practically vanishes. Thus it is fully justified to use for ϵ_{in} in (4.42) the corresponding bulk value ϵ_M. This gives the continuous line of Fig. 4.9, in excellent agreement with the numerical values. However the use of an average size dependent $\epsilon_{\text{in}}(R)$ as in [166] makes little difference except for very small clusters.

In Fig. 4.10, we plot $\delta\Sigma - \delta\Sigma_{\text{bulk}} = \delta\Sigma_{\text{surf}}$ for a (100) quantum well of width d. As for spherical nanocrystals, we find that $\delta\Sigma_{\text{surf}}$ is close to the macroscopic value calculated using the image-charge method presented in Sect. 3.5.2, assuming an effective mass wave function $\cos(\pi z/d)$ for the electron and the hole. From (3.76) with $\epsilon_1 = \epsilon_{\text{in}}$ and $\epsilon_2 = 1$, we have

$$\delta\Sigma_{\text{surf}} \approx \frac{2e^2}{4\pi\epsilon_0 \epsilon_{\text{in}} d} \left[0.219 \left(\frac{\epsilon_{\text{in}} - 1}{\epsilon_{\text{in}} + 1} \right) - \ln\left(\frac{2}{\epsilon_{\text{in}} + 1} \right) \right] . \tag{4.43}$$

4.4.3 Conclusions

From the previous sections, we see that one can write to an excellent degree of accuracy and for any dimensionality

$$\varepsilon_g^{\text{qp}} = \varepsilon_g^0 + \delta\Sigma \approx \varepsilon_g^0 + \delta\Sigma_{\text{bulk}} + \delta\Sigma_{\text{surf}} . \tag{4.44}$$

It applies to any type of one-particle calculations, in particular to LDA where $\delta\Sigma_{\text{bulk}} \approx 0.65$ eV [225, 226]. This is supported by recent results [64, 132, 133] which consistently show that $(\varepsilon_g^0)_{\text{TB}}$ obtained by the best tight binding methods agrees with $(\varepsilon_g^0)_{\text{LDA}} + (\delta\Sigma_{\text{bulk}})_{\text{LDA}}$ and with the EPM value $(\varepsilon_g^0)_{\text{EPM}}$.

In Fig. 4.11, we plot the variations of $\delta\Sigma_{\text{surf}}$ when going from 0D to 3D [235]. As a gedanken experiment, we can consider a continuous change in the shape of a nanostructure with an ellipsoid surface of equation $x^2/a^2 +$

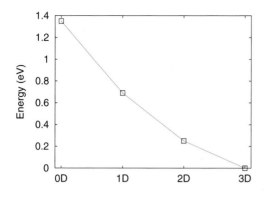

Fig. 4.11. Variation of $\delta\Sigma - \delta\Sigma_{\text{bulk}}$ in Si nanostructures of increasing size, going from a sphere (0D) of diameter d (d = 2 nm) to a cylindrical wire (1D) of diameter d to a well (2D) of width d and then to the bulk. The lines are only guides for the eyes

$y^2/b^2 + z^2/c^2 = 1$, going from a sphere of diameter d ($a = b = c = d/2$), to a cylindrical wire ($a \to \infty$), to a well ($b \to \infty$) and to the bulk ($c \to \infty$). It is important to note that the size of the nanostructure is always increasing, and thus we expect that the influence of the surfaces must decrease. This is verified in Fig. 4.11 where $\delta\Sigma_{\text{surf}}$ is continuously decreasing as must be the case for the self interaction energy with surface polarization charges.

From this, one can deduce the variations of the gap correction $\Delta - \Delta_{\text{bulk}}$ from those of $\delta\Sigma - \delta\Sigma_{\text{bulk}} = \delta\Sigma_{\text{surf}}$. From (4.37) and using the fact that $\Delta_{\text{bulk}} = \delta\Sigma_{\text{bulk}}$, we obtain that :

$$\Delta - \Delta_{\text{bulk}} = \delta\Sigma_{\text{surf}} - \left[(\varepsilon_g^{\text{qp}})_{\text{LDA}} - (\varepsilon_g^0)_{\text{LDA}}\right] . \quad (4.45)$$

For reasons discussed previously $(\varepsilon_g^{\text{qp}})_{\text{LDA}} = (\varepsilon_g^0)_{\text{LDA}}$ in extended systems (1D, 2D, 3D), while at 0D the situation is different because $(\varepsilon_g^{\text{qp}})_{\text{LDA}}$ fully includes the surface contribution $\delta\Sigma_{\text{surf}}$. This leads to the curve of Fig. 4.12, calculated for $d = 2$ nm. Contrary to $\delta\Sigma - \delta\Sigma_{\text{bulk}}$ the behavior of the gap correction is strongly non monotonic with dimensionality.

The main result here is that the correction to the exchange–correlation potential is much larger in wires and wells than in spherical dots, whereas

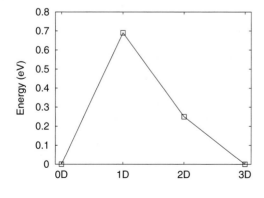

Fig. 4.12. Variation of $\Delta - \Delta_{\text{bulk}}$ in Si nanostructures of increasing size, going from a sphere (0D) of diameter d (d = 2 nm) to a cylindrical wire (1D) of diameter d to a well (2D) of width d and then to the bulk. The lines are only guides for the eyes

the system size is constantly increasing. Thus $\Delta - \Delta_{\text{bulk}}$ does not follow a simple scaling law with respect to the size of the nanostructure. These results point out a fundamental difference between finite and infinite systems as regards their description in a Kohn–Sham approach. It is particularly interesting to note that a similar conclusion can be drawn for calculations of optical spectra in time-dependent density functional theory using the adiabatic LDA (TDLDA) for the exchange–correlation kernel f_{xc}. It turns out that in general TDLDA yields good results in finite systems [36, 48], with spectra considerably improved compared to bare spectra based on time-independent Kohn–Sham LDA eigenvalues (e.g. [240]), but for solids TDLDA results are close to those obtained in a simple random-phase approximation [36, 48, 50] (see Sect. 1.2.7). Transition energies in finite systems calculated in TDLDA include the time-dependent response to the density variation due the electron–hole excitation, which is infinitesimally small in extended systems [48, 241].

We can now discuss the origin of the evolution of $\Delta - \Delta_{\text{bulk}}$. Taking into account that the variations of the self-energy correction $\delta\Sigma - \delta\Sigma_{\text{bulk}}$ have a simple dependence on the system size which can be understood from macroscopic electrostatics, the behavior of $\Delta - \Delta_{\text{bulk}}$ reflects that the decomposition between the different terms in the Kohn–Sham effective potential is arbitrary [233]. This is particularly clear when going from 0D to 1D (i.e. varying a from d to infinity) since the same physical quantity, the self-polarization energy, which is provided by the Hartree self-consistent term at 0D has to be totally included in the true V_{xc} at 1D. Thus any attempt to include surface self-polarization terms in the exchange–correlation energy, for example using a more sophisticated description of the exchange–correlation hole in the vicinity of a surface would require for V_{xc} an ultra-non-local functional of the electron density.

Our previous considerations show how one might improve current LDA calculations of nanostructures in a simple way. One possibility would be to use a standard LDA calculation for the nanostructure, then add the bulk correction $\delta\Sigma_{\text{bulk}}$ as a scissor operator and finally determine the surface correction $\delta\Sigma_{\text{surf}}$. To calculate this macroscopic surface polarization term, one could discretize the system into cells (of the order of the atomic cell or more), evaluate the charge in such a cell, calculate self-consistently the corresponding W_{surf} and determine the average self-energy correction from (4.39) or (4.40).

As a final point, we would like to underline the fact that the macroscopic surface contribution also occurs for metallic nanostructures, even when using a free electron approximation. In that case the full screening hole is repelled to the surface ($\epsilon_{\text{in}}^{-1} \to 0$). It is this term which is at the origin of the Coulomb blockade effect and which is usually treated in terms of capacitance in association with the electrodes (Chap. 8).

4.5 Quantitative Treatment of Excitons

We now consider numerical calculations of the excitonic gap performed via direct resolution of (1.125) and (1.126) [186, 187, 234]. Again, we consider silicon crystallites as a test case. We start from (4.10) and express the excitonic gap $\varepsilon_g^{\text{exc}}$ as the difference between the quasi-particle gap $\varepsilon_g^{\text{qp}}$ and E_{coul}, the attractive interaction between these two quasi-particles. We have

$$\varepsilon_g^{\text{exc}} = \varepsilon_g^{\text{qp}} - E_{\text{coul}} = \varepsilon_g^0 + \delta\Sigma - E_{\text{coul}}, \tag{4.46}$$

where $E_{\text{coul}} = -\langle H_{\text{eh}} \rangle$ of (4.10) and $\varepsilon_g^{\text{qp}}$ is written as the sum of the independent particle value ε_g^0 and the self-energy correction $\delta\Sigma$. We shall see that there is strong cancellation between the two large quantities $\delta\Sigma - \delta\Sigma_{\text{bulk}} = \delta\Sigma_{\text{surf}}$ and E_{coul}, such that $\varepsilon_g^{\text{exc}} \approx \varepsilon_g^0 + \delta\Sigma_{\text{bulk}}$. This justifies why the single particle calculations yield accurate results for $\varepsilon_g^{\text{exc}}$. We also show that E_{coul} like $\delta\Sigma$ is dominated to a large extent by surface polarization charges, and we discuss on this basis the amount of cancellation between $\delta\Sigma_{\text{surf}}$ and E_{coul}.

4.5.1 Numerical Calculations

These calculations proceed in two steps:

- by calculating the separate electron and hole quasi-particle energies via the GW method discussed in the previous section
- by determining the attractive Coulomb interaction between these quasi-particles via the resolution of (1.125) and (1.126).

Such work has already been achieved with success from an ab initio point of view for bulk semiconductors [242], Na_4 clusters [186] and small silicon clusters saturated by hydrogen atoms (up to $Si_{14}H_{20}$) [187]. However the computation is very time consuming and cannot be extended to nanocrystals. This is why the use of the tight binding formulation allows again to treat much larger clusters and get more information about the trends of $\varepsilon_g^{\text{exc}}$ with size.

Let us then rewrite (1.125) and (1.126) which can be derived from the Bethe–Salpeter equations [187]. We first consider the case of triplet states for which we can drop the exchange terms (their influence will be discussed later). As shown in the derivation of (1.125), the excitonic states can be obtained from a diagonalization of an Hamiltonian, expressed in a basis of single particle excited states ψ_{vc} as

$$\langle \psi_{vc} | H_{\text{eff}} | \psi_{v'c'} \rangle = (\varepsilon_c^{\text{qp}} - \varepsilon_v^{\text{qp}}) \delta_{vv'} \delta_{cc'} - \langle cv' | v_{\text{SCH}} | c'v \rangle, \tag{4.47}$$

where the screened Hartree potential is obtained from (1.126). In (4.47), we have made apparent that ε_c and ε_v of (1.125) are the quasi-particle energies $\varepsilon_c^{\text{qp}}$ and $\varepsilon_v^{\text{qp}}$ obtained by the GW method. As for GW, these matrix elements are calculated in a tight binding framework. For the frequency dependence of (1.126), a single plasmon pole approximation is used together with a first order expansion of the correction with respect to the static approximation

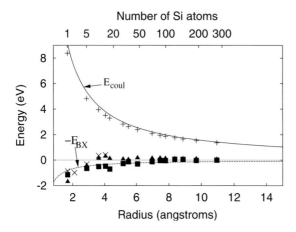

Fig. 4.13. Exciton Coulomb energy E_{coul} versus size in Si nanocrystals: full GW + Bethe–Salpeter calculation (+), classical electrostatics calculation with effective mass wave functions (*straight line*). Difference between the self-energy correction $\delta\Sigma - \delta\Sigma_{\text{bulk}} = \delta\Sigma_{\text{surf}}$ and E_{coul}: full tight binding calculation (■), second LDA approximation (▲) and ab initio results (×) from [187]. E_{BX} is the classical binding energy of the exciton given by (4.25)

[187, 234]. The matrix equation (4.47) is then diagonalized increasing the number of electron hole states till convergence is reached (this usually requires 10 electron and hole states). The lowest eigenvalue obtained in this way thus corresponds to the triplet exciton gap $\varepsilon_g^{\text{exc}}$. The corresponding results are given on Fig. 4.13. However, for reasons which will become clear later, it is more interesting to plot E_{coul} versus size, taken from (4.46) as the difference $\varepsilon_g^{\text{qp}} - \varepsilon_g^{\text{exc}}$. On the same figure, the computed E_{coul} is also compared with the result of the classical electrostatic argument [166] discussed in Sect. 4.2.4, where the effective interaction for the electron and hole at distance r_{eh} is the sum of two terms: a direct screened interaction plus the interaction of one particle with the polarization charge induced by the other. Taking the average of this with respect to the electron and hole distribution in the effective mass approximation gives (4.21) to (4.24). With $\epsilon_{\text{out}} = 1$, this leads to

$$E_{\text{coul}} \approx \left\langle \frac{e^2}{4\pi\epsilon_0 \epsilon_{\text{in}} r_{\text{eh}}} \right\rangle + \frac{\alpha_{\text{eh}}}{4\pi\epsilon_0 R}(\epsilon_{\text{in}} - \epsilon_{\text{out}}) = \left(\frac{0.79}{\epsilon_{\text{in}}} + 1\right) \frac{e^2}{4\pi\epsilon_0 R} , \quad (4.48)$$

which we plot on Fig. 4.13. The values for E_{coul} are extremely well approximated by the classical law.

Figure 4.13 also shows the difference $(\delta\Sigma - \delta\Sigma_{\text{bulk}}) - E_{\text{coul}}$ for the two extreme values of $\delta\Sigma - \delta\Sigma_{\text{bulk}}$ obtained in [234] (see Fig. 4.12). The latter are compared with the same quantity obtained from the full ab initio GW calculation [187] for SiH_4, Si_5H_{12}, $Si_{10}H_{16}$ and $Si_{14}H_{20}$. The tight binding values fall in the same range as the ab initio values, especially those arising from the second LDA model. A striking feature displayed by Fig. 4.13 is that the quantities E_coul and $\delta\Sigma - \delta\Sigma_\mathrm{bulk}$, while being pretty large, compensate each other to a large degree and, for clusters with $R > 0.6$ nm, the two quantities are practically identical so that their contributions to the excitonic gap cancel each other.

4.5.2 Interpretation of the Results

From (4.46), the compensation between E_coul and $\delta\Sigma - \delta\Sigma_\mathrm{bulk}$ leads to

$$\varepsilon_\mathrm{g}^\mathrm{exc} \approx \varepsilon_\mathrm{g}^0 + \delta\Sigma_\mathrm{bulk} \,. \tag{4.49}$$

This means that $\varepsilon_\mathrm{g}^\mathrm{exc}$ is, to an excellent approximation, directly given by the single-particle gap $(\varepsilon_\mathrm{g}^0)_\mathrm{TB}$ in full tight binding (where $\delta\Sigma_\mathrm{bulk} = 0$) and $(\varepsilon_\mathrm{g}^0)_\mathrm{LDA} + 0.65$ eV in LDA calculations. This result not only justifies the use of single particle calculations to get the excitonic gap but also explains the agreement between empirical and LDA results once these are shifted by the bulk correction 0.65 eV [64]. Of course the cancellation is not strictly exact but for $R > 0.6$ nm it is verified to better than 0.2 eV on Fig. 4.13. One can also notice that (4.49) is likely to hold true to some extent for other semiconductor crystallites. It has been checked in [234] that this is indeed the case for Ge and even for C for which, at $R = 0.8$ nm the deviation from perfect cancellation is 0.8 eV still small compared to the gap value (≈ 12 eV in this case).

An important point to consider is the accuracy of tight binding predictions. The most important source of errors is probably the short range contribution in the GW part. In this regard one measure of the uncertainty in the calculations is the dispersion of the results for $\delta\Sigma - \delta\Sigma_\mathrm{bulk}$ between the three approximations used to include this short range term. The corresponding error is ± 0.2 eV at $R = 0.6$ nm but decreases very rapidly with size to become practically negligible at $R = 0.8$ nm. Another interesting point is illustrated on Fig. 4.13 which shows that the results with the Thomas–Fermi approximation (second LDA model) agree well with the ab initio calculations for small crystallites. As they also provide a fairly accurate bulk value $\delta\Sigma_\mathrm{bulk} = 0.75$ eV, this certainly means that they must remain practically exact over the whole range of sizes, strengthening the conclusion concerning the amount of cancellation between E_coul and $\delta\Sigma - \delta\Sigma_\mathrm{bulk}$.

Finally the cancellation has been confirmed by recent ab initio calculations using the quantum Monte-Carlo method [243], in quantitative agreement with the tight binding predictions.

To explore the generality of this effect it is necessary to come back to the general analysis of Sect. 4.1, in particular to (4.11) which contains all effects discussed above. Let us then first compare it to the numerical results of Fig. 4.13. For the macroscopic surface contribution, it is certainly a quite accurate approximation to use $\sin(\pi r/R)/r$ wave functions leading to α_1 and α_eh given by (4.22) and (4.23). Furthermore, in the strong confinement limit and for a non degenerate excitonic state, the term $\langle T - \frac{e^2}{4\pi\epsilon_0\epsilon_{in}r_{eh}}\rangle$ can be analyzed by first order perturbation theory, for which $\sin(\pi r/R)/r$ wave functions are certainly pretty good. Thus the binding energy of the exciton E_{BX} (which corresponds to our $\delta\Sigma - \delta\Sigma_{bulk} - E_{coul}$) given by (4.25) should provide an already accurate answer. It is exactly what is demonstrated in Fig. 4.13 where (4.25) is represented for $\epsilon_{out} = 1$. This curve provides a fair average of the numerical results except that it does not contain their oscillations which are in fact due to interference effects characteristic of the degeneracies of the low energy excitonic states. Thus probably the most accurate economical method would be to use (4.11), evaluate the surface correction (term $\alpha_1 - \alpha_{eh}$) with effective mass wave functions and then calculating directly $\langle T - \frac{e^2}{4\pi\epsilon_0\epsilon_{in}r_{eh}}\rangle$ as was done in Sect. 4.3. However, in all cases of interest, the surface correction is likely to be quite small and should not affect the excitonic splittings determined in this section.

A last point concerns the influence of the surface polarization contribution on the exchange terms. From the previous discussions, the corresponding induced potential varies slowly in space and should not affect the exchange terms that contain products $u_c(\boldsymbol{r})u_v^*(\boldsymbol{r})$ which strongly oscillate in space.

As a general conclusion, the full numerical calculations tend to support semi-empirical methods which, starting from ε_g^{qp}, diagonalize the screened electron–hole Hamiltonian in the basis of the states corresponding to single electron–hole excitations. However the use of static screening might become wrong if ε_g/ω_s becomes too large. Finally, in all these problems, the use of bulk dielectric screening within the cluster is appropriate for reasons given in Chap. 3 and gives results in better agreement with numerical calculations.

4.5.3 Comparison with Experiments

Since the discovery of the photoluminescence of porous silicon in 1990 [244], much work has been carried out to study the optical properties of silicon nanocrystals. After many debates, it seems that there are several origins to the luminescence depending on the nature of the materials, their synthesis and their treatment. In most cases, the luminescence is either due to electron–hole recombination between quantum confined levels in the nanocrystals or involves defect states at the interface between the nanocrystals and their surrounding oxide layer. Recently, experiments performed on silicon nanocrystals produced by laser pyrolysis of silane have shown that, in this case, the photoluminescence could be most probably associated with quantum confinement effects in a wide range of size [245, 246]. A size-selective deposition of the nanocrystals on quartz substrates allows to measure the photoluminescence arising from an extremely narrow distribution of sizes. Figure 4.14 presents the results for three distinct samples. Note that the estimation of the size takes into account the shrinking of the crystalline core as a result of the surface oxidation [247]. We also plot for comparison the gap versus size obtained from tight binding calculations [64] ((ε_g^0)$_{TB}$) and the gap corrected

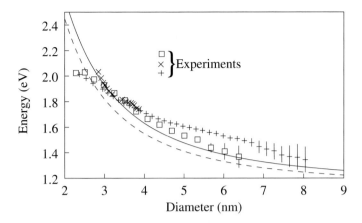

Fig. 4.14. Correlation between average diameter and photoluminescence peak energy measured on three different samples [247]. Tight binding prediction for the gap versus size without (*straight line*) or with (*dashed line*) the correction -E_{BX} given in (4.25)

by E_{BX}. We have seen in Sect. 2.3 that these calculations predict bandgaps in good agreement with other methods. The agreement between theory and experiments in Fig. 4.14 is good for sizes around 3 nm. In the range 4-7.5 nm, photoluminescence peak energies are higher than the predicted ones, especially for one sample (+) for reasons which are not yet understood. The discrepancy in this range is larger than the difference between the different theoretical predictions (see Fig. 2.10 in Chap. 2). For particles with diameters below 3 nm, the saturation of the photoluminescence energy around 2 eV could be due to the appearance of an oxide-related surface state within the gap of the nanocrystals [248].

4.6 Charging Effects and Multi-excitons

We discuss here situations frequently observed experimentally where injection of carriers and excitations across the gap result in charging effects and multi-excitons. We describe calculations with different levels of sophistication and compare them to the results of tunneling spectroscopy and photoluminescence experiments.

4.6.1 Charging Effects: Single Particle Tunneling Through Semiconductor Quantum Dots

We consider the tunneling spectroscopy experiments on InAs nanocrystals performed by Banin et al. [249] using a scanning tunneling microscope. They reveal rich features due to the interplay between quantum confinement and

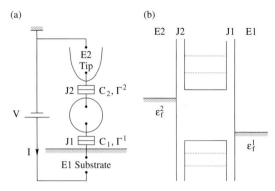

Fig. 4.15. Typical double barrier tunnel junction. (a) It consists of two metallic electrodes E1 and E2 (e.g. a substrate and the tip of a scanning tunneling microscope) coupled to a nanostructure by tunnel junctions J1 and J2 with capacitances C_1 and C_2 and tunneling rates Γ^1 and Γ^2. (b) E1 and E2 are characterized by Fermi energies ε_f^1 and ε_f^2.

charging effects. In this context, we start by describing the calculations [138, 173] which allow a detailed understanding of the experimental data.

In [138, 173], the energy levels ε_i^e and ε_i^h of spherical InAs nanocrystals have been calculated with a sp^3 tight binding model with second nearest neighbors interactions, as described in Sects. 2.4 and 2.5. The lowest conduction level (1S$_e$) is s-like, twofold degenerate, and the next level (1P$_e$) is p-like, sixfold degenerate (see Fig. 2.12 in Sect. 2.5). The highest two valence levels are found fourfold degenerate. The calculations of the transport properties use an extension of the theory of Averin et al. [250] (see Sect. 8.2.2) and consider a standard double barrier tunnel junction [251] (Fig. 4.15). The system consists of two metallic electrodes E1 and E2 weakly coupled to a semiconductor quantum dot by two tunnel junctions J1 and J2 with capacitances C_1 and C_2. The metallic electrodes E1 and E2 are characterized by their Fermi energies $\varepsilon_f^1 = \varepsilon_f - e\varphi$ and $\varepsilon_f^2 = \varepsilon_f$, where φ is the bias voltage. The total energy of the quantum dot charged with n electrons and p holes with respect to the neutral state can be approximated by [138, 251] (Sect. 3.5.1):

$$E(\{n_i\}, \{p_i\}, \varphi) = \sum_i n_i \varepsilon_i^e - \sum_i p_i \varepsilon_i^h + \eta e\varphi q + \frac{1}{2} U q^2 \ . \tag{4.50}$$

ε_i^e and ε_i^h are the conduction and valence energy levels in the quantum dot, n_i and p_i are electron and hole occupation numbers ($n = \sum_i n_i$, $p = \sum_i p_i$), and $q = p - n$. In terms of the junction capacitances C_1 and C_2, $U = e^2/(C_1 + C_2)$ is the charging energy (Sect. 3.5.1), and $\eta = C_1/(C_1 + C_2)$ is the part of the bias voltage φ that drops across junction J2 in the neutral quantum dot. Tunneling of an electron via the energy level ε_i^e occurs at transition energy (see also Sect. 8.2.2)

$$\varepsilon_i^e(q|q-1,\varphi) = E(n_i = 1, \{p_j\}, \varphi) - E(n_i = 0, \{p_j\}, \varphi) \,, \tag{4.51}$$

$$= \varepsilon_i^e - \eta e\varphi + U(-q + \frac{1}{2}) \,. \tag{4.52}$$

Symmetrically, hole tunneling occurs at $\varepsilon_i^h(q+1|q,\varphi)$. The current is calculated using the orthodox theory presented in Sect. 8.2.2 where one defines tunneling rates [138, 250] through the junctions (Fig. 4.15). Both electrons and holes are treated at the same time incorporating the electron–hole recombination rate $R(n,p)$ from the charge state (n,p) to the charge state $(n-1, p-1)$ into the master equations [138, 173]. At $T \to 0$ K, the $I(\varphi)$ curve looks like a staircase [250] (Sect. 8.2.2). It exhibits a step each time ε_f^1 or ε_f^2 crosses a transition energy. A new charge state then becomes available in the quantum dot (addition step), or a new channel ε_i^e or ε_i^h is opened for tunneling to a given, already available charge state (excitation step).

This behavior is apparent in the results of Banin et al. [249]. The differential conductance $G(\varphi) = dI(\varphi)/d\varphi$ is shown in Fig. 4.16 for an InAs nanocrystal 6.4 nm in diameter. The tip was retracted from the quantum dot so that C_1/C_2 is maximum and η is close to unity. A zero-current gap is observed around $\varphi = 0$, followed by a series of conductance peaks for $\varphi < 0$ and $\varphi > 0$.

To compare with the interpretation of Banin et al. [249], two types of calculations have been performed [138, 173]:

- using the capacitive model of (4.50) with the calculated tight binding level structure, U and η being considered as fitting parameters chosen to optimize the agreement with the position of the peaks in the $G(\varphi)$ curve
- a full self-consistent treatment on a system with a realistic geometry described in Fig. 4.17 and in [138, 173]. The ground state energy $E_0(n, p, \varphi)$ is self-consistently computed for a set of charge states (n, p) and several voltages φ_i. This is done in the Hartree approximation corrected from the unphysical self-interaction term [138] (the method is described in Sect. 3.2.2). The electrostatic potential inside the quantum dot is computed with a finite difference method. The tunneling rates (Sect. 8.2.2) are taken as adjustable parameters but the position of the calculated conductance peaks does not depend on their value. Details on the calculation are given in [138].

The calculated $G(\varphi)$ curves are compared to the experimental one on Fig. 4.16. The agreement with experiment is extremely good with practically a one to one correspondence between the calculated and experimental peaks over a range of 3.5 V. The negative bias voltages side is clearly improved in the self-consistent calculation.

For $\varphi > 0$, the first group of two peaks is assigned to the tunneling of electrons filling the $1S_e$ level [249], the splitting between the two peaks corresponding to the charging energy. Similarly, the next group of six peaks mainly corresponds to the tunneling of electrons through the $1P_e$ level, and there is some contributions from the tunneling of holes. There are also two excitation peaks X_1 and X_2 on Fig. 4.16 (tunneling through the $1P_e$ level in the charge states $n = 0$ and $n = 1$) that are hardly visible on the experimental $G(\varphi)$ curve.

For $\varphi < 0$, the first two peaks can be unambiguously assigned to the tunneling of holes filling the highest valence level. However, the next group of peaks is a very intricate structure involving single-hole charging peaks and tunneling of electrons through the $1S_e$ level. This disagrees with the interpretation of Banin et al. [249] based on single-hole transitions. In particular, the strong increase of the current below -1.25 V is mainly related to the tunneling of electrons through the $1P_e$ level.

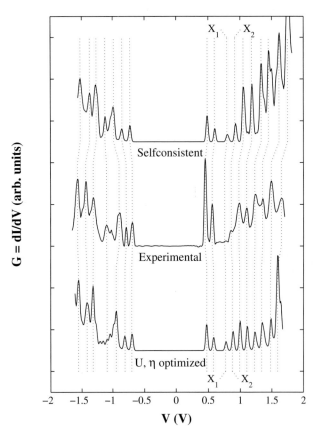

Fig. 4.16. Comparison between calculated [138, 173] and experimental [249] differential conductance $G(\varphi)$ curves for a 6.4 nm diameter InAs nanocrystal. The optimized parameters for the capacitive model are $U = 100$ meV and $\eta = 0.9$ ($C_1 = 1.44$ aF, $C_2 = 0.16$ aF). The calculated peaks are broadened with a Gaussian of width $\sigma = 15$ meV

4.6 Charging Effects and Multi-excitons   137

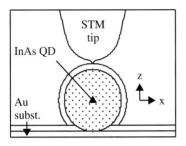

Fig. 4.17. A scanning tunneling microscope is used to probe the electronic structure of a 6.4 nm InAs nanocrystal. The quantum dot is linked to the gold substrate by a 5 Å thick hexane dithiol layer and is surrounded by a 5 Å thick layer of molecular ligands (dielectric constant = 2.6). The radius of curvature of the Pt-Ir tip is 2.5 nm, and the tip nanocrystal distance is 5 Å

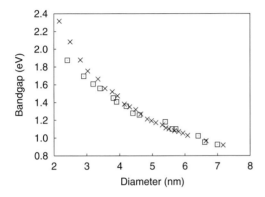

Fig. 4.18. Comparison between calculated [138, 173] (×) and experimental [249] (□) bandgap energies ε_g^0 of InAs nanocrystals versus size

The experimental spectra have been measured for different nanocrystal sizes allowing to deduce the one-particle bandgap $\varepsilon_g^0 = \varepsilon_1^e - \varepsilon_1^h$ (Fig. 4.18) and the charging energy U (Fig. 4.19) [249]. Figures 4.18 and 4.19 show that the self-consistent tight binding values agree extremely well with experimental ones in the whole range of sizes. This result strongly supports the above interpretation of the tunneling spectra and validates the predictions of the tight binding calculations even for large confinement energies.

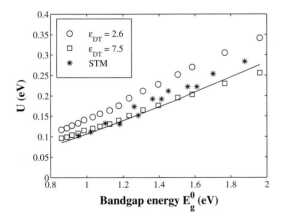

Fig. 4.19. Comparison between the calculated (tight binding) and experimental (STM = scanning tunneling microscopy results of [249]) charging energies U versus the bandgap energy for the geometry shown in Fig. 4.17. The dielectric constant of the hexane dithiol layer is either $\varepsilon_{DT} = 2.6$ or $\varepsilon_{DT} = 7.5$. Charging energy U given by (3.72) with $\epsilon_{out} = 6$ (*straight line*)

Such calculations point to the importance of the medium surrounding the nanocrystal for an accurate determination of the charging energies. This is clearly apparent in the analytical expressions (3.67) to (3.72) obtained for spherical nanocrystals of dielectric constant ϵ_{in} embedded in an external medium of dielectric constant ϵ_{out} and calculated with simple effective mass envelope functions. One finds that the self-energy of a carrier directly depends on $\epsilon_{in} - \epsilon_{out}$ and can thus reverse sign with this quantity. This is not quite the case of the electronic Coulomb repulsion which is always positive but which, from (3.72), is proportional to $e^2/(\epsilon_{out} R)$ which can vary between 0 ($\epsilon_{out} \to +\infty$) and e^2/R ($\epsilon_{out} \to 0$). Such conclusions have been confirmed by more refined calculations [175, 252]. In these works, the charging energies was calculated with single particle wave functions obtained from an empirical pseudopotential method. The results are completely in line with the simple formula of Sect. 3.5.1. As with the tight binding approach just described, the authors have calculated the single particle energies of neutral and charged clusters for InAs nanocrystals. They have also compared their results with those of Banin et al. [249] and they have shown that the best agreement with experiments occurs for $\epsilon_{out} = 6$, as shown in Fig. 4.19. However care should be taken when comparing the predictions of idealized situations (like spherical quantum dots embedded in an homogeneous medium) with the experimental geometry, as done in [138, 173], which is equivalent to calculating the corresponding capacitances for each particular population of the quantum dot (see the discussion in [138]).

4.6.2 Multi-excitons

Several papers [224, 253] have dealt recently with the theory of photoluminescence spectra for several types of quantum dots corresponding to direct gap compound semiconductors. It is found that the low-power luminescence spectra consist of sharp lines [253] with energy separation of a few meV which must correspond to the formation of single excitons. However, when increasing the photo-excitation intensity, new features appear which have been associated with exciton-exciton interactions.

The way of dealing theoretically with such a problem is to work in the spirit of the configuration interaction method, as briefly described in Sect. 1.1.4 and used in Sect. 4.3.1. For a single exciton, we have seen that the obvious technique is to write the wave function as

$$|\Psi_{exc}\rangle = \sum_{cv} a_{vc} |\psi_{vc}\rangle, \qquad (4.53)$$

where the functions ψ_{vc} are Slater determinants corresponding to single electron–hole excitations. One has then to diagonalize a matrix, with diagonal terms equal to $\varepsilon_c - \varepsilon_v$ and non diagonal terms given by the screened Hartree and exchange interactions, its size being determined by the product $N_e \times N_h$ of the number of single particle electron (N_e) and hole (N_h)

states included in the treatment. This way of doing corresponds to a first order configuration interaction, except that the use of screened interactions corresponds to the folding of higher plasmon-like excitations.

As detailed in [224], this procedure can be generalized to excited states corresponding to N interacting excitons by using a basis of Slater determinants corresponding to N electron–hole excitations, i.e.

$$|\Psi_{\text{exc}}(N)\rangle = \sum_{\substack{c_1...c_N \\ v_1...v_N}} a_{v_1...v_N,c_1...c_N} |\psi_{v_1...v_N,c_1...c_N}\rangle . \tag{4.54}$$

This matrix is obtained from the common rules of single particle and two-particle operators between Slater determinants. The diagonal terms are $\sum_{i=1}^{N}(\varepsilon_{c_i} - \varepsilon_{v_i})$ and the non diagonal terms are screened inter-particle interactions. Again, such an expansion corresponds to a first-order configuration interaction procedure. Obviously, the complexity of the treatment increases rapidly with N (numbers are given in [224]) but up to $N = 3$ excitons have been treated in this way. As detailed in this reference, the multi-exciton spectra correspond to transitions between N and $N - 1$ excitonic states as is shown schematically on Fig. 4.20. A detailed description of the corresponding transitions for CdSe has been worked out.

A fairly complete and realistic interpretation of the experimental data for various quantum dots is described in [253, 254]. It is based on the previously described technique coupled to the assumption that radiative transitions from $N \to N - 1$ only occur after decay to the ground state of the N excitonic system. Clear evidence for the existence of multi-excitonic transitions as well as charged excitons is provided in this work.

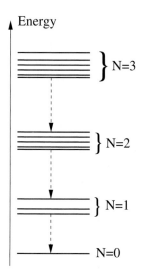

Fig. 4.20. Schematic representation of multi-excitonic transitions in quantum dots

It is clear from this that the emission of light from single exciton and multi-excitons in a nanostructure occurs at different energies. One can take advantage of this effect to make single photon sources [255]. If several electron–hole pairs are excited by a light pulse, only the last one leads to the emission of a photon at the single-exciton frequency. Thus, if the other frequencies are filtered out, only one photon is emitted for each pulse.

To end up this section, a word of warning should be said concerning the validity of configuration interaction treatments. As discussed in Sect. 1.1.4 for the single exciton, this is only applicable to low-energy configurations, i.e. $\varepsilon_g/\omega_s \ll 1$ (ω_s being the plasmon energy). If so the static screening limit is valid to lowest order in ε_g/ω_s. It is reasonable to think that a similar derivation could be done for the N exciton system but only if $N\varepsilon_g/\omega_s \ll 1$ which is more and more difficult to realize as N increases. If applicable, dynamical corrections to dielectric properties will become increasingly important.

4.7 Conclusion

It seems to us that the central results of this chapter come from the analysis of the quantitative calculations in Sects. 4.4 and 4.5. They are the following:

- quasi-particle energies can be obtained from refined single particle calculations for the neutral system (LDA + gap correction, tight binding or empirical pseudopotential methods) plus a self-energy due to the surface polarization charge which can be obtained from a simple calculation
- the lowest triplet exciton energy is equal to the gap calculated in the same single-particle treatment plus a small correction due to the average direct electron–hole attraction reduced by an even smaller surface polarization term.

Finally, the fine structure of the excitonic spectra can be accurately obtained from a first-order configuration interaction treatments with screened interactions, dynamical screening becoming important when ε_g/ω_s is not negligible.

5 Optical Properties and Radiative Processes

In this chapter, we deal with the optical properties of nanostructures. In a first part, we start with a general formulation of the optical transition probabilities taking into account specific problems related to the small size of the systems. In a second part, we consider the electron–phonon coupling using macroscopic or microscopic formulations and we analyze its consequences on the optical line-shape. The last two sections are devoted to the description of the optical properties in nanostructures of semiconductors with direct or indirect bandgap.

5.1 General Formulation

In this section, we describe the general basis to calculate the optical properties of nanostructures when their electronic structure is known. If the formalism of the electron-photon interaction in condensed matter is well-known and has been subject of considerable body of literature, our aim here is to insist on difficulties which are specific to nano-size objects. Taking into account that the diversity of possible physical situations does not allow to make a synthetic review of the problems, we consider in the following the particular case of semiconductor nanocrystals embedded in a dielectric matrix (Fig. 5.1), which is a common experimental situation. If the host material is a good insulator with a large optical gap, the study of the optical properties of the composite material within the insulator gap allows to probe transitions between quantized levels of the nanocrystals. The main tools of optical characterization are absorption and photoluminescence experiments. Note that in this section we will not discuss the optical properties of metal nanoparticles which can be described using the classical Mie theory [256] or using more elaborated approaches [257].

5.1.1 Optical Absorption and Stimulated Emission

We consider an optical absorption experiment made on the sample depicted in Fig. 5.1. A beam of monochromatic light and of intensity I_0 is irradiated perpendicularly to the sample surface which is supposed to be flat and one measures the intensity I_t of the transmitted light. The electromagnetic field

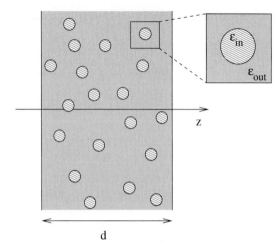

Fig. 5.1. Semiconductor nanocrystals of dielectric constant ϵ_{in} embedded in a matrix of dielectric constant ϵ_{out}. An optical absorption experiment consists of measuring the attenuation of the light passing through the sample along the axis z

inside the nanocrystals induces the transition of electrons to excited levels by absorption of a photon (Fig. 5.2a). The transition of electrons from excited states to states with lower energy can be non radiative, which is described in the next chapter, or radiative by stimulated emission (Fig. 5.2b) or spontaneous emission (Fig. 5.2c). The physics of the spontaneous emission in nanostructures will be briefly described in Sects. 5.1.2 and 5.1.3.

In absorption, and in stimulated emission, the electronic transitions between the energy levels directly result from the interaction of the electrons with the electromagnetic field in the system. These effects can be treated using a semi-classical model, which we present hereafter. The probability of transition of an electron is thus proportional to the intensity of the electromagnetic field inside the nanocrystal. But the field inside a particle is not equal to the field in the surrounding medium of different (and usually smaller) dielectric constant. Local-field effects [258–260] due to the dielectric

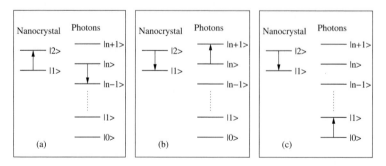

Fig. 5.2. Mechanisms of optical transitions in a two-level system in interaction with a quantized electromagnetic field: (**a**) absorption, (**b**) stimulated emission, (**c**) spontaneous emission

confinement (see Chap. 3) may strongly influence the optical properties of the quantum dots. In the general case, there is no simple analytical form to describe the distribution of the field in the system. However, because the size of the nanocrystals is small compared to the photon wavelength ($> 100\mu m$), we can define a macroscopic electromagnetic field as an average over a volume large compared to the heterogeneities. With respect to this average field, the composite material is seen as an homogeneous and isotropic material characterized by an effective dielectric constant ϵ_M [155, 157]. Thus it remains to calculate the optical absorption for the macroscopic material as a function of ϵ_M and to relate ϵ_M to the electronic structure of the nanocrystals.

Macroscopic Optical Properties. We consider that the homogeneous material is dielectric ($\boldsymbol{D} = \epsilon_0 \boldsymbol{E} + \boldsymbol{P} = \epsilon_M \epsilon_0 \boldsymbol{E}$), absorbing and non magnetic ($\mu = 1$). The electric field inside the medium is given by a plane wave of the form $\boldsymbol{E} = \boldsymbol{E}_0 \exp[i(kz - \omega t)]$. The dielectric constant is complex and frequency dependent ($\epsilon_M(\omega) = \epsilon'_M(\omega) + i\epsilon''_M(\omega)$). The wave-vector k is complex, and is related to a complex refractive index $N = n_{op} + iK_{op}$ by $k = (\omega/c)N$ with $N^2 = \epsilon_M$. n_{op} is the (real) refractive index of the sample and K_{op} is the extinction coefficient which are related to ϵ'_M and ϵ''_M by

$$\epsilon'_M = n_{op}^2 - K_{op}^2 ,$$
$$\epsilon''_M = 2 n_{op} K_{op} . \tag{5.1}$$

The propagating wave becomes

$$\boldsymbol{E} = \boldsymbol{E}_0 \exp\left(-\frac{\omega K_{op}}{c} z\right) \exp\left[i\left(\frac{n_{op}\omega}{c} z - \omega t\right)\right] . \tag{5.2}$$

Thus the wave is damped and ϵ''_M determines the absorption through K_{op}. In the absorption experiment, it can be shown [158, 261] that the intensity of the transmitted light is given by

$$I_t = \frac{I_0 (1-R)^2 \exp(-\alpha d)}{1 - R^2 \exp(-2\alpha d)} , \tag{5.3}$$

where d is the thickness of the sample, α is the absorption coefficient

$$\alpha = \frac{2\omega K_{op}}{c} , \tag{5.4}$$

and R is the sample reflectance:

$$R = \frac{(n_{op} - 1)^2 + K_{op}^2}{(n_{op} + 1)^2 + K_{op}^2} . \tag{5.5}$$

Local-Fields. The macroscopic dielectric constant can be calculated using the Maxwell–Garnett effective medium theory described in [157, 262]. One difficulty is to describe the distribution of the field in the system. However, for some geometrical shapes of crystallites, such as ellipsoids or spheres, the field inside the nanocrystal is uniform (using the fact that the size of the particles is much smaller than the wavelength of the electromagnetic wave) and proportional to the field outside [155, 158]. It is then possible to define a local-field factor F, such that $\boldsymbol{E}_{\text{in}} = F\boldsymbol{E}_{\text{out}}$. For simplicity, we consider in the following that all the nanocrystals are spherical and identical. In that case, we have [155, 158]

$$F = \frac{3\epsilon_{\text{out}}}{\epsilon_{\text{in}} + 2\epsilon_{\text{out}}}, \tag{5.6}$$

where ϵ_{in} is the frequency-dependent dielectric constant of a nanocrystal (to be defined) and ϵ_{out} the dielectric constant of the host material (Fig. 5.1). The average electric displacement $\overline{\boldsymbol{D}}$ and the average electric field $\overline{\boldsymbol{E}}$ in the composite medium are given by

$$\overline{\boldsymbol{D}} = p\boldsymbol{D}_{\text{in}} + (1-p)\boldsymbol{D}_{\text{out}},$$
$$\overline{\boldsymbol{E}} = p\boldsymbol{E}_{\text{in}} + (1-p)\boldsymbol{E}_{\text{out}}, \tag{5.7}$$

where p is the volume fraction of nanocrystals in the composite medium. Using $\boldsymbol{E}_{\text{in}} = F\boldsymbol{E}_{\text{out}}$, $\boldsymbol{D}_{\text{in}} = \epsilon_{\text{in}}\epsilon_0 \boldsymbol{E}_{\text{in}}$, $\boldsymbol{D}_{\text{out}} = \epsilon_{\text{out}}\epsilon_0 \boldsymbol{E}_{\text{out}}$ and $\epsilon_M = \overline{\boldsymbol{D}}/(\epsilon_0\overline{\boldsymbol{E}})$, we find

$$\frac{\epsilon_M - \epsilon_{\text{out}}}{\epsilon_M + 2\epsilon_{\text{out}}} = p\frac{\epsilon_{\text{in}} - \epsilon_{\text{out}}}{\epsilon_{\text{in}} + 2\epsilon_{\text{out}}}. \tag{5.8}$$

For small values of p, the dielectric function becomes:

$$\epsilon_M \approx \epsilon_{\text{out}}\left(1 + 3p - 9p\frac{\epsilon_{\text{out}}}{\epsilon_{\text{in}} + 2\epsilon_{\text{out}}}\right). \tag{5.9}$$

Near the optical threshold, ϵ_{out} is real and constant, $\epsilon_M'' \ll \epsilon_M'$, so that $n_{\text{op}} \approx \sqrt{\epsilon_M'}$ and the absorption coefficient in the composite medium is

$$\alpha \approx \frac{\omega}{cn_{\text{op}}}\epsilon_M'' \approx \frac{\omega}{cn_{\text{op}}}p\left(\frac{3\epsilon_{\text{out}}}{\epsilon_{\text{in}}' + 2\epsilon_{\text{out}}}\right)^2 \epsilon_{\text{in}}'' = \frac{\omega}{cn_{\text{op}}}pF^2\epsilon_{\text{in}}''. \tag{5.10}$$

Thus, in this limit of $p \ll 1$, we recover the usual expression of the absorption coefficient $\omega/(cn_{\text{op}})\epsilon_{\text{in}}''$ multiplied by the volume concentration p of nanocrystals and by the square of the local-field factor F, which is a quite intuitive result. However, in the general case, the relation between ϵ_M'' and ϵ_{in}'' in (5.8) is more complex: we will come back to this point later.

Density Matrix Formulation of the Dielectric Constant. It remains to calculate ϵ_{in}. One conceptual difficulty is how to define a dielectric constant in a nano-size object. We have already discussed this point in the static limit in Chap. 3. We have shown that it is sometimes interesting to define an average quantity over the nanocrystal volume. Here we will work in the same spirit, extending this approximation to non-zero frequency. Once again, we use the fact that the dimensions of the nanostructures are small compared the electromagnetic wavelength. In particular, we will replace $\exp(i\boldsymbol{k}\cdot\boldsymbol{r})$ by 1 in the following. The electric field $\boldsymbol{E}$ inside the nanocrystal polarizes the system

5.1 General Formulation

leading to a total dipolar momentum $\mathcal{P}$. Thus we can define the average polarization as $\boldsymbol{P} = \mathcal{P}/\Omega$ where Ω is the volume of the nanocrystal and we can deduce the dielectric constant using $\boldsymbol{P} = (\epsilon_{\text{in}} - 1)\epsilon_0 \boldsymbol{E}$.

The dipolar momentum $\mathcal{P}$ is induced by the transitions of electrons between the discrete levels of the nanocrystal. To characterize this system, we calculate the statistical density matrix ρ_{ij} in the basis of the eigenstates of the electronic Hamiltonian H_0 in absence of electromagnetic field. In this formulation, a diagonal element ρ_{ii} gives the probability to find the system in the state $|i\rangle$. To simplify the problem, we first consider the case of a nanocrystal with only two electronic levels, the ground state $|1\rangle$ of energy ε_1 and the excited state $|2\rangle$ of energy ε_2 (Fig. 5.2). In the presence of the electric field $\boldsymbol{E}(t) = \boldsymbol{E}_0 \exp(-i\omega t)$, the Hamiltonian becomes $H = H_0 + W(t)$ where the perturbation is written in the usual dipolar form [57, 261, 263] as

$$W(t) = e\boldsymbol{r} \cdot \boldsymbol{E}_0 e^{-i\omega t} . \tag{5.11}$$

The matrices of H_0 and W are

$$H_0 = \begin{vmatrix} \varepsilon_1 & 0 \\ 0 & \varepsilon_2 \end{vmatrix} ,$$

$$W = \begin{vmatrix} 0 & e\langle \boldsymbol{r}_{12}\rangle \cdot \boldsymbol{E}_0 e^{-i\omega t} \\ e\langle \boldsymbol{r}_{12}\rangle \cdot \boldsymbol{E}_0 e^{-i\omega t} & 0 \end{vmatrix} . \tag{5.12}$$

We have supposed that the system is centro-symmetric ($\langle j|\boldsymbol{r}|j\rangle = 0$, $j = 1, 2$) and that $\langle \boldsymbol{r}_{12}\rangle = \langle 1|\boldsymbol{r}|2\rangle$ is real.

The evolution of the system due to the perturbation is given by the Schrödinger equation which in the density matrix formulation becomes [79, 263]

$$i\hbar \frac{d\rho}{dt} = [H, \rho] . \tag{5.13}$$

It leads to the following equations for each element $\rho_{ij} = \langle i|\rho|j\rangle$ of the density matrix

$$\frac{d\rho_{11}}{dt} = -\frac{i}{\hbar} W_{12}(\rho_{21} - \rho_{12}) , \tag{5.14}$$

$$\frac{d\rho_{22}}{dt} = -\frac{i}{\hbar} W_{12}(\rho_{12} - \rho_{21}) , \tag{5.15}$$

$$\frac{d\rho_{12}}{dt} = -\frac{i}{\hbar} W_{12}(\rho_{22} - \rho_{11}) + i\omega_{21}\rho_{12} , \tag{5.16}$$

$$\frac{d\rho_{21}}{dt} = -\frac{i}{\hbar} W_{12}(\rho_{11} - \rho_{22}) - i\omega_{21}\rho_{21} , \tag{5.17}$$

where $\omega_{21} = (\varepsilon_2 - \varepsilon_1)/\hbar$. Combining (5.14) and (5.15), we verify that $d(\rho_{11} + \rho_{22})/dt = 0$ resulting from the conservation of the electronic population in the system ($\rho_{11} + \rho_{22} = 1$). When the perturbation is switched off ($W_{12} = 0$), the populations ρ_{ii} should tend after some time T towards their equilibrium values f_i given by the Fermi-Dirac statistics, which is clearly not the case in

(5.14) and (5.15). The reason is that there are other physical processes which are not included in the model (e.g. electron–phonon coupling, spontaneous emission) coming from interactions of the electronic system with its environment. In the case of random interactions, these effects are usually described by a relaxation time T leading to extra terms in the master equations:

$$\frac{d\rho_{11}}{dt} = -\frac{i}{\hbar}W_{12}(\rho_{21} - \rho_{12}) - \frac{\rho_{11} - f_1}{T}, \tag{5.18}$$

$$\frac{d\rho_{22}}{dt} = -\frac{i}{\hbar}W_{12}(\rho_{12} - \rho_{21}) - \frac{\rho_{22} - f_2}{T}, \tag{5.19}$$

$$\frac{d\rho_{12}}{dt} = -\frac{i}{\hbar}W_{12}(\rho_{22} - \rho_{11}) + i\omega_{21}\rho_{12} - \frac{\rho_{12}}{\tau}, \tag{5.20}$$

$$\frac{d\rho_{21}}{dt} = -\frac{i}{\hbar}W_{12}(\rho_{11} - \rho_{22}) - i\omega_{21}\rho_{21} - \frac{\rho_{21}}{\tau}. \tag{5.21}$$

In (5.20) and (5.21), τ defines the decoherence time of the non diagonal elements of the density matrix due to random interactions.

These equations allow to describe a very rich physics such as the saturation of the absorption at high optical excitation [264, 265]. Here we will only consider weak excitations such that $\Delta\rho = \rho_{11} - \rho_{22} \approx f_1 - f_2$. In the permanent regime, the populations ρ_{ii} become constant and the non diagonal elements oscillate with a frequency ω [264, 265]. Thus we can write $\rho_{12} = \rho_{12}^0 \exp(-i\omega t)$ and $\rho_{21} = \rho_{21}^0 \exp(-i\omega t)$. From (5.21) we obtain that

$$\rho_{21}^0 i(\omega - \omega_{21}) - \frac{\rho_{21}^0}{\tau} - \frac{i}{\hbar}W_{12}^0[f_1 - f_2] = 0 \tag{5.22}$$

which gives

$$\rho_{21}^0 = \frac{W_{12}^0[f_1 - f_2]}{\hbar\omega - \hbar\omega_{21} + \frac{i\hbar}{\tau}}. \tag{5.23}$$

Similarly we deduce from (5.20) that

$$\rho_{12}^0 = -\frac{W_{12}^0[f_1 - f_2]}{\hbar\omega + \hbar\omega_{21} + \frac{i\hbar}{\tau}}. \tag{5.24}$$

The induced dipolar momentum $\mathcal{P}$ is given by the statistical average of the operator $-e\mathbf{r}$:

$$\mathcal{P} = \text{Tr}[\rho(-e\mathbf{r})]. \tag{5.25}$$

We deduce the polarization in the nanocrystal

$$\mathbf{P} = -\frac{e\langle \mathbf{r}_{12}\rangle}{\Omega}\left(\rho_{21}^0 + \rho_{12}^0\right)e^{-i\omega t} \tag{5.26}$$

and the dielectric constant using $\epsilon_{\text{in}} = 1 + \mathbf{P}/(\epsilon_0 \mathbf{E})$,

$$\epsilon_{\text{in}} = 1 - \frac{e^2|\langle 1|\mathbf{r}\cdot\mathbf{e}|2\rangle|^2}{\epsilon_0 \Omega}\left\{\frac{1}{\hbar\omega - \hbar\omega_{21} + \frac{i\hbar}{\tau}} - \frac{1}{\hbar\omega + \hbar\omega_{21} + \frac{i\hbar}{\tau}}\right\}[f_1 - f_2], \tag{5.27}$$

where $\mathbf{e}$ is the polarization vector of the electric field.

The imaginary part of the dielectric constant has a resonance when $\omega = \omega_{21}$, i.e. when the electron undergoes a transition $|1\rangle \rightarrow |2\rangle$ following the absorption of a photon of energy $\hbar\omega_{21}$ (Fig. 5.2a) or a transition $|2\rangle \rightarrow |1\rangle$ following a stimulated emission of a photon of energy $\hbar\omega_{21}$ (Fig. 5.2b). In the case of a system with an arbitrary number of levels, one simply needs to sum in (5.27) over all the possible transitions $|i\rangle \rightarrow |j\rangle$ of one electron. We can now derive the expression of the absorption coefficient. For small values of the volume fraction p, (5.10) gives

$$\alpha(\omega) = pF^2 \sum_{ij} \frac{\omega e^2 |\langle i|\mathbf{r}\cdot\mathbf{e}|j\rangle|^2 \pi}{cn_{op}\epsilon_0 \Omega} \delta(\hbar\omega - \hbar\omega_{ji})[f_i - f_j] = C(\omega)[f_i - f_j] \tag{5.28}$$

which is written in the limit $\tau \rightarrow +\infty$ using the well-known relation

$$\lim_{\tau \rightarrow +\infty} \mathrm{Im}\left(\frac{1}{\hbar\omega - \hbar\omega_{21} + \frac{i\hbar}{\tau}}\right) = -\pi\delta(\hbar\omega - \hbar\omega_{21}) \,. \tag{5.29}$$

We can also rewrite (5.28) in the following form

$$\alpha(\omega) = C(\omega)f_i[1 - f_j] - C(\omega)f_j[1 - f_i] \,, \tag{5.30}$$

where the first term correspond to the transition $|i\rangle \rightarrow |j\rangle$ (absorption) and the second one to the transition $|j\rangle \rightarrow |i\rangle$ (stimulated emission). The strength of the optical coupling between two levels $|i\rangle$ and $|j\rangle$ is often described by a quantity without dimension, the oscillator strength

$$f_{ji} = \frac{2m_0}{\hbar}\omega_{ji}|\langle i|\mathbf{r}\cdot\mathbf{e}|j\rangle|^2 \tag{5.31}$$

which verifies the Thomas–Reiche–Kuhn sum rule

$$\sum_j f_{ji} = 1 \,. \tag{5.32}$$

It is sometimes interesting to write (5.27), (5.28) or (5.31) in terms of the matrix elements of the momentum $\mathbf{p}$ instead of those of $\mathbf{r}$ using the relation [263–265]

$$\langle j|\mathbf{p}\cdot\mathbf{e}|i\rangle = im_0\omega_{ji}\langle j|\mathbf{r}\cdot\mathbf{e}|i\rangle \,, \tag{5.33}$$

which is obtained via the commutator relation $\mathbf{p} = m_0[\mathbf{r}, H]/(i\hbar)$.

Equation (5.28) shows that the absorption spectrum is the sum of narrow lines at each transition energy $\hbar\omega_{ji}$. However, in a real medium, size dispersion of the nanocrystals often broadens the lines (inhomogeneous broadening), giving rise for example to a Gaussian profile for each peak. But there are also intrinsic sources of broadening (homogeneous broadening) such as the random interactions described previously by the decoherence time τ. When τ is finite, it is easy to see from (5.29) that each delta function $\delta(\hbar\omega - \hbar\omega_{ji})$ in (5.28) must be replaced by a Lorentzian:

$$\mathcal{L}(\hbar\omega) = \frac{1}{\hbar}\frac{1/(\pi\tau)}{(\omega - \omega_{ji})^2 + (1/\tau)^2} \,. \tag{5.34}$$

More complex line-shapes due to electron–phonon coupling will be discussed in Sect. 5.2.

It is also important to remind that (5.28) is not always valid. In the general case, the absorption coefficient is given by $\omega/(cn_{\mathrm{op}})\epsilon''_{\mathrm{M}}$ and the resonances of ϵ''_{M} may be slightly shifted with respect to the transition frequencies ω_{ji}, i.e. the resonances of ϵ_{in}. These shifts are proportional to the oscillator strengths [266] and values of the order of few meV have been predicted [262]. Splittings of the lines have also been predicted in the case of optically anisotropic materials [262]. In (5.28), we have also implicitly assumed that the refractive index n_{op} is homogeneous, which may not be the case at resonance when a high oscillator strength gives an important contribution to the dielectric constant [260].

At high light intensities, the optical response becomes non-linear. The saturation of the absorption leads for example to the intensity dependence of field penetration [258, 260, 262]. An interesting consequence which has been predicted is an intrinsic optical bistability [258, 260, 262].

5.1.2 Luminescence

In the previous section, we have seen that the optical absorption and the stimulated emission are induced by the electromagnetic field. In contrast, the spontaneous emission occurs even when there is no photon in the system. It is not described in the previous calculation because the electromagnetic field is treated classically [263–265]. In the following, we shall relate the spontaneous emission to the absorption through the Einstein relationships, which will allow to take into account the local-field effects in a simple manner.

We consider the composite material as a system of two levels in thermal equilibrium in an optical cavity of volume V. The number of photons per unit of energy $\hbar\omega$ in the cavity is given by the Planck formula for the black body:

$$N = \frac{8\pi(\hbar\omega)^2 n_{\mathrm{op}}^3 V}{h^3 c^3} \frac{1}{\exp\left(\frac{\hbar\omega}{kT}\right) - 1} \,. \tag{5.35}$$

The effective rate of transition $|1\rangle \to |2\rangle$ per photon and per unit time, i.e. the balance between the absorption and the stimulated emission, is given by the product of the absorption coefficient by the velocity of the light

$$P_{12} = \alpha(\omega) \frac{c}{n_{\mathrm{op}}} \,. \tag{5.36}$$

The effective number of transitions $|1\rangle \to |2\rangle$ in the cavity per unit of energy is NP_{12}. At equilibrium, they must be compensated by spontaneous transitions $|2\rangle \to |1\rangle$ whose number per unit of energy is proportional to the mean occupancy f_2 of the level $|2\rangle$

$$NP_{12} = Af_2 \,. \tag{5.37}$$

This relation must be verified at any temperature. At equilibrium, we have $f_2/f_1 = \exp(-\hbar\omega/kT)$. Using (5.28), (5.35) and (5.36), we obtain

$$A = \frac{8\pi\hbar^2\omega^2 n_{\rm op}^2 V}{h^3 c^2} C(\hbar\omega) . \tag{5.38}$$

To calculate the spontaneous recombination rate $\Gamma_{\rm sp} = 1/\tau_{\rm sp}$, we must divide A by the number of nanocrystals in the volume V ($= pV/\Omega$) and we must integrate over the energy

$$\Gamma_{\rm sp} = \frac{1}{\tau_{\rm sp}} = \frac{\Omega}{pV} \int A d(\hbar\omega) \tag{5.39}$$

leading to

$$\Gamma_{\rm sp} = \frac{1}{\tau_{\rm sp}} = \frac{\omega_{21}^3 F^2 e^2 |\langle 1|\mathbf{r}\cdot\mathbf{e}|2\rangle|^2 n_{\rm op}}{\pi c^3 \epsilon_0 \hbar} . \tag{5.40}$$

In this calculation, we have implicitly assumed that the absorption is isotropic. Thus, one usually prefers the following expression

$$\Gamma_{\rm sp} = \frac{1}{\tau_{\rm sp}} = \frac{\omega_{21}^3 F^2 e^2 r_{12}^2 n_{\rm op}}{3\pi c^3 \epsilon_0 \hbar} , \tag{5.41}$$

where $r_{12}^2 = |\langle 1|x|2\rangle|^2 + |\langle 1|y|2\rangle|^2 + |\langle 1|z|2\rangle|^2$.

We note the presence of the square of the local-field factor F in the spontaneous emission rate.

5.1.3 Nanostructures in Optical Cavities and Photonic Crystals

The spontaneous emission can be described in quantum theory if one includes the quantization of the electromagnetic field. In this theory of the light, the vacuum state has a non zero energy carried by virtual photons. These photons provoke fluctuations which are at the origin of the spontaneous emission, basically with the same mechanism as when real photons provoke stimulated emission. Therefore, the local-field factor appears naturally in (5.41).

According to this model, the spontaneous emission rate can be modified by changing the density of optical modes in the system. This effect, which can be achieved by placing the emitter in an optical cavity, was first proposed by Purcell [267]. If the optical transition is resonant with a cavity mode, the spontaneous emission is enhanced in this mode [268]. Similarly, if the transition is non resonant, the emission rate is decreased. For example, the spontaneous emission of a single atom in a cavity can be enhanced [269] or inhibited [270].

Since they can be thought as artificial atoms, semiconductor quantum dots and nanocrystals present the same kind of properties in optical cavities [255, 271–273], which opens the road to very interesting applications, such as high-efficiency light-emitting diodes. Also, single dots in a cavity can produce efficient sources of single photons under pulsed optical excitation, for application to quantum cryptography or quantum computing [273] (see Sect. 4.6.2 for a brief description of the mechanism).

Various types of semiconductor cavities have been investigated, producing from one to three-dimensional confinement of the light. For example, micro-disks containing InAs quantum boxes have been fabricated, showing high quality factors [274]. Another structure that has attracted considerable attention is the photonic crystal [275–277]. A photonic crystal is a periodic structure made of dielectric materials with a length-scale of the order of the optical wavelength. Light is scattered in the crystal, and a photonic bandgap may appear in the optical frequencies, exactly like the electronic bandgap in semiconductors. A point defect in the photonic crystal can be designed to introduce a defect mode in the optical gap. In that case, light trapped in the defect mode is localized around the defect which acts as an optical cavity. Light emission from quantum dots coupled to defect modes has been demonstrated [277, 278].

5.1.4 Calculation of the Optical Matrix Elements

In this section, we briefly discuss how the optical matrix elements are calculated. In ab initio approaches including all the electrons (valence and core ones), one has just to calculate the matrix elements of r or p, as shown in Sect. 5.1.1. With pseudopotential or tight binding methods, the situation is slightly more complex, as described below.

Optical Matrix Elements in the Pseudopotential Formalism. In presence of an electric field, the total pseudo Hamiltonian is equal to $H_{\mathrm{ps}} + W$ where $H_{\mathrm{ps}} = T + V_{\mathrm{ps}}$, T is the kinetic energy, V_{ps} is the pseudopotential (Sect. 1.3.2) and W is the coupling term (Sect. 5.1.1). W can still be treated in perturbation and the oscillator strength is proportional to the square of the matrix element $\langle i|\boldsymbol{e}\cdot\boldsymbol{r}|j\rangle$ where $|i\rangle$ and $|j\rangle$ are pseudo wave-functions. In the case of extended systems with periodic boundary conditions, the matrix elements of r are ill defined and one is forced to work with the matrix elements of p [182, 197] using the commutator relation with the Hamiltonian. However, H_{ps} is a non-local operator (Sect. 1.3.2), and the commutator relation becomes

$$\frac{m_0}{i\hbar}[\boldsymbol{r}, H_{\mathrm{ps}}] = \boldsymbol{p} + \frac{m_0}{i\hbar}[\boldsymbol{r}, V_{\mathrm{nl}}] \,, \tag{5.42}$$

where $V_{\mathrm{nl}} = V_{\mathrm{ps}} - V$ is the non-local part of the pseudopotential ($= \sum_c (E - E_c)|c\rangle\langle c|$, following the notations of Sect. 1.3.2). Thus, we have

$$i m_0 \omega_{ij} \langle i|\boldsymbol{e}\cdot\boldsymbol{r}|j\rangle = \boldsymbol{e}\cdot\langle i|\boldsymbol{p} + \frac{m_0}{i\hbar}[\boldsymbol{r}, V_{\mathrm{nl}}]|j\rangle \,, \tag{5.43}$$

which requires to calculate the matrix elements

$$\langle i|[\boldsymbol{r}, V_{\mathrm{nl}}]|j\rangle = \int \varphi_i^*(\boldsymbol{r}) V_{\mathrm{nl}}(\boldsymbol{r}, \boldsymbol{r}')(\boldsymbol{r} - \boldsymbol{r}') \varphi_j(\boldsymbol{r}') \mathrm{d}\boldsymbol{r}\mathrm{d}\boldsymbol{r}' \,, \tag{5.44}$$

where $\varphi_i(\boldsymbol{r}) = \langle \boldsymbol{r}|i\rangle$.

Optical Matrix Elements in the Tight Binding Formalism. The main difficulty in tight binding comes from the fact that the basis of atomic orbitals is incomplete and is not explicitly defined since only the matrix elements of the Hamiltonian are parametrized. One possible approach is to consider the matrix elements of the momentum $\boldsymbol{p}$ as free parameters that must be determined empirically. For instance, they can be calculated using orbitals of the free atoms [279].

Another approach is to express the matrix elements of $\boldsymbol{p}$ in terms of the Hamiltonian matrix elements using the commutator relation $\boldsymbol{p} = m_0/(i\hbar)[\boldsymbol{r}, H]$. To this end, we consider atomic orbitals $|\alpha, \boldsymbol{R}\rangle$ centered at position $\boldsymbol{R}$ and characterized by a label α. We have:

$$\langle \alpha, \boldsymbol{R}|\boldsymbol{p}|\beta, \boldsymbol{R}'\rangle = \frac{m_0}{i\hbar}\left[\langle \alpha, \boldsymbol{R}|\boldsymbol{r}H|\beta, \boldsymbol{R}'\rangle - \langle \alpha, \boldsymbol{R}|H\boldsymbol{r}|\beta, \boldsymbol{R}'\rangle\right] . \quad (5.45)$$

Then we assume that the basis is approximately complete to write:

$$\langle \alpha, \boldsymbol{R}|\boldsymbol{r}H|\beta, \boldsymbol{R}'\rangle \approx \sum_{\gamma, \boldsymbol{R}''} \langle \alpha, \boldsymbol{R}|\boldsymbol{r}|\gamma, \boldsymbol{R}''\rangle\langle \gamma, \boldsymbol{R}''|H|\beta, \boldsymbol{R}'\rangle . \quad (5.46)$$

If only intra-atomic matrix elements of $\boldsymbol{r}$ are retained, we obtain

$$\langle \alpha, \boldsymbol{R}|\boldsymbol{p}|\beta, \boldsymbol{R}'\rangle = \frac{m_0}{i\hbar}\Big\{(\boldsymbol{R} - \boldsymbol{R}')\langle \alpha, \boldsymbol{R}|H|\beta, \boldsymbol{R}'\rangle$$
$$+ \sum_{\gamma}[\langle \gamma, \boldsymbol{R}|H|\beta, \boldsymbol{R}'\rangle d_{\alpha\gamma} - \langle \alpha, \boldsymbol{R}|H|\gamma, \boldsymbol{R}'\rangle d_{\gamma\beta}]\Big\} , \quad (5.47)$$

where the terms $d_{\alpha\gamma} = \langle \alpha, \boldsymbol{R}|\boldsymbol{r}|\gamma, \boldsymbol{R}\rangle$ describe intra-atomic polarizations. The simplest approximation consists to set all these intra-atomic terms to zero since there are no adjustable parameters beyond those for the Hamiltonian [57, 280–282]. In semiconductors, this is often justified because the polarizability of the atom gives a small contribution to the total polarizability. But it is clear that the intra-atomic terms cannot be exactly equal to zero because, if it was the case, in the limit of a periodic system of well separated atoms, all the optical matrix elements should vanish [283, 284]. Many authors have suggested to include intra-atomic matrix elements [283, 285, 286], but these models are not gauge invariant [284, 287]. However, our experience in this field shows that this approach, using (5.47) with intra-atomic terms calculated from free atom orbitals, leads to results in good agreement with experiments in many situations, even for the treatment of semiconductor nanostructures.

5.2 Electron–Phonon Coupling and Optical Line-Shape

The electron–lattice interaction can profoundly affect the optical line-shape in absorption or in photoluminescence, especially when the transitions involve localized electronic states, for example in molecules or with defects in insulators or semiconductors [288, 289]. In contrast band-to-band transitions in bulk semiconductors are usually less affected by electron–phonon coupling. Electron–lattice interaction usually plays an important role in the optical properties of nanostructures, and its importance generally increases when going to small sizes. In this section we discuss the origin of the broadening of the optical line-shape due to electron–lattice interaction in a nanostructure. Since the number N of atoms in a nanostructure is large (typically larger than 10^2 atoms), we assume that the electron–phonon coupling occurs with $3N$ delocalized modes of vibration. The case of coupling to localized modes, for example in presence of defects, is treated in detail in [288, 289].

5.2.1 Normal Coordinates

We consider first the motion of the N atoms in the system. Let their equilibrium position be $\mathbf{R}_n$ and their instantaneous displacement be $\mathbf{s}_n$ ($n = 1...N$). In the harmonic approximation [291–293], one expands the potential energy V about the equilibrium situation to second order in displacements which leads to the Hamiltonian

$$H = \sum_n \frac{p_n^2}{2M_n} + \frac{1}{2} \sum_{ni,n'i'} \frac{\partial^2 V}{\partial R_{ni} \partial R_{n'i'}} s_{ni} s_{n'i'} \,, \tag{5.48}$$

where the index i distinguishes the three coordinates of $\mathbf{s}_n$, M_n is the mass of the atom n and $\mathbf{p}_n$ is the momentum conjugate to $\mathbf{s}_n$. If we consider the classical problem, the Lagrangian equations of motion deduced from H are

$$\omega^2 u_{ni} = \sum_{n'i'} D_{ni,n'i'} u_{n'i'} \tag{5.49}$$

if we are looking for solutions periodic in time of the form

$$s_{ni}(t) = \frac{1}{\sqrt{M_n}} u_{ni} \exp(-i\omega t) \,. \tag{5.50}$$

The $3N \times 3N$ matrix D in (5.49) is the dynamical matrix defined by

$$D_{ni,n'i'} = \frac{1}{\sqrt{M_n M_{n'}}} \frac{\partial^2 V}{\partial R_{ni} \partial R_{n'i'}} \,. \tag{5.51}$$

The $3N$ eigenvalues ω_j of (5.49) and their corresponding normalized eigenvectors $\mathbf{u} = \mathbf{e}^{(j)}$ define the normal modes. If we express the displacements in terms of normal coordinates

$$s_{ni} = \frac{1}{\sqrt{M_n}} \sum_j Q_j e_{ni}^{(j)} \,, \tag{5.52}$$

the Hamiltonian resolves into a sum of $3N$ independent harmonic oscillators

$$H = \frac{1}{2} \sum_j \left(P_j^2 + w_j^2 Q_j^2 \right) \,, \tag{5.53}$$

where, in quantum mechanics, P_j is the conjugate momentum of Q_j such that $[Q_j, P_{j'}] = i\hbar\delta_{jj'}$. The Hamiltonian can be rewritten in the usual second-quantized form

$$H = \sum_j \hbar\omega_j \left(a_j^+ a_j + \frac{1}{2} \right), \tag{5.54}$$

where $a_j = \frac{1}{\sqrt{2\hbar\omega_j}} (\omega_j Q_j + iP_j)$. The eigenstates $|\chi_n\rangle$ of H are products of the eigenstates of the $3N$ independent harmonic oscillators

$$|\chi_n\rangle = \prod_{j=1}^{3N} |n_j\rangle, \tag{5.55}$$

where n_j is the occupation number in the mode j. The corresponding total energy is

$$E_n = \sum_j \hbar\omega_j \left(n_j + \frac{1}{2} \right). \tag{5.56}$$

5.2.2 Calculation of Phonons in Nanostructures

The quantitative calculation of the frequencies ω_j is in principle quite straightforward using atomistic approaches. For small systems like molecules (< 50 atoms), and for periodic systems with a small unit cell, ab initio methods are applicable. Those based on density functional theory are in general quite efficient [290], with predictions in good agreement with experiments (see Chap. 1). For larger systems, semi-empirical methods must be applied (e.g. valence force-field methods). For nanocrystals and quantum dots, one can use the model inter-atomic potentials designed to describe the lattice dynamics in bulk crystals [291–293]. These potentials usually include Coulomb interactions between ions in ionic crystals and short-range interactions with various degrees of sophistication. They can be directly transferred to the case of the nanocrystals with the appropriate boundary conditions. Examples of such calculations are given in [294] for quantum wells and [295–297] for quantum dots. To calculate the full spectrum of frequencies ω_j, it is necessary to diagonalize completely the $3N \times 3N$ dynamical matrix D, which becomes rapidly impossible for large sizes. Thus, to go beyond, continuum models must be used for acoustic [298–302] or optical [303–306] phonons.

In a continuum model, the acoustic properties may be described in terms of elastic vibrations, solutions of the standard Navier–Stokes equations. In the case of an elastically isotropic sphere under stress-free boundary conditions, the solutions have been studied by Lamb [298]. The modes of vibration can be classified in two categories, namely, the torsional modes and spheroidal modes. The former ones are purely transversal, whereas the latter ones are mixed modes of transverse and longitudinal characters. The optical modes in polar semiconductors are usually described by a dielectric model [303–306]. An example will be described in Sect. 5.2.6.

In isolated quantum dots, the phonon density of states is theoretically discrete (see Sect. 2.1.5) and the band edges are shifted in energy with respect to the bulk situation. As a consequence, coherent optical phonons have been reported in quantum dots of CdSe [307], InP [147], and PbS [308]. In contrast, the acoustic modes of quantum dots are typically strongly damped due to the coupling with the surrounding host [302, 309] even if coherent acoustic phonons have been observed in PbS quantum dots using femto-second optical techniques [309]. An important consequence of the lack of translational periodicity is a mixing of the transverse optical and longitudinal optical modes [305, 306, 309]. Another difference with the bulk situation is the existence of surface modes whose frequencies strongly depend on the boundary conditions (chemical passivation, connection to other materials). These modes are superpositions of many bulk phonon states from different bands and different points of the Brillouin zone [296].

5.2.3 Configuration Coordinate Diagram

In the following, we consider optical transitions from initial states Ψ_{in} of energy E_{in} to final states $\Psi_{fn'}$ of energy $E_{fn'}$. We start with the adiabatic approximation to separate the electronic and nuclear motions. If there is no degeneracy, we can split the total wavefunctions into an electronic and a vibrational part (Born–Oppenheimer approximation)

$$\Psi_{in} = \phi_i(x,Q)\chi_{in}(Q) ,$$
$$\Psi_{fn'} = \phi_f(x,Q)\chi_{fn'}(Q) , \qquad (5.57)$$

where Q denotes all the normal displacements Q_j of the atoms, x represents the electronic coordinates including spin, and n, n' indicate the occupation numbers of the various lattice modes. The electronic parts ϕ depend parametrically on Q. To discuss qualitatively the physical properties associated with the electron–lattice interaction, it is often interesting to draw the configuration coordinate diagram that represents the electronic energies $\varepsilon_i(Q)$ and $\varepsilon_f(Q)$ for the initial and final states, respectively, as function of the normal displacement Q_j (Fig. 5.3). Following the Born–Oppenheimer approximation, these energies are the potential energies for nuclear motion. Near each minimum we can use the harmonic approximation, assuming that the modes and their frequencies are the same in both initial and final states:

$$\varepsilon_i(Q) = \varepsilon_i(Q^{(i)}) + \sum_{j=1}^{3N} \frac{\omega_j^2}{2}\left(Q_j - Q_j^{(i)}\right)^2 ,$$
$$\varepsilon_f(Q) = \varepsilon_f(Q^{(f)}) + \sum_{j=1}^{3N} \frac{\omega_j^2}{2}\left(Q_j - Q_j^{(f)}\right)^2 . \qquad (5.58)$$

5.2 Electron–Phonon Coupling and Optical Line-Shape

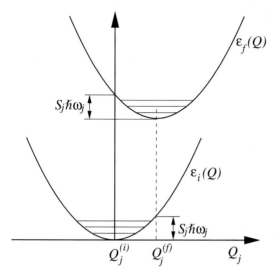

Fig. 5.3. Configuration coordinate diagram showing phonon states of the initial state i and of the final states f in optical absorption. The situation in luminescence is reversed

Thus the electronic energies of the initial and final states just differ by a term which is linear in the displacements Q_j (linear electron–phonon coupling), and the system oscillates about different mean positions in the two states. The change in mean position between the initial and final states is a measure of the electron–lattice interaction. It can be defined by a dimensionless factor V_j for each mode j:

$$V_j = \sqrt{\frac{\omega_j}{2\hbar}} \left(Q_j^{(i)} - Q_j^{(f)} \right) . \tag{5.59}$$

More commonly, one defines the Huang–Rhys factor [310] $S_j = V_j^2$. The importance of this factor is apparent in Fig. 5.3. When an electronic transition occurs at $Q = Q_j^{(i)}$ with no change in nuclear configuration following the so-called Franck–Condon principle [311, 312], the system relaxes to $Q = Q_j^{(f)}$ through phonon emission, the energy released being $S_j \hbar \omega_j = \omega_j^2 \left(Q_j^{(i)} - Q_j^{(f)} \right)^2 /2$. The difference between the energies in the final and initial states can be written

$$\varepsilon_f(Q) - \varepsilon_i(Q) = \varepsilon_f(Q^{(f)}) - \varepsilon_i(Q^{(i)}) + d_{\mathrm{FC}} + \sum_{j=1}^{3N} \sqrt{2\hbar \omega_j^3} V_j \, Q_j , \tag{5.60}$$

where

$$d_{\mathrm{FC}} = \sum_{j=1}^{3N} S_j \hbar \omega_j \tag{5.61}$$

is the so-called Franck–Condon energy corresponding to the total relaxation energy of the lattice. The terms linear in Q_j in (5.60) give the electron–phonon coupling Hamiltonian in a quantum mechanical description (the Hamiltonian is usually written in terms of the operators a_j and a_j^+). Finally, the vibrational parts of the wave functions (5.55) become

$$|\chi_{in}\rangle = \prod_{j=1}^{3N} |in_j\rangle \,, \quad |\chi_{fn'}\rangle = \prod_{j=1}^{3N} |fn'_j\rangle \,, \tag{5.62}$$

where the harmonic oscillator wave functions are centered on the $Q_j^{(i)}$ and $Q_j^{(f)}$, respectively.

5.2.4 General Expression for the Optical Transition Probabilities

We can now calculate the optical matrix elements that determine the optical transitions in absorption or in photoluminescence. For example, from (5.28) and (5.33), the absorption coefficient is given by

$$\alpha(h\nu) = A \sum_{i,n,f,n'} p(i,n) |\langle \Psi_{in}| \boldsymbol{e} \cdot \boldsymbol{p} |\Psi_{fn'}\rangle|^2 \delta[h\nu - (E_{fn'} - E_{in})] \,, \tag{5.63}$$

where $p(i,n)$ is the occupation probability of the state Ψ_{in}, A contains all other factors and

$$E_{fn'} - E_{in} = h\nu_0 + \sum_{j=1}^{3N} (n'_j - n_j) \hbar \omega_j \,, \tag{5.64}$$

where $h\nu_0 = \varepsilon_f(Q^{(f)}) - \varepsilon_i(Q^{(i)})$. Since we assumed that there is no localized mode, the coupling to anyone mode is weak and a first order expansion of the optical matrix elements is sufficient:

$$\langle \Psi_i | \boldsymbol{e} \cdot \boldsymbol{p} | \Psi_f \rangle \approx \langle \phi_i | \boldsymbol{e} \cdot \boldsymbol{p} | \phi_f \rangle_{Q=Q^{(f)}} \langle \chi_{in} | \chi_{fn'} \rangle$$
$$+ \sum_j \left(\frac{\partial}{\partial Q_j} \langle \phi_i | \boldsymbol{e} \cdot \boldsymbol{p} | \phi_f \rangle \right)_{Q=Q^{(f)}} \left\langle \chi_{in} \left| Q_j - Q_j^{(f)} \right| \chi_{fn'} \right\rangle \,. \tag{5.65}$$

In the case where the direct transition $\phi_i \to \phi_f$ is allowed (a common situation in direct gap semiconductors), the first term in (5.65) is usually sufficient. Basically, it is equivalent to the Condon approximation [311] which asserts that $\langle \phi_i | \boldsymbol{e} \cdot \boldsymbol{p} | \phi_f \rangle$ is essentially independent of Q. Then the effect of the phonons is just a shift and a broadening of the optical line-shape. When the direct transition is not allowed, the second term in (5.65) must be evaluated: it describes phonon-assisted transitions, where the coupling to phonons makes that the optical transition becomes possible. Phonon-assisted processes are particularly important in indirect gap semiconductors.

5.2 Electron–Phonon Coupling and Optical Line-Shape

Direct Transitions. We now evaluate the first term in (5.65), the zero-order one. The calculation is considerably simplified by the factorization of $\langle \chi_{in}|\chi_{fn'}\rangle$ which from (5.62) is given by

$$\langle \chi_{in}|\chi_{fn'}\rangle = \prod_{j=1}^{3N} \langle in_j|fn'_j\rangle . \tag{5.66}$$

The overlap between displaced harmonic oscillators can be calculated exactly as a function of the coefficients V_j and S_j, but their expression is quite heavy [288, 289, 310, 313]. However, since the Huang–Rhys factors S_j are of order N^{-1}, we only need to keep the terms up to first order in S_j:

$$\langle in_j|fn_j\rangle = 1 - (n_j + 1/2)S_j ,$$
$$\langle in_j|fn_j + 1\rangle = \sqrt{n_j+1}\,V_j ,$$
$$\langle in_j|fn_j - 1\rangle = -\sqrt{n_j}\,V_j . \tag{5.67}$$

All the other terms are of higher order. The change in quantum number in the transition can only be equal to 0, +1, or −1. We now evaluate the intensity of one transition in which $p+r$ modes are excited by +1, and r modes by −1. We label by the index l the first set of modes, by k the second, and j the modes with no change in quantum number. Using (5.67), the intensity of the transition is given by [310]

$$\left(\prod_{l=1}^{p+r}(n_l+1)S_l\right)\left(\prod_{k=1}^{r}n_k S_k\right) \prod_{j=1}^{3N-(p+2r)}[1-(2n_j+1)S_j] . \tag{5.68}$$

The thermal average of (5.68) can be simply obtained by replacing n_j by

$$\bar{n}_j = \left[\exp\left(\frac{\hbar\omega_j}{kT}\right) - 1\right]^{-1} . \tag{5.69}$$

The expression (5.68) allows in principle to calculate the intensity of all possible direct transitions. We will see in the following that it can be done using microscopic electronic structure calculations for small systems (e.g. < 1000 atoms in tight binding). To go beyond, further approximations are required. A simple case occurs when all phonon frequencies can be approximated by a single one ω. Then one can sum the intensities of all transitions corresponding to fixed values $p+r$ and r and one can express the result in the form

$$W_{p+r,r} = \frac{1}{r!(p+r)!}\left[\sum_{l=1}^{3N}(\bar{n}_l+1)S_l\right]^{p+r}\left[\sum_{k=1}^{3N}\bar{n}_k S_k\right]^{r}\prod_{j=1}^{3N}[1-(2\bar{n}_j+1)S_j] , \tag{5.70}$$

where $p+2r$ has been neglected compared to $3N$. Since the summation on l and k extends over the $3N$ phonon modes, (5.70) contains unphysical terms corresponding to products with equal values of l and k but their contribution is negligible. All the transitions occur at the same energy $h\nu_0 + p\hbar\omega$ and the total intensity W_p is then

$$W_p = \sum_{r=0}^{\infty} \frac{1}{r!(p+r)!} [(\bar{n}+1)S]^{p+r} [\bar{n}S]^r \exp\left[-(2\bar{n}+1)S\right], \quad (5.71)$$

where $S = \sum_{j=1}^{3N} S_j$ is the total Huang–Rhys factor and $\bar{n}$ is given by (5.69) at the frequency ω. This expression is usually rewritten in terms of the Bessel functions I_p with imaginary argument of order p [310] as

$$W_p = \left(\frac{\bar{n}+1}{\bar{n}}\right)^{p/2} \exp\left[-(2\bar{n}+1)S\right] I_p\left\{2S[\bar{n}(\bar{n}+1)]^{1/2}\right\},$$

$$= \exp\left[\frac{p\hbar\omega}{2kT} - S\coth\left(\frac{\hbar\omega}{2kT}\right)\right] I_p\left\{\frac{S}{\sinh(\hbar\omega/2kT)}\right\}. \quad (5.72)$$

Note that this expression is valid for all value of p, even negative, that may be the case at $T \neq 0\mathrm{K}$ [289]. The low-temperature limit can be obtained by keeping only the $r = 0$ term in (5.71). The opposite limit can be determined from an asymptotic expansion of the Bessel function

$$I_p(z) = \frac{1}{\sqrt{2\pi}\,(p^2+z^2)^{1/4}} \exp\left(\sqrt{p^2+z^2}\right) \left(\frac{z}{p+\sqrt{p^2+z^2}}\right)^p, \quad (5.73)$$

which is valid if $\sqrt{p^2+z^2} \gg 1$. When $S \gg 1$ and $|p-S| \ll S$, the expression of W_p becomes

$$W_p \approx \frac{\exp\left(-\frac{(p-S)^2}{2S\coth(\hbar\omega/2kT)}\right)}{\sqrt{2\pi S \coth(\hbar\omega/2kT)}}, \quad (5.74)$$

which shows that in the strong coupling limit the line-shape is a Gaussian centered on $p = S$ if we treat p as a continuous variable. In this approximation, and in the high-temperature limit ($kT \gg \hbar\omega$), the line-shape function has a simple form:

$$W(h\nu) = \int W_p\, \delta\left[h\nu - (h\nu_0 + p\hbar\omega)\right] \mathrm{d}p,$$

$$= \frac{1}{\sqrt{4\pi kT S\hbar\omega}} \exp\left(-\frac{(h\nu - h\nu_0 - S\hbar\omega)^2}{4kT S\hbar\omega}\right). \quad (5.75)$$

We now show that this expression can be obtained from a purely classical treatment [289]. To this end, we consider that the optical transitions are vertical in the configuration coordinate diagram (Fig. 5.4). Thus, at a given Q, the line-shape is a Dirac function $\delta(h\nu - \varepsilon_f(Q) + \varepsilon_i(Q))$. To obtain the total line-shape in optical absorption, we have simply to perform an average over the ground state, using classical Boltzmann statistics (over the excited

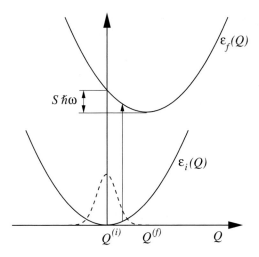

Fig. 5.4. Configuration coordinate diagram (*straight line*) showing the classical view of the optical line-shape in optical absorption. The line-shape is given by the average over the ground state (*dashed line*) of the vertical transitions from the initial state i to the final states f

state in luminescence). Using (5.60) in the case of a single phonon frequency, we thus write

$$W(h\nu) = \frac{\int_{-\infty}^{+\infty} \mathrm{d}P\mathrm{d}Q \; \delta(h\nu - h\nu_0 - S\hbar\omega + IQ)\exp\left[-\left(\frac{P^2}{2} + \frac{\omega^2 Q^2}{2}\right)/kT\right]}{\int_{-\infty}^{+\infty} \mathrm{d}P\mathrm{d}Q \exp\left[-\left(\frac{P^2}{2} + \frac{\omega^2 Q^2}{2}\right)/kT\right]}, \tag{5.76}$$

where P the classical momentum and $I = V\sqrt{2\hbar\omega^3}$, with $V^2 = S$. This expression can be readily calculated and gives

$$W(h\nu) = \frac{1}{I}\sqrt{\frac{\omega^2}{2\pi kT}} \exp\left[-\frac{\omega^2}{2I^2 kT}(h\nu - h\nu_0 - S\hbar\omega)^2\right], \tag{5.77}$$

which can be rewritten in the final form (5.75).

In the case where the coupling occurs with extended phonons centered on two frequencies ω_1 and ω_2 (e.g. acoustic and optical phonons), the same treatment can be applied to each frequency separately, leading to optical line-shape functions of the form

$$W_1(h\nu) = \sum_{p_1} W_{p_1}\delta\left[h\nu - (h\nu_0 + p_1\hbar\omega_1)\right],$$

$$W_2(h\nu) = \sum_{p_2} W_{p_2}\delta\left[h\nu - (h\nu_0 + p_2\hbar\omega_2)\right]. \tag{5.78}$$

The total line-shape function can be written

$$W(h\nu) = \sum_{p_1,p_2} W_{p_1}W_{p_2}\delta\left[h\nu - (h\nu_0 + p_1\hbar\omega_1 + p_2\hbar\omega_2)\right],$$

$$= \int W_1(h\nu - E)W_2(E + h\nu_0)\mathrm{d}E, \tag{5.79}$$

which is just the convolution of the two functions. The generalization to a larger number of modes is straightforward. In the case where ω_1 corresponds to a single localized mode, $W_1(h\nu)$ must be calculated following the rules established for point defects [288, 289, 314]. However, in the high temperature limit $(kT \gg \hbar\omega)$, $W_1(h\nu)$ is still given by (5.75).

Phonon-Assisted Transitions. We deal now with the evaluation of the second term in (5.65) corresponding to phonon-assisted transitions. For the sake of generality, we also consider the first term since in some cases there may be a competition between direct and phonon-assisted processes (see for example Sect. 5.4). Some general formula will be established in the limit of the coupling to $3N$ extended phonon modes. But first we consider the simpler limit of a vanishing electron–phonon coupling, i.e. $V_j, S_j \to 0$. The oscillators in both initial and final states are centered on the same position and we can make use of the fact that $\langle \chi_{in} | \chi_{fn'} \rangle = \delta_{n,n'}$. Using the transformation

$$Q_j - Q_j^{(f)} = \sqrt{\frac{\hbar}{2\omega_j}} \left(a_j^{+(f)} + a_j^{(f)} \right), \tag{5.80}$$

where $a_j^{+(f)}$ and $a_j^{(f)}$ are the creation and annihilation operators for the final state, respectively, we see in (5.65) that the change in quantum number in the transition can only be equal to 0 (no-phonon transition), $+1$ (phonon emission) or -1 (phonon absorption) because

$$a_j^{+(f)} |fn_j\rangle = \sqrt{n_j + 1} |fn_j + 1\rangle ,$$
$$a_j^{(f)} |fn_j\rangle = \sqrt{n_j} |fn_j - 1\rangle . \tag{5.81}$$

We deduce the contribution of each process to the absorption coefficient:

no-phonon: $A\, p(i)\, |\langle \phi_i | \boldsymbol{e} \cdot \boldsymbol{p} | \phi_f \rangle|^2\, \delta(h\nu - h\nu_0)$,

phonon emission: $A\, p(i)\, \dfrac{\hbar}{2\omega_j} \left| \boldsymbol{A}_j^{(if)} \right|^2 (\bar{n}_j + 1)\, \delta(h\nu - h\nu_0 - \hbar\omega_j)$,

phonon absorption: $A\, p(i)\, \dfrac{\hbar}{2\omega_j} \left| \boldsymbol{A}_j^{(if)} \right|^2 \bar{n}_j\, \delta(h\nu - h\nu_0 + \hbar\omega_j)$, (5.82)

with

$$\boldsymbol{A}_j^{(if)} = \left(\frac{\partial}{\partial Q_j} \langle \phi_i | \boldsymbol{e} \cdot \boldsymbol{p} | \phi_f \rangle \right)_{Q=Q^{(f)}} . \tag{5.83}$$

We come now to the general formulation of the problem. We consider that the electrons can be coupled to all modes ($S_j \neq 0$), with $S_j \propto N^{-1}$. We need to calculate

$$\langle \phi_i | \boldsymbol{e} \cdot \boldsymbol{p} | \phi_f \rangle_{Q=Q^{(f)}} \langle \chi_{in} | \chi_{fn'} \rangle + \sum_j \boldsymbol{A}_j^{(if)} \sqrt{\frac{\hbar}{2\omega_j}} \left\langle \chi_{in} | a_j^{+(f)} + a_j^{(f)} | \chi_{fn'} \right\rangle . \tag{5.84}$$

5.2 Electron–Phonon Coupling and Optical Line-Shape

We multiply and divide the second term by $\langle \chi_{in} | \chi_{fn'} \rangle$ which allows to factorize it in the whole expression. Since this term has been completely evaluated in (5.68), it only remains to calculate

$$\sum_j A_j^{(if)} \sqrt{\frac{\hbar}{2\omega_j}} \frac{\langle \chi_{in} | a_j^{+(f)} + a_j^{(f)} | \chi_{fn'} \rangle}{\langle \chi_{in} | \chi_{fn'} \rangle} \tag{5.85}$$

that simplifies into

$$\sum_j A_j^{(if)} \sqrt{\frac{\hbar}{2\omega_j}} \frac{\langle in_j | a_j^{+(f)} + a_j^{(f)} | fn_j' \rangle}{\langle in_j | fn_j' \rangle} . \tag{5.86}$$

We now evaluate the intensity of one transition in which $p + r$ modes are excited by $+1$, and r modes by -1. We label by the index l the first set of modes, by k the second, and j the modes with no change in quantum number. Using (5.67), we finally obtain that the intensity of the transitions is given by

$$\left| \langle \phi_i | \boldsymbol{e} \cdot \boldsymbol{p} | \phi_f \rangle_{Q=Q^{(f)}} + \sum_{l=1}^{p+r} A_l^{(if)} \sqrt{\frac{\hbar}{2\omega_l}} \frac{1 - (\bar{n}_l + \tfrac{1}{2})S_l}{V_l} \right.$$

$$\left. - \sum_{k=1}^{r} A_k^{(if)} \sqrt{\frac{\hbar}{2\omega_k}} \frac{1 - (\bar{n}_k + \tfrac{1}{2})S_k}{V_k} + \sum_{j=1}^{3N-(p+2r)} A_j^{(if)} \sqrt{\frac{\hbar}{2\omega_j}} \frac{V_j}{1 - (\bar{n}_j + \tfrac{1}{2})S_j} \right|^2$$

$$\times \left(\prod_{l=1}^{p+r} (\bar{n}_l + 1) S_l \right) \left(\prod_{k=1}^{r} \bar{n}_k S_k \right) \left(\prod_{j=1}^{3N-(p+2r)} [1 - (2\bar{n}_j + 1) S_j] \right) . \tag{5.87}$$

In spite of its apparent complexity, this formula can be easily implemented in electronic structure calculations, allowing to determine the absorption coefficient by injecting (5.87) in (5.63). The problem can be considerably simplified when phonon modes involved in the assisted transitions and those strongly coupled to electrons are well separated in energy. In that case, we can calculate the intensities of the phonon-assisted transitions using (5.82) and convolute the resulting spectrum to account for the coupling, for example using (5.72). An example of calculations using this simplified procedure is described in Sect. 5.4.1.

5.2.5 Calculation of the Coupling Parameters

The theory developed in the previous sections cover many cases of interest. It remains now to see how the parameters involved in the expressions of the optical line-shape can be calculated in a practical way, in particular the coefficients V_j and the related Huang–Rhys factors [310] $S_j = V_j^2$ that determine the electron–phonon coupling. From (5.58) and (5.59), we have

$$V_j = \frac{1}{2\hbar \omega_j^3} \frac{\partial (\varepsilon_f(Q) - \varepsilon_i(Q))}{\partial Q_j} . \tag{5.88}$$

The coefficients can be obtained with ab initio approaches by calculating the derivative of the total energy in the initial and final states with respect to the normal displacements. For obvious reasons, this can be done using density functional theory only for small systems [315]. For larger systems like small nanocrystals, non-self-consistent semi-empirical methods can be used. In tight binding, the one-electron state energies can be calculated as a function of the atomic displacements if the tight binding parameters are made dependent on the atomic positions, for example using the Harrison's rules [57] given in (1.149). This is the approach used in [295] for Si nanocrystals. Note that the coefficients $\boldsymbol{A}_j^{(if)}$ for phonon-assisted transitions can be calculated similarly using (5.83).

An interesting limit is to consider that the electrons mainly couple to lattice deformations characterized by long wavelengths compared to the size of the unit cell, which is a reasonable approximation in large quantum dots. In that limit, one can use continuum models that we present in the following sections. We first consider the optical modes, and next the acoustic ones.

5.2.6 Fröhlich Coupling: Optical Modes

In polar materials, the dominant electron–phonon coupling is the Fröhlich interaction: the optical vibrations induce a macroscopic polarization $\boldsymbol{P}$ which, in the bulk, induces in a coupling between electrons and longitudinal optical (LO) phonons. The resulting electron–phonon coupling Hamiltonian can be calculated with the well-known Fröhlich continuum model [316]. But a large number of works have shown that one cannot use the Hamiltonian based on bulk phonons to treat heterostructures [317–319] and nanostructures [303, 304, 320] because optical phonons can be strongly influenced by the presence of interfaces which gives rise to confinement of optical phonons as well as interface modes, the so-called surface optical (SO) modes [321]. Due to its practical importance in a large number of problems, we derive in the following the coupling Hamiltonian in the case of spherical quantum dots including LO and SO modes, following [303, 304]. The case of quantum wells is treated in [318]. Note that these models do not take into account the coupling between LO and transverse optical (TO) modes imposed by the boundary conditions and neglect the dispersion of the phonon branches. Improved models can be found in [305, 306].

We consider a semiconductor sphere of radius R and dielectric constant $\epsilon_{\text{in}}(\omega)$ embedded in a medium of dielectric constant ϵ_{out}. In the particular case of a crystal with two oppositely charged ions in the unit cell, we define by $\boldsymbol{s}_\pm$ the instantaneous displacement of the ion of effective charge $\pm e^*$ [292]. In the limit of long wavelength, the two atoms vibrate opposite to each other while the motion in adjacent cell is practically identical [291]. The displacements induce a dipole in each cell given by $e^*(\boldsymbol{s}_+ - \boldsymbol{s}_-) \equiv e^*\boldsymbol{s}$ (Fig. 5.5). The displacement is treated as a continuous variable $\boldsymbol{s}(\boldsymbol{r}, t)$. The total polarization is given by [322]

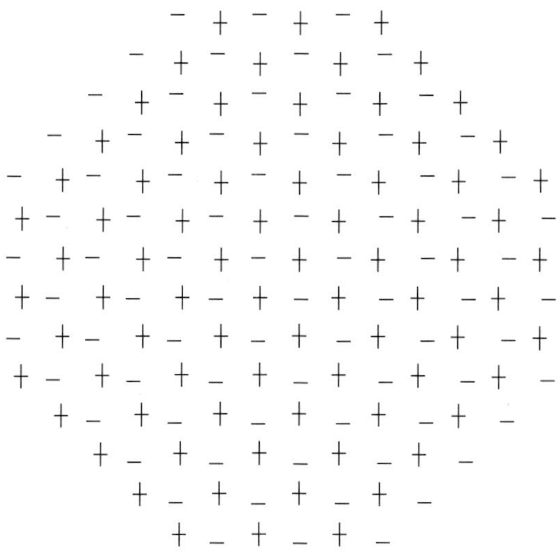

Fig. 5.5. Lattice deformation induced by an electron injected in a quantum dot with a cubic lattice

$$\boldsymbol{P} = [\epsilon_{\text{in}}(\omega) - 1]\epsilon_0 \boldsymbol{E} = \frac{1}{\Omega_0}\left(e^*\boldsymbol{s} + \alpha \boldsymbol{E}_{\text{loc}}\right) , \tag{5.89}$$

where $\alpha \boldsymbol{E}_{\text{loc}}$ is the contribution coming from the polarization of the two ions induced by the local-field $\boldsymbol{E}_{\text{loc}}$, α being the sum of the polarizability of the two atoms and Ω_0 is the volume of the unit cell. In the case of cubic lattices, we can use the Lorentz relation

$$\boldsymbol{E}_{\text{loc}} = \boldsymbol{E} + \frac{\boldsymbol{P}}{3\epsilon_0} , \tag{5.90}$$

where $\boldsymbol{E}$ is the macroscopic field. The equations of motion are given by

$$\begin{aligned} M_+ \frac{\partial^2 \boldsymbol{s}_+}{\partial t^2} &= -k(\boldsymbol{s}_+ - \boldsymbol{s}_-) + e^*\boldsymbol{E}_{\text{loc}} , \\ M_- \frac{\partial^2 \boldsymbol{s}_-}{\partial t^2} &= +k(\boldsymbol{s}_+ - \boldsymbol{s}_-) - e^*\boldsymbol{E}_{\text{loc}} , \end{aligned} \tag{5.91}$$

where k describes the short-range force. With a reduced mass $\bar{M} = M_+M_-/(M_+ + M_-)$, it leads to

$$\bar{M}\frac{\partial^2 \boldsymbol{s}}{\partial t^2} = -k\boldsymbol{s} + e^*\boldsymbol{E}_{\text{loc}} . \tag{5.92}$$

In the bulk semiconductor, the solutions of (5.90) and (5.92) give the longitudinal (LO) and transverse (TO) optical modes of frequencies ω_{LO} and ω_{TO}, respectively (in the long wavelength limit). The equations solved in the static case and in the limit of high frequency lead to the following relations [291, 292]

$$-\frac{k}{\bar{M}} + \frac{e^{*2}}{\bar{M}(3\epsilon_0\Omega_0 - \alpha)} = -\omega_{TO}^2 \,, \quad \frac{\alpha}{\Omega_0 - \frac{\alpha}{3\epsilon_0}} = \epsilon_0(\epsilon_{in}(\infty) - 1) \,,$$

$$\frac{1}{\sqrt{\Omega_0 \bar{M}}} \frac{e^*}{1 - \frac{\alpha}{3\epsilon_0 \Omega_0}} = \omega_{TO}\sqrt{\epsilon_0(\epsilon_{in}(0) - \epsilon_{in}(\infty))} \,, \quad \frac{\omega_{LO}^2}{\omega_{TO}^2} = \frac{\epsilon_{in}(0)}{\epsilon_{in}(\infty)} \,, \quad (5.93)$$

the last relationship being known as the Lyddane–Sachs–Teller relation. The dielectric constant is given by

$$\epsilon_{in}(\omega) = \epsilon_{in}(\infty) + \frac{[\epsilon_{in}(0) - \epsilon_{in}(\infty)]\omega_{TO}^2}{\omega_{TO}^2 - \omega^2} = \epsilon_{in}(\infty)\frac{\omega^2 - \omega_{LO}^2}{\omega^2 - \omega_{TO}^2} \,. \quad (5.94)$$

The phonon modes in the sphere are determined by the classical equations in absence of external field:

$$\boldsymbol{D} = \epsilon_{in}(\omega)\epsilon_0 \boldsymbol{E} = \epsilon_0 \boldsymbol{E} + \boldsymbol{P} \,,$$
$$\boldsymbol{\nabla} \cdot \boldsymbol{D} = 0 \,, \quad \boldsymbol{E} = -\boldsymbol{\nabla}\phi \,. \quad (5.95)$$

They lead to the following equation for the electrostatic potential ϕ:

$$\epsilon_{in}(\omega)\Delta\phi = 0 \,. \quad (5.96)$$

LO Modes. The first type of solutions corresponds to $\epsilon_{in}(\omega) = 0$ that gives the internal LO modes of frequency ω_{LO}. Using spherical coordinates, the potential may be written in terms of the spherical Bessel functions j_l defined in (2.19) and of the spherical harmonics Y_{lm}:

$$\phi(\boldsymbol{r}) = \sum_{k,l,m} c_{k,l,m} \, j_l(kr) Y_{lm}(\theta,\varphi) \,. \quad (5.97)$$

The continuity of $\boldsymbol{D}$ implies that $\boldsymbol{D} = 0$ outside the sphere and therefore $\phi = 0$ for $r > R$. From the continuity of ϕ, we deduce that the allowed values of k are given by

$$k = X_{nl}/R \,, \quad (5.98)$$

where the coefficients X_{nl} are the zeros of the spherical Bessel functions (see Sect. 2.1.4). It remains to connect the potential with the atomic displacement $\boldsymbol{s}$. Since $\boldsymbol{D} = 0$, (5.90) becomes

$$\boldsymbol{E}_{loc} = -\frac{2\boldsymbol{P}}{3\epsilon_0} \,. \quad (5.99)$$

Putting this into (5.89) and using $\boldsymbol{E} = -\boldsymbol{\nabla}\phi = -\boldsymbol{P}/3\epsilon_0$ we find

$$\boldsymbol{\nabla}\phi = \frac{e^*}{\epsilon_0 \Omega_0 + 2\alpha/3}\boldsymbol{s} \,, \quad (5.100)$$

which with (5.97) gives an expression for $\boldsymbol{s}(\boldsymbol{r})$. The next step is to connect the displacements $\boldsymbol{s}_{n+}$ and $\boldsymbol{s}_{n-}$ of the two atoms in the cell n with $\boldsymbol{s}(\boldsymbol{r})$ in order to establish the quantized form of the operators. From (5.52) we have

$$\boldsymbol{s}_{n\pm} = \frac{1}{\sqrt{M_\pm}} \sum_{k,l,m} \boldsymbol{e}_{n\pm}^{(klm)} Q_{k,l,m} \,, \quad (5.101)$$

where the vectors $\boldsymbol{e}^{(klm)}$ are mutually orthogonal and $Q_{k,l,m}$ are the normal displacements. From (5.91) we deduce that $\sqrt{M_+}\boldsymbol{e}^{(klm)}_{n+} + \sqrt{M_-}\boldsymbol{e}^{(klm)}_{n-} = 0$ and therefore

$$\boldsymbol{s}_n = \boldsymbol{s}_{n+} - \boldsymbol{s}_{n-} = \frac{\sqrt{M_+}}{\bar{M}} \sum_{k,l,m} \boldsymbol{e}^{(klm)}_{n+} Q_{k,l,m} = \frac{\sqrt{M_-}}{\bar{M}} \sum_{k,l,m} \boldsymbol{e}^{(klm)}_{n-} Q_{k,l,m} ,$$

$$= \frac{1}{\sqrt{\bar{M}}} \sum_{k,l,m} \boldsymbol{u}^{(klm)}_n Q_{k,l,m} , \tag{5.102}$$

where we have defined new vectors $\boldsymbol{u}^{(klm)}$. We easily verify that these vectors must be also mutually orthogonal. With (5.97) and (5.100) we can write

$$\boldsymbol{u}^{(klm)}_n = u_0 \boldsymbol{\nabla}\left(j_l(kr_n) Y_{lm}(\theta_n, \phi_n)\right) , \tag{5.103}$$

where r_n, θ_n, ϕ_n are the spherical coordinates of the cell n. We note that with this definition the normal displacements $Q_{k,l,m}$ and the vectors $\boldsymbol{u}^{(klm)}$ are complex. We verify that

$$\sum_n \boldsymbol{u}^{(klm)*}_n \boldsymbol{u}^{(k'l'm')}_n \approx \frac{|u_0|^2}{\Omega_0} \int_V \boldsymbol{\nabla}\left(j_l(kr) Y^*_{lm}(\theta,\phi)\right) \boldsymbol{\nabla}\left(j_{l'}(kr) Y_{l'm'}(\theta,\phi)\right) \mathrm{d}\boldsymbol{r}$$

$$= \delta_{kk'} \delta_{ll'} \delta_{mm'} , \tag{5.104}$$

if

$$u_0 = \sqrt{\frac{\Omega_0 B_k^2}{k^2}} \quad \text{with} \quad B_k^{-2} = \frac{R^3}{2} j^2_{l+1}(kR) . \tag{5.105}$$

To obtain this result, we use the well-known relation

$$\int_V \boldsymbol{\nabla} f \boldsymbol{\nabla} g \mathrm{d}\boldsymbol{r} = -\int_V f \boldsymbol{\nabla}^2 g \mathrm{d}\boldsymbol{r} + \int_S f \frac{\partial g}{\partial n} \mathrm{d}S , \tag{5.106}$$

where the integration is made over the sphere (V) or its surface (S), and n is the normal to the surface of the sphere. In our case, f and g are of the form $j_l(kr) Y_{lm}(\theta,\phi)$, the surface integral vanishes and $\boldsymbol{\nabla}^2 g = -k^2 g$. In accordance with (5.102), we obtain

$$\boldsymbol{s}(\boldsymbol{r}) = \sum_{k,l,m} \sqrt{\frac{\Omega_0 B_k^2}{\bar{M} k^2}} \boldsymbol{\nabla}\left(j_l(kr) Y_{lm}(\theta,\phi)\right) Q_{k,l,m} . \tag{5.107}$$

For simplicity, we choose the phase factors of the spherical harmonics such that $Y_{l-m} = Y^*_{lm}$. Therefore $\boldsymbol{s}(\boldsymbol{r})$ is real if $Q_{k,l,-m} = Q^*_{k,l,m}$. Then, using the orthogonality of the normal modes, we derive the classical Hamiltonian for the free phonons

$$H = \frac{1}{2} \sum_{k,l,m} \left[P^*_{k,l,m} P_{k,l,m} + \omega^2_{\mathrm{LO}} Q^*_{k,l,m} Q_{k,l,m}\right] , \tag{5.108}$$

where $P_{k,l,m} = \partial Q_{k,l,m}/\partial t$. The transition to a quantum mechanical description is now straightforward. We have just to interpret $P_{k,l,m}$ and $Q_{k,l,m}$ as operators that fulfill the commutation relations:

$$[Q_{k,l,m}, P_{k',l',m'}] = i\hbar\, \delta_{kk'}\delta_{ll'}\delta_{mm'}\ . \tag{5.109}$$

If we introduce the operators

$$a^+_{klm} = \sqrt{\frac{1}{2\hbar\omega_{\text{LO}}}}\left(\omega_{\text{LO}} Q^*_{k,l,m} - iP_{k,l,m}\right)\ ,$$

$$a_{klm} = \sqrt{\frac{1}{2\hbar\omega_{\text{LO}}}}\left(\omega_{\text{LO}} Q_{k,l,m} + iP^*_{k,l,m}\right)\ , \tag{5.110}$$

the Hamiltonian takes the usual form

$$H = \sum_{k,l,m} \hbar\omega_{\text{LO}}\left(a^+_{klm} a_{klm} + 1/2\right)\ . \tag{5.111}$$

We can now calculate the electron–LO–phonon Hamiltonian. If $\rho(\mathbf{r})$ is the charge density, the Hamiltonian reads

$$H^{\text{LO}}_{\text{e-ph}} = \int_V \phi(\mathbf{r})\rho(\mathbf{r})d\mathbf{r}\ . \tag{5.112}$$

Equations (5.100) and (5.107) give the expression

$$\phi(\mathbf{r}) = \sum_{k,l,m} \frac{e^*}{\epsilon_0 \Omega_0 + 2\alpha/3}\sqrt{\frac{\Omega_0 B_k^2 \hbar}{2\bar{M} k^2 \omega_{\text{LO}}}} \times [a_{klm} j_l(kr) Y_{lm}(\theta,\phi) + \text{H.c.}]\ , \tag{5.113}$$

where H.c. means Hermitian conjugate. Using (5.93), we obtain finally

$$H^{\text{LO}}_{\text{e-ph}} = \sum_{k,l,m}\sqrt{\frac{B_k^2 \hbar\omega_{\text{LO}}}{2\epsilon_0 k^2}}\left(\frac{1}{\epsilon_{\text{in}}(\infty)} - \frac{1}{\epsilon_{\text{in}}(0)}\right)^{1/2}$$

$$\times \left[a_{klm}\int_V \rho(\mathbf{r}) j_l(kr) Y_{lm}(\theta,\phi) d\mathbf{r} + \text{H.c.}\right]\ . \tag{5.114}$$

SO Modes. The surface (SO) modes are obtained following the same approach. They correspond to $\Delta\phi = 0$ in (5.96). The solutions are of the form

$$\phi(\mathbf{r}) = A_{l,m} r^l Y_{lm}(\theta,\phi)\ \text{for}\ r < R\ ,$$
$$\phi(\mathbf{r}) = A_{l,m} R^{2l+1} r^{-l-1} Y_{lm}(\theta,\phi)\ \text{for}\ r > R\ , \tag{5.115}$$

where the coefficient $A_{l,m} R^{2l+1}$ in the second line has been determined by the condition of continuity of ϕ across the interface. The continuity of the normal component of $\mathbf{D}$ implies that

$$l\epsilon_{\text{in}}(\omega_l) = -(l+1)\epsilon_{\text{out}}\ , \tag{5.116}$$

which with (5.94) leads to

$$\omega_l^2 - \omega_{\text{TO}}^2 = \frac{l[\epsilon_{\text{in}}(0) - \epsilon_{\text{in}}(\infty)]}{l\epsilon_{\text{in}}(\infty) + (l+1)\epsilon_{\text{out}}}\omega_{\text{TO}}^2\ . \tag{5.117}$$

Thus, for a given couple (l, m), there are several LO modes and only one SO mode. The frequencies of the LO modes depend on the quantum number l. We must now calculate the electrostatic potential with respect to the displacement s. Looking for solutions of the form $s = s_0 e^{i\omega t}$, (5.92) gives

$$\boldsymbol{E}_{\text{loc}} = \frac{k - \omega^2 \bar{M}}{e^*} \boldsymbol{s} . \tag{5.118}$$

Eliminating $\boldsymbol{P}$ between (5.89) and (5.90), we obtain

$$\boldsymbol{E} = \left(1 - \frac{\alpha}{3\epsilon_0 \Omega_0}\right) \frac{\bar{M}}{e^*} \left[\frac{k}{\bar{M}} - \omega^2 - \frac{e^{*2}}{\bar{M}(3\epsilon_0 \Omega_0 - \alpha)}\right] \boldsymbol{s} , \tag{5.119}$$

which using (5.93) leads to

$$\boldsymbol{E} = -\boldsymbol{\nabla}\phi = \sqrt{\frac{\bar{M}}{\Omega_0}} \frac{\omega_{\text{TO}}^2 - \omega^2}{\omega_{\text{TO}} \sqrt{\epsilon_0 [\epsilon_{\text{in}}(0) - \epsilon_{\text{in}}(\infty)]}} \boldsymbol{s} . \tag{5.120}$$

Following (5.102), we start now with

$$\boldsymbol{s}_n = \frac{1}{\sqrt{\bar{M}}} \sum_{l,m} \boldsymbol{v}_n^{(lm)} Q_{l,m}^{\text{SO}} ,$$

$$\boldsymbol{v}_n^{(lm)} = v_0 \boldsymbol{\nabla}[r^l Y_{lm}(\theta, \phi)] . \tag{5.121}$$

The relation

$$\sum_n \boldsymbol{v}_n^{(lm)*} \boldsymbol{v}_n^{(l'm')} = \delta_{ll'} \delta_{mm'} \tag{5.122}$$

is verified if

$$v_0 = \frac{1}{R^l} \sqrt{\frac{\Omega_0}{lR}} . \tag{5.123}$$

To establish this result, we also used (5.106) with f and g of the form $r^l Y_{lm}(\theta, \phi)$, but this time it is the volume integral that vanishes. Combining (5.120), (5.121), (5.117) and (5.123), we obtain finally

$$H_{\text{e-ph}}^{\text{SO}} = \sum_{l,m} \sqrt{\frac{\hbar l}{2\epsilon_0 \omega_l R}} \sqrt{\frac{\epsilon_{\text{in}}(0) - \epsilon_{\text{in}}(\infty)}{l \epsilon_{\text{in}}(\infty) + (l+1)\epsilon_{\text{out}}}} \omega_{\text{TO}}$$

$$\times \left[a_{lm}^{\text{SO}} \int_V \rho(\boldsymbol{r}) \left(\frac{r}{R}\right)^l Y_{lm}(\theta, \phi) \mathrm{d}\boldsymbol{r} + \text{H.c.}\right] , \tag{5.124}$$

where a_{lm}^{SO} is the annihilation operator for the SO mode of quantum numbers (l, m).

The coupling Hamiltonians (5.114) and (5.124) have been extensively used to study polarons in spherical quantum dots using variational techniques [303, 304, 320, 324] and within second-order perturbation theory [325–329], and also the polaron bound to an impurity in quantum dots [320, 330, 331]. The coupling in quantum dots of ternary alloy semiconductors such as

$CdS_{1-x}Se_x$ has been studied in [332]. The influence of the electron–optical–phonon coupling on the optical properties of direct gap semiconductor nanostructures will be discussed in Sect. 5.3. But before proceeding, we consider the behavior of an electron (free polaron) in a small quantum dot [331].

Electron–Phonon Interaction Energy of an Electron in a Small Quantum Dot. In the limit where the kinetic energy of the electron predominates, the interaction with phonons may be regarded as a perturbation, the quantum dot radius being much smaller than the effective Bohr radius of the electron. In the envelope function approximation, we can use the wavefunction (2.18) for the ground state in a spherical quantum well and the charge density is given by

$$\rho(r) = -\frac{e}{2\pi R} \left(\frac{\sin\left(\frac{\pi r}{R}\right)}{r} \right)^2 . \tag{5.125}$$

This radial charge distribution only couples to the $l = 0$ LO modes. The total electron–phonon interaction energy, the Franck–Condon energy (5.61), is then equal to

$$d_{\mathrm{FC}} = \sum_k S_{k,0,0} \, \hbar\omega_{\mathrm{LO}} , \tag{5.126}$$

where $S_{k,0,0}$ are the Huang–Rhys factors for the coupling to the LO modes with $l = 0$ and $m = 0$. These factors are easily deduced by comparing the term linear in the normal displacements in (5.60) and $H_{\mathrm{e-ph}}^{\mathrm{LO}}$ in (5.114). We obtain:

$$S_{k,0,0} \, \hbar\omega_{\mathrm{LO}} = \frac{B_k^2}{2\epsilon_0 k^2} \left(\frac{1}{\epsilon_{\mathrm{in}}(\infty)} - \frac{1}{\epsilon_{\mathrm{in}}(0)} \right) \left[\int_V \rho(r) j_0(kr) Y_{00}(\theta,\phi) \mathrm{d}\mathbf{r} \right]^2 . \tag{5.127}$$

Using the fact that $k = n\pi/R$ and $B_k^2 = 2k^2/R$ when $l = 0$, we are left with [320, 331]

$$d_{\mathrm{FC}} = \frac{e^2}{4\pi\epsilon_0 R} \left(\frac{1}{\epsilon_{\mathrm{in}}(\infty)} - \frac{1}{\epsilon_{\mathrm{in}}(0)} \right) C , \tag{5.128}$$

with

$$C = \frac{4}{\pi^2} \sum_{n=1}^{\infty} \left[\int_0^\pi \frac{\sin^2(x) \, \sin(nx)}{nx} \mathrm{d}x \right]^2 \tag{5.129}$$

that may be written in closed analytical form [331] as

$$C = [1 - \mathrm{Si}(2\pi) + \mathrm{Si}(4\pi)/2]/2 = 0.3930 , \tag{5.130}$$

where Si is the integral sine. In fact, the interaction energy (5.128) can be obtained using a much simpler approach in this limit of small quantum dots.

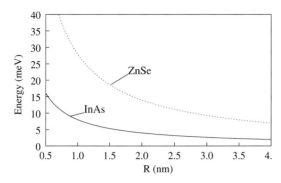

Fig. 5.6. Electron–phonon interaction energy $d_{\rm FC}$ as a function of the radius of the quantum dot R. For comparison, the energy for the bulk polaron is 1.6 meV in InAs and 14.8 meV for ZnSe

The kinetic energy of the electron is so high that the lattice can only respond to the average charge density given by $\rho(r)$. Thus we have

$$d_{\rm FC} = \left(\frac{1}{2}\int_V \rho(r)\phi(r)\mathrm{d}\mathbf{r}\right)_\infty - \left(\frac{1}{2}\int_V \rho(r)\phi(r)\mathrm{d}\mathbf{r}\right)_0 , \qquad (5.131)$$

where $\phi(r)$ is the electrostatic potential induced by $\rho(r)$. The factor $1/2$ comes from the fact that it is a self-energy (see Sect. 3.1.1). We calculate the self-energy with the dielectric constant $\epsilon_{\rm in}(\omega)$ and we make the difference between the high frequency limit ($\omega \to \infty$) and the static one ($\omega \to 0$) to keep only the response of the phonons. Using the Gauss theorem, we are left with

$$d_{\rm FC} = \frac{e^2}{4\pi\epsilon_0 R}\left(\frac{1}{\epsilon_{\rm in}(\infty)} - \frac{1}{\epsilon_{\rm in}(0)}\right)\int_0^\pi \frac{(u - \sin(2u)/2)^2}{2\pi u^2}\mathrm{d}u , \qquad (5.132)$$

where the integral is also equal to C.

The perturbation calculation is justified if $d_{\rm FC} \ll E_c$ where E_c is the confinement energy. We plot in Fig. 5.6 the electron–phonon interaction energy $d_{\rm FC}$ for InAs and ZnSe, for comparison. It is also interesting to compare with the free polaron in the bulk semiconductor where the relaxation energy is given by [291]

$$d_{\rm FC}^{\rm bulk} = \alpha_0 \hbar\omega_{\rm LO} \text{ with } \alpha_0 = \frac{e^2}{8\pi\epsilon_0\hbar\omega_{\rm LO}}\left(\frac{2m^*\omega_{\rm LO}}{\hbar}\right)^{1/2}\left(\frac{1}{\epsilon_{\rm in}(\infty)} - \frac{1}{\epsilon_{\rm in}(0)}\right), \qquad (5.133)$$

m^* being the electron effective mass. The coupling constant α_0 is typically less than unity in III–V and II–VI semiconductors (e.g. $\alpha_0 \approx 0.06$ in InAs).

5.2.7 Coupling to Acoustic Modes

We deal now with the coupling to longitudinal acoustic modes in the limit of long wavelengths. These modes correspond to compression waves associated with local variations of volume, i.e. of the lattice constant. Such a lattice relaxation takes place in particular after excitation of an electron–hole pair. The relaxation occurs because an electron has been transferred from a bonding-like (valence) to an antibonding-like (conduction) state, that tends to weaken the bonds. The amplitude of the distortion is connected to the electron–hole density. In quantum dots, the confinement increases this density and therefore the relaxation is enhanced.

We consider the particular case of a spherical nanocrystal of radius R in which an electron–hole pair has been excited. This induces a deformation which we describe once again by a continuous displacement field $s(r)$. We need to find a functional $E(\{s\})$ for the total energy of the system that will be minimum for the true displacement field. We write

$$E(\{s\}) = E_0(\{s\}) + E_{\text{exc}}(\{s\}) , \tag{5.134}$$

where E_0 is the ground state energy and E_{exc} is the excitonic energy. The lattice deformation is characterized classically by the strain parameters [158]

$$e_{ij} = \frac{1}{2}\left[\frac{\partial s_i}{\partial x_j} + \frac{\partial s_j}{\partial x_i}\right] , \tag{5.135}$$

where $i, j = x, y, z$. The ground state energy is given by a constant term plus the total elastic energy

$$E_0(\{s\}) = E_0(\{s = 0\}) + \int_V U(r)\mathrm{d}r ,$$

with $U(r) = \dfrac{C_{11}}{2}(e_{xx}^2 + e_{yy}^2 + e_{zz}^2) + 2C_{44}(e_{yz}^2 + e_{zx}^2 + e_{xy}^2)$
$$+ C_{12}(e_{zz}e_{yy} + e_{zz}e_{xx} + e_{xx}e_{yy}) , \tag{5.136}$$

where C_{11}, C_{12}, and C_{44} are the elastic constants of the bulk semiconductor. We assume that these constants are not influenced by the confinement. The energy $E_{\text{exc}}(\{s\})$ of the electron–hole pair can be easily calculated in the effective mass approximation following Sect. 2.2.1. In this approximation, each band edge (conduction band for the electron and valence band for the hole) is interpreted as a potential energy that depends locally on the elastic strains through the deformation potentials [106]. Thus, for a given displacement field $s(r)$, we can calculate the exciton energy and we can iterate the procedure to minimize the total energy $E(\{s\})$ with respect to $s(r)$, for example using a discrete spatial grid. A simpler approach can be used in the limit of strong confinement where the electron–hole wavefunction is not influenced by the lattice relaxation. In this approximation, the electron–hole density is given by (Sect. 2.1.4)

$$n(r) = \frac{1}{2\pi R}\left(\frac{\sin(\pi r/R)}{r}\right)^2 , \tag{5.137}$$

where r is the distance from the center of the crystallite. We can write to first order in perturbation that

$$E_{\text{exc}}(\{s\}) = \int_V E_{\text{g}}(\boldsymbol{r}) n(\boldsymbol{r}) \mathrm{d}\boldsymbol{r} \,, \tag{5.138}$$

where $E_{\text{g}}(\boldsymbol{r})$ is the local bandgap energy taking into account the shift of the band edges induced by the deformation. If we further assume that the displacement field is radial, i.e. $\boldsymbol{s}(\boldsymbol{r}) = s(r)\boldsymbol{e}_r$, we have

$$e_{ii} = \left(\frac{x_i}{r}\right)^2 \left(\frac{\partial s}{\partial r}\right) + \frac{s}{r}\left(\frac{r^2 - x_i^2}{r^2}\right),$$

$$e_{ij} = \frac{x_i x_j}{r^2}\left[\left(\frac{\partial s}{\partial r}\right) - \frac{s}{r}\right], \quad i \neq j \,. \tag{5.139}$$

Then the total energy becomes a simple functional of $s(r)$ and can be minimized to get the stable configuration in the excited state. An example of such a calculation is given in [171] for Si nanocrystals where the function $s(r)$ is written as a Fourier sum whose coefficients are adjusted to get the minimum of the energy functional.

Case of an Isotropic System. In order to get a rough estimate of the relaxation energy, we consider further approximations. In the case of an isotropic material, we have $C_{44} = (C_{11} - C_{12})/2$. Thus using (5.139), we obtain

$$U(r) = \frac{C_{11}}{2}\left(\frac{\partial s}{\partial r}\right)^2 + (C_{11} + C_{12})\left(\frac{s}{r}\right)^2 + 2C_{12}\left(\frac{\partial s}{\partial r}\right)\left(\frac{s}{r}\right). \tag{5.140}$$

In the limit of isotropic conduction and valence bands, the bandgap is just a function of the deformation potential $a = a_{\text{c}} - a_{\text{v}}$ that describes the response of the conduction (c) and valence (v) band edges to a uniform compression defined by $e_{xx} + e_{yy} + e_{zz}$. We have using (5.139)

$$E_{\text{g}}(r) = E_{\text{g}0} + \delta E_{\text{g}} = E_{\text{g}0} + a\left[\left(\frac{\partial s}{\partial r}\right) + 2\left(\frac{s}{r}\right)\right], \tag{5.141}$$

where $E_{\text{g}0}$ is the bandgap energy in absence of lattice relaxation. Using the symmetry of the problem, we look for solutions of the form $s(r) = v_0 v(x)$ where $x = r/R$. We have to minimize the total energy with respect to v_0 and $v(x)$ and then to calculate the relaxation energy $d_{\text{FC}} = E(\{s=0\}) - E(\{s\})$. After minimization with respect to v_0, we obtain that

$$d_{\text{FC}} = \frac{a^2}{4\pi R^3 C_{11}} g(\alpha) \,, \tag{5.142}$$

where $\alpha = C_{12}/C_{11}$ and the function g is given by

$$g(\alpha) = \left\{ \frac{\left[\int_0^1 \sin^2(\pi x)\left(\frac{\partial v}{\partial x} + 2\frac{v}{x}\right) \mathrm{d}x\right]^2}{\int_0^1 x^2 \left\{\frac{1}{2}\left(\frac{\partial v}{\partial x}\right)^2 + (1+\alpha)\left(\frac{v}{x}\right)^2 + 2\alpha\left(\frac{\partial v}{\partial x}\right)\left(\frac{v}{x}\right)\right\} \mathrm{d}x} \right\}_{\min}, \tag{5.143}$$

where the minimum is taken with respect to the function v, which can be found numerically. Figure 5.7 shows that the corresponding $g(\alpha)$ remains

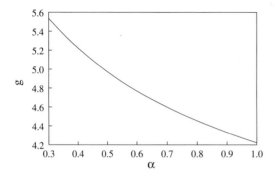

Fig. 5.7. Plot of $g(\alpha)$ with respect to $\alpha = C_{12}/C_{11}$

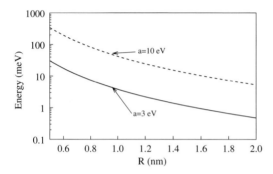

Fig. 5.8. Relaxation energy d_{FC} as a function of the nanocrystal radius R for $C_{11} = 1.5 \times 10^{12}$ dynes/cm^2 and two values of the deformation potential a

of the order of 5 when α is in the range 0.3–1.0 that covers almost all the experimental situations (α is usually close to 0.5). Figure 5.8 presents the variation with size of the relaxation energy for two values of the deformation potential a. We see that the relaxation energy can be substantial in small nanocrystals ($R < 2$ nm) with a large deformation potential for the bandgap. It is interesting to notice that the relaxation energy varies like $1/R^3$ for the coupling to acoustic phonons whereas it behaves like $1/R$ with optical phonons (previous section).

Finally, one must note that the coupling to acoustic phonons in polar materials also arises from the piezoelectric effect because the lattice strain produces a polarization. Details on this mechanism can be found in [202].

5.2.8 The Importance of Non-adiabatic Transitions

Up to now, we have always assumed that the systems could be described in the adiabatic approximation which means that the electrons are in a stationary state for each instantaneous position of the nuclei and that the total wave functions can be written in the Born–Oppenheimer approximation (Sect. 5.2.3), i.e. under the form

$$\Psi_{in} = \phi_i(x, Q)\chi_{in}(Q) , \tag{5.144}$$

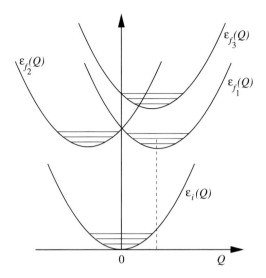

Fig. 5.9. Configuration coordinate diagram in the case of degenerate or almost degenerate excited states

where i enumerates all possible electronic configurations. In this approximation, the electron–phonon coupling does not give rise to transitions between different electronic configurations (e.g. $i \to j$). However, when there are degenerate states or states with energy spacing comparable to phonon energies, non-adiabatic processes become important [333–335]. Figure 5.9 describes a situation where degeneracies or near-degeneracies occur in the excited state (but they may also occur in the ground state). Then the electron–phonon coupling induces the mixing of states belonging to different electronic configurations. Nanosystems with electronic degeneracy can be described in close analogy with impurities in semiconductors [288, 289]. In particular, we can apply the Jahn–Teller theorem [336] which states that there is always a lattice distortion which lowers the energy and reduces the symmetry of the system (it does not apply to the Kramers' degeneracy). The Jahn–Teller effect can also result from the coupling between nearly degenerate states. Furthermore, the asymmetric state has lower symmetry than the Hamiltonian, so there are several equivalent but distinct distortions [288, 289] (Fig. 5.9). Transitions between these degenerate configurations give rise to the dynamic Jahn–Teller effect. In quantum dots, degeneracies arise in particular from the complex structure of the valence band which is P-like. Thus non-adiabatic processes play an important role in excitonic transitions [333].

The usual way to include non-adiabatic processes is to calculate the vibronic states corresponding to coupled electronic and nuclear motion. One can follow the general Born–Oppenheimer treatment and write the total vibronic wave function Ψ as a combination of Born–Oppenheimer products of the general form given by (5.144):

$$\Psi(x,Q) = \sum_{i,n} c_{in}\phi_i(x,Q)\chi_{in}(Q) \,. \tag{5.145}$$

In practice, the basis can be limited to Born–Oppenheimer products which are close in energy, i.e. those corresponding to degenerate or nearly degenerate electronic configurations. Then one must calculate the matrix of the total Hamiltonian in this basis. The total Hamiltonian is usually written as

$$H = H_\mathrm{e} + H_\mathrm{ph} + H_\mathrm{int} \;, \tag{5.146}$$

where H_e is the electronic part, H_ph is the phonon part of the form (5.54), and H_int is the electron–phonon Hamiltonian usually approximated by its first order terms. Fomin and coworkers [333] have performed such calculations using a continuous model for the coupling to optical modes and using a $\boldsymbol{k} \cdot \boldsymbol{p}$ model for the electronic Hamiltonian. Applied to CdSe quantum dots, the calculations predict photoluminescence spectra in good agreement with experiments.

The inclusion of non-adiabatic processes in microscopic theories (e.g. tight binding for the electronic part plus valence force field for the phonons) is in principle possible but, to our knowledge, this has not yet been done.

5.3 Optical Properties of Heterostructures and Nanostructures of Direct Gap Semiconductors

In this section, we deal with the optical absorption of systems with reduced dimensionality based on direct gap semiconductors, going from the bulk to quantum wells and quantum dots. We mainly describe the systems in the effective mass approximation for the envelope functions, considering one-particle and excitonic transitions. We only consider the effects of the electronic structure, discarding all the proportionality constants such as the local-field factor discussed in Sect. 5.1.1. We write the absorption coefficient from (5.28) and (5.33) as

$$\alpha(\omega) \propto \frac{1}{\omega} \sum_{i,f} |\langle i|\boldsymbol{e} \cdot \boldsymbol{p}|f\rangle|^2 \delta(\varepsilon_f - \varepsilon_i - \hbar\omega) \;, \tag{5.147}$$

where the sum is over the (final) empty states $|f\rangle$ and the (initial) occupied states $|i\rangle$ ($T \to 0$K). Following (2.1), the wavefunction of the initial state has the following form

$$\Psi_i(\boldsymbol{r}) = u_{b_i}(\boldsymbol{r})\phi_i(\boldsymbol{r}) \;, \tag{5.148}$$

where $u_{b_i}(\boldsymbol{r})$ is the periodic part of the Bloch functions at the zone center for the band b_i and $\phi_i(\boldsymbol{r})$ is the envelope function. A similar expression holds for the final state. The optical matrix element is [86]

$$\langle i|\boldsymbol{e} \cdot \boldsymbol{p}|f\rangle \approx \boldsymbol{e} \cdot \langle u_{b_i}|\boldsymbol{p}|u_{b_f}\rangle \int_\Omega \phi_i^* \phi_f \mathrm{d}\boldsymbol{r} + \delta_{b_i b_f} \boldsymbol{e} \cdot \int_\Omega \phi_i^* \boldsymbol{p} \phi_f \mathrm{d}\boldsymbol{r} \tag{5.149}$$

with

$$\langle u_{b_i}|\boldsymbol{p}|u_{b_f}\rangle = \int_{\Omega_0} u_{b_i}\boldsymbol{p} u_{b_f}\,d\boldsymbol{r}\;, \qquad (5.150)$$

where Ω_0 denotes the volume of the elementary cell of the semiconductor. In (5.149), we have used the fact that the envelope functions are slowly variable functions on the length-scale of the unit cell.

In the following, we consider two categories of optical transitions:

- interband transitions that occur between states originating from different bands ($b_i \equiv$ valence, $b_f \equiv$ conduction) where the optical matrix element reduces to the first term in (5.149)
- intraband transitions ($b_f = b_i$) involving the dipole matrix elements between envelope functions, the second term in (5.149).

5.3.1 Interband Transitions

Interband Transitions in Bulk Semiconductors. In this introductory part, we just briefly recall essential aspects of the optical absorption in bulk semiconductors. We concentrate on the behavior near the optical threshold since in this energy range the differences with quantum confined systems are the most important. The conduction and valence bands are described by single parabolic bands of effective masses m_e^* and m_h^*, respectively. The valence band represents either the heavy hole band, the light hole one or the split-off one (Sect. 1.3.3) which are treated separately. We completely neglect the intricate anisotropic dispersion in the valence band.

From the Bloch theorem, we easily check that the optical matrix element between two states with different wave vectors is equal to zero. This means that the transition is vertical within the Brillouin zone, i.e. it occurs at fixed $\boldsymbol{k}$ because, physically, the wave vector of the light is much smaller than the dimension of the first Brillouin zone [261, 263–265]. Thus the absorption coefficient becomes

$$\alpha(\omega) \propto \frac{1}{\omega}\sum_{\boldsymbol{k}}|M_{\mathrm{vc}}(\boldsymbol{k})|^2 \delta\left(\varepsilon_{\mathrm{c}} + \frac{\hbar k^2}{2m_e^*} - \varepsilon_{\mathrm{v}} + \frac{\hbar k^2}{2m_h^*} - \hbar\omega\right)\;, \qquad (5.151)$$

where $M_{\mathrm{vc}}(\boldsymbol{k})$ is the optical matrix element, ε_{c} and ε_{v} are the energies of the bottom of the conduction band and of the top of the valence band, respectively ($\varepsilon_{\mathrm{c}} - \varepsilon_{\mathrm{v}} = \varepsilon_{\mathrm{g}}$). A common approximation is to discard the $\boldsymbol{k}$ dependence of the optical matrix element near the threshold. M_{vc} is related in (2.32) to the parameter P of the Kane Hamiltonian which takes similar values in all semiconductors ($2m_0 P^2/\hbar^2 \approx 17\text{--}23$ eV). With a constant M_{vc}, $\alpha(\omega)$ becomes proportional to the joint density of states

$$\alpha(\omega) \propto \frac{1}{\omega}|M_{\mathrm{vc}}|^2 \sum_{\boldsymbol{k}} \delta\left(\varepsilon_{\mathrm{g}} + \frac{\hbar k^2}{2m^*} - \hbar\omega\right)\;, \qquad (5.152)$$

where m^* is the reduced mass ($1/m^* = 1/m_e^* + 1/m_h^*$). This sum reduces to the density of states of a 3D electron gas, so that

5 Optical Properties

$$\alpha(\omega) \propto \frac{1}{\omega}|M(\omega)|^2 \tag{5.153}$$

with

$$|M(\omega)|^2 = |M_{\text{vc}}|^2 \frac{\Omega}{4\pi^2}\left(\frac{2m^*}{\hbar^2}\right)^{3/2}(\hbar\omega - \varepsilon_{\text{g}})^{1/2} . \tag{5.154}$$

The absorption process leads to the formation of an electron–hole pair. So far we have assumed that there is no interaction between the quasiparticles. In Chap. 4, we have seen that the two quasiparticles attract each other via a screened Coulomb potential which give rise to localized gap states. Therefore, the absorption spectrum, instead of starting at the energy gap can show lines at $\varepsilon_{\text{g}} - \frac{m^* e^4}{2\hbar^2(4\pi\epsilon_0\epsilon_{\text{M}})^2}\frac{1}{n^2}$, where ϵ_{M} is the static dielectric constant. From Sect. 4.2.1, we know that the wave function of the excited states may be written in the following form

$$\Psi_{\text{exc}} = \sum_{\boldsymbol{k}_{\text{e}},\boldsymbol{k}_{\text{h}}} a(\boldsymbol{k}_{\text{e}},\boldsymbol{k}_{\text{h}})\Phi(\boldsymbol{k}_{\text{e}},\boldsymbol{k}_{\text{h}}) , \tag{5.155}$$

where the function $\Phi(\boldsymbol{k}_{\text{e}},\boldsymbol{k}_{\text{h}})$ is a Slater determinant which describes a situation in which a valence band electron of wave vector $\boldsymbol{k}_{\text{h}}$ has been excited to a conduction band state of wave vector $\boldsymbol{k}_{\text{e}}$. The Fourier transform

$$F(\boldsymbol{r}_{\text{e}},\boldsymbol{r}_{\text{h}}) = \sum_{\boldsymbol{k}_{\text{e}},\boldsymbol{k}_{\text{h}}} a(\boldsymbol{k}_{\text{e}},\boldsymbol{k}_{\text{h}}) \exp\left(\mathrm{i}(\boldsymbol{k}_{\text{e}} \cdot \boldsymbol{r}_{\text{e}} - \boldsymbol{k}_{\text{h}} \cdot \boldsymbol{r}_{\text{h}})\right) \tag{5.156}$$

defines a two-particle envelope function solution of the effective mass equation (4.15).

It is interesting to compare the strength for exciton absorption to the one for one particle transitions. For many particle states, the optical matrix element is given by

$$M_{\text{exc}} = \langle \Psi_0 | \boldsymbol{e} \cdot \sum_i \boldsymbol{p}_i | \Psi_{\text{exc}} \rangle , \tag{5.157}$$

where $\sum_i \boldsymbol{p}_i$ is the sum of the one-electron momenta and Ψ_0 is the ground state. From (5.155), we obtain

$$M_{\text{exc}} = \sum_{\boldsymbol{k}_{\text{e}},\boldsymbol{k}_{\text{h}}} a(\boldsymbol{k}_{\text{e}},\boldsymbol{k}_{\text{h}})\langle \boldsymbol{k}_{\text{h}}|\boldsymbol{e} \cdot \boldsymbol{p}|\boldsymbol{k}_{\text{e}}\rangle . \tag{5.158}$$

We have seen that one-particle matrix element is non-zero only if $\boldsymbol{k}_{\text{e}} = \boldsymbol{k}_{\text{h}} = \boldsymbol{k}$. This matrix element is identical to $M_{\text{vc}}(\boldsymbol{k})$ defined above and we take it to be constant over the small range of $\boldsymbol{k}$ involved. We then get

$$M_{\text{exc}} = M_{\text{vc}} \sum_{\boldsymbol{k}} a(\boldsymbol{k},\boldsymbol{k}) . \tag{5.159}$$

From (5.156), we see that

$$\sum_{\boldsymbol{k}} a(\boldsymbol{k},\boldsymbol{k}) = \int F(\boldsymbol{r},\boldsymbol{r})\mathrm{d}\boldsymbol{r} , \tag{5.160}$$

5.3 Optical Properties of Heterostructures and Nanostructures

where we take $r_e - r_h = r$. We thus obtain

$$|M_{\text{exc}}|^2 = |M_{\text{vc}}|^2 \left| \int F(r,r) dr \right|^2 . \qquad (5.161)$$

Using the wave function (4.17) for the lowest excitonic state, the optical matrix element for the 3D exciton is given by

$$|M_{\text{exc}}^{(3)}|^2 = |M_{\text{vc}}|^2 \frac{\Omega}{\pi a^3} , \qquad (5.162)$$

where a is the exciton Bohr radius. This result can be compared to $|M(\omega)|^2$ in (5.154). In fact, it is better to compare with $|M(\omega)|^2$ integrated up to an energy $\hbar\omega$. One thus gets

$$\frac{|M_{\text{exc}}^{(3)}|^2}{\int |M(\omega)|^2 d(\hbar\omega)} = 6\pi \left[\frac{\hbar^2}{2m^* a^2 (\hbar\omega - \varepsilon_g)} \right]^{3/2} = 6\pi \left[\frac{\varepsilon_{1s}}{\hbar\omega - \varepsilon_g} \right]^{3/2} , \qquad (5.163)$$

where ε_{1s} is the exciton binding energy in the 1s state. With a typical value $\varepsilon_{1s} \approx 10$ meV, the ratio (5.163) is of the order of 20% for $\hbar\omega - \varepsilon_g \approx 200$ meV showing an important concentration of the oscillator strength in a single line.

Interband Transitions in Quantum Wells. We start with transitions between single particle states. We have seen in Sect. 2.1.1 that the confinement in the z direction leads to the formation of subbands starting at discrete energies and having a free dispersion in the x and y directions (Fig. 5.10). The envelope function (2.3) for the initial state (hole state) is

$$\phi_i(r) = \frac{1}{\sqrt{S}} \exp\left[i k^{(h)} \cdot \rho \right] \chi_n^{(h)}(z) , \qquad (5.164)$$

with the same expression holding for the final state (electron state $\equiv$ e). Once again, only vertical transitions are possible ($k^{(h)} = k^{(e)}$). Thus the optical matrix element (5.149) becomes approximately:

$$e \cdot \langle u_v | p | u_c \rangle \langle \chi_n^{(h)} | \chi_m^{(e)} \rangle . \qquad (5.165)$$

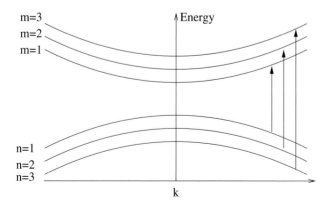

Fig. 5.10. Main interband transitions in a quantum well

The first term determines the polarization selection rules which depend on the valence band (v) under consideration, i.e. heavy holes, light holes and split-off (details can be found in [86]). For simplicity, we will neglect this dependence in the following, replacing $\langle u_v | p | u_c \rangle$ by the average value M_{vc}. The second term in (5.165) gives selection rules on the envelope function quantum numbers. In the case of symmetric wells, we can deduce from the parity of the wave functions that the overlap $\langle \chi_n^{(h)} | \chi_m^{(e)} \rangle$ is non zero only if $n + m$ is even. In the case of infinitely deep barriers, the envelope functions are given by (2.6) both for the electron and the hole. Thus the transition is allowed only if $n = m$. In the general case, it is observed that transitions with $n \neq m$ are always much less efficient than those with $m = n$.

We come now to the shape of the absorption spectrum. From (5.147) and (2.3), we get:

$$\alpha(\omega) \propto \frac{1}{\omega} |M_{vc}|^2 \sum_{k,m,n} \left| \langle \chi_n^{(h)} | \chi_m^{(e)} \rangle \right|^2 \delta \left(\varepsilon_g + \varepsilon_m^{z(e)} + \varepsilon_n^{z(h)} + \frac{\hbar k^2}{2m^*} - \hbar\omega \right). \tag{5.166}$$

The sum over k of the delta functions corresponds to the 2D joint density of states which, following Sect. 2.1.1, leads to:

$$\alpha(\omega) \propto \frac{1}{\omega} |M_{vc}|^2 \frac{2m^* S}{\pi \hbar^2} \sum_{m,n} \left| \langle \chi_n^{(h)} | \chi_m^{(e)} \rangle \right|^2 \Theta \left(\varepsilon_g + \varepsilon_m^{z(e)} + \varepsilon_n^{z(h)} - \hbar\omega \right). \tag{5.167}$$

Thus the absorption coefficient has a staircase-like shape as expected from the 2D density of states (Fig. 5.11). As discussed in Sect. 2.1.1, the absorption threshold is blue-shifted with respect to the 3D situation due to quantum confinement effects.

Let us consider now the excitonic transition. We use the same approach as in the 3D case. But the problem is complicated by the confinement in the z direction and by the fact that there is still free motion of the center of mass $\mathbf{R}_\perp$ in the other directions. One possible envelope function for the lowest exciton state as given in (4.18) is

$$F(\mathbf{r}_e, \mathbf{r}_h) = \frac{e^{i\mathbf{k}\cdot\mathbf{R}_\perp}}{\sqrt{S}} \chi_1^{(e)}(z_e) \chi_1^{(h)}(z_h) \sqrt{\frac{2}{\pi\lambda^2}} e^{-\rho/\lambda}, \tag{5.168}$$

where $\rho = \sqrt{(x_e - x_h)^2 + (y_e - y_h)^2}$ and λ is a variational parameter. Using for $\chi_1^{(h)}$ and $\chi_1^{(e)}$ the solution (2.6) of the infinite square well and with $\mathbf{k} = 0$, one readily obtains from (5.161):

$$|M_{exc}^{(2)}|^2 = |M_{vc}|^2 \frac{2S}{\pi\lambda^2}. \tag{5.169}$$

5.3 Optical Properties of Heterostructures and Nanostructures

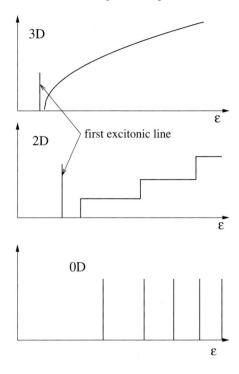

Fig. 5.11. Comparison of the ideal optical absorption spectra in a bulk semiconductor (3D), in a quantum well (2D) and in a quantum dot (0D)

Thus the relative strength of optical absorption between the quantum well and bulk exciton is, for the same volume $\Omega = SL_z$ of material, given by

$$\left|\frac{M_{\text{exc}}^{(2)}}{M_{\text{exc}}^{(3)}}\right|^2 = \frac{2a^3}{L_z \lambda^2} \ . \tag{5.170}$$

For strong confinement, λ tends to its 2D limiting value $a/2$ (Sect. 4.2.2) so that this ratio becomes equal to

$$\left|\frac{M_{\text{exc}}^{(2)}}{M_{\text{exc}}^{(3)}}\right|^2 = \frac{8a}{L_z} \ , \tag{5.171}$$

where L_z is of the order of the inter-atomic spacing. This ratio can thus become, for GaAs, as large as 300 showing that the 2D confined exciton has much more oscillator strength than the bulk one. This is confirmed by experimental observations [337].

Interband Transitions in Quantum Dots. We have seen in Sects. 2.1.3 and 2.1.4 that the confinement splits the bands into series of discrete energy levels. The optical transitions will also become discrete. In the limit of strong confinement, the kinetic energy terms dominate the electron–hole attraction (Chap. 4) and the latter can thus be included in first order perturbation theory. At the lowest order, the envelope function for an excitonic state is just the product of uncorrelated electron and hole envelope functions. Under the assumption of infinitely high barriers, the electron and hole eigenfunctions are identical. Since the optical matrix element in (5.149) contains the overlap between the envelope functions of the electron and the hole, we deduce the well-known selection rule that optical transitions can only take place between states having the same quantum numbers, regardless of the shape of the quantum dot [260]. In reality, this rule may be lifted due to a different penetration of the electron and hole wave-functions in the barriers. Another complication comes from the intricate nature of the hole states (see the discussion in Sect. 2.5.3). Therefore a precise evaluation of the oscillator strengths usually requires more elaborate calculations such as tight binding or empirical pseudopotentials.

The confinement in quantum dots makes that the bulk oscillator strength is concentrated into discrete lines and small volumes. To estimate this enhancement, we extend the simplest effective mass model used in the previous sections. We then consider a spherical nanocrystal with infinite potential at the boundary $r = R$. The envelope function for the lowest excitonic state takes the form

$$F(\mathbf{r}_e, \mathbf{r}_h) = N \frac{\sin(\pi r_e/R)}{r_e} \frac{\sin(\pi r_h/R)}{r_h}, \tag{5.172}$$

where N is a normalizing factor. One can now evaluate $|M_{exc}^{(0)}|^2$ for this 0D case from the general expression (5.161)

$$|M_{exc}^{(0)}|^2 = |M_{vc}|^2. \tag{5.173}$$

Thus the strength of this exciton relative to the bulk one for the same volume of material given here by $\Omega = \frac{4}{3}\pi R^3$ is equal to

$$\left|\frac{M_{exc}^{(0)}}{M_{exc}^{(3)}}\right|^2 = \frac{3}{4}\left(\frac{a}{R}\right)^3, \tag{5.174}$$

which is the result of Kayanuma [338]. We easily verify that the enhancement is still larger than in quantum wells.

The optical matrix element in (5.173) does not depend on the nanocrystal volume, which is confirmed by tight binding or pseudopotential calculations. Thus (5.28) shows that the absorption coefficient of a composite material made of nanocrystals embedded in a dielectric matrix is proportional to p/Ω which corresponds to the concentration of nanocrystals. A similar result holds for the stimulated emission, meaning that lasing will be favored in materials with a high concentration of nanocrystals. Indeed, lasers based on nanocrystals have been obtained only recently using compact arrays of nanocrystals where the stimulated emission strength is high enough to overcome the non radiative effects such as the Auger recombination [339] (see Sect. 7.2). This is an interesting result because it is known that semiconductor quantum dots promise the lowest lasing threshold for semiconductor media.

To conclude on the efficiency of the optical transitions in quantum dots, it is important to consider the exciton fine structure. In spherical nanocrystals, the exciton ground state is eightfold degenerate, a factor 2 coming from the spin S_e of the electron and a factor 4 coming from the total momentum ($J_h = 3/2$) of the hole (Sect. 2.5.3). However, the electron–hole exchange interaction, which increases with decreasing size, lifts the degeneracy (see Sect. 4.3.2). The excitonic levels can be labeled by their total angular momentum $J = S_e + J_h$. The electron–hole exchange term splits the manifold into two states with $J = 1$ and $J = 2$. The optical transition from the lowest energy state with $J = 2$ is forbidden and the optically allowed state $J = 1$ is shifted toward high energy. Thus, for example, this leads to an increase of the decay time of the photoluminescence, depending on the thermal occupation of the two levels. At low temperature, it may lead to very long lifetimes, in the microsecond or even in the millisecond range. These effects are discussed in detail in Sect. 4.3.2.

We conclude that the oscillator strength in quantum dots made of direct gap semiconductors is high but, in small quantum dots, the electron–hole exchange splitting may become large enough to reduce the radiative recombination rate.

5.3.2 Intraband Transitions

Intraband Transitions in Bulk Semiconductors. Here we deal with the intraband optical transitions, for example in the case of doped semiconductors. Only one type of carriers is involved in the transitions, which allows to probe the electronic structure either in the valence or in the conduction band. In bulk materials, this is usually treated as a free electron gas. Since transitions must be vertical in k space, direct transitions between two states that necessarily have different wave vectors are not possible. Electron–phonon interactions or scattering on defects must therefore be involved in the transitions. Since there is a continuous succession of states, the transition of the carrier can be seen as an acceleration induced by the electromagnetic field and the absorption of light by quasi-free carriers can be consequently treated as a transport problem [291]. We will see that the situation is different in quantum confined systems because the confinement potential breaks the k selection rules.

Intraband Transitions in Quantum Wells. For carriers confined in the z direction and free to move in x and y directions, one must consider two situations for the optical absorption depending the polarization of the light. With polarizations along x and y, the problem is the 2D analogue of the free-carrier absorption in bulk semiconductors. Direct transitions are not allowed and scattering mechanisms induced by phonons are necessary. With a z polarization corresponding to a wave propagating within the well, direct transitions are allowed between different subbands (Fig. 5.12), provided that

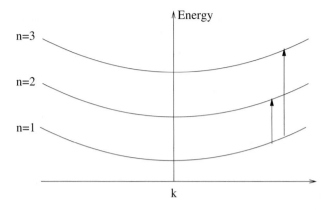

Fig. 5.12. Intraband optical transitions in a quantum well

the in-plane wavevector of the carrier is conserved [86]. Using from (5.164) the wavefunctions $\phi_i \propto \exp\left(i\boldsymbol{k}^{(i)}.\boldsymbol{\rho}\right) \chi_n(z)$ and $\phi_f \propto \exp\left(i\boldsymbol{k}^{(f)}.\boldsymbol{\rho}\right) \chi_m(z)$ for the initial and final states, respectively, the optical matrix element (5.149) becomes

$$\delta_{\boldsymbol{k}^{(i)},\boldsymbol{k}^{(f)}} \langle \chi_n | p_z | \chi_m \rangle \ . \tag{5.175}$$

In order to estimate the magnitude of the transitions, we consider the case of a well with infinite barriers. Using (2.6), we easily obtain the total oscillator strength for all vertical transitions from the subband n to the subband m:

$$f_{nm} = \frac{2^6}{\pi^2} \frac{(nm)^2}{(m^2 - n^2)^3} \text{ if } n - m \text{ is odd },$$

$$f_{nm} = 0 \text{ otherwise } . \tag{5.176}$$

Let us consider now that only the subband $n = 1$ is populated in the initial state. We see from (5.176) that the amplitude of the optical matrix element quickly decreases with m. The absorption spectrum is a sum of delta functions $\delta\left(\varepsilon_m^z - \varepsilon_1^z - \hbar\omega\right)$ due to the exact parallelism of the subbands in the effective mass approximation (Fig. 5.12). Using more elaborate calculations, the bands would not be perfectly parabolic and each line in the optical spectrum would acquire a finite width.

Intraband transitions in quantum wells also occur between bound states to the continuum of states above the barrier potential. They are a possible loss mechanism in quantum well lasers. They are used to make infrared detectors [265] and they are treated in detail in [86, 92, 265].

Intraband Transitions in Quantum Dots. Due to the confinement potential in the three directions of space, intraband transitions in quantum dots may be possible for any polarization of the light without the assistance of scattering mechanisms. The absorption spectrum is made of sharp lines with an intensity given by the optical matrix element between the envelope functions in the effective mass approximation. In the case of cubic dots, the matrix elements can be easily calculated from (5.176). In the case of spherical dots or nanocrystals, we have seen in Sect. 2.1.4 that the electron and hole states are those of an artificial atom. Thus atomic-like selection rules can be applied to the optical transitions. In particular, the transitions may be allowed only if the quantum number l differs by $+1$ or -1 between the initial and final states (this rule can be established from the general properties of the spherical harmonics [340]):

$$\Delta l = \pm 1 \Rightarrow \text{ transition may be allowed.} \tag{5.177}$$

Intraband transitions in nanocrystals are particularly interesting because they usually occur in the infrared spectral range and they involve only one type of carriers which allows to study separately the dynamics of electrons and holes [341]. They also strongly depend on the charge state of the quantum dot which can be changed deliberately [142, 342, 343] or not [344]. Several works on quantum dots charged with one electron (n-type) reported optical transitions from the lowest state in the conduction band to the next higher state which is the analogue of 1S → 1P transitions in atoms [142, 342, 345].

Recent work [140] on ZnO nanocrystals charged with electrons have confirmed the $\Delta l = \pm 1$ selection rule (Fig. 5.13). On the theoretical side, tight binding calculations described in Sect. 2.5.1 (see Fig. 2.11 for the one-particle energies) show that transitions with $\Delta l \neq \pm 1$ are characterized by an oscillator strength smaller by several order of magnitudes compared to the allowed ones. On the experimental side, near-infrared spectroscopy has been performed for gradually increasing number of electrons up to ten. When the number of electrons is close to one, the main transition is 1S → 1P. When there is an average of five electrons, the 1S and 1P states are populated and the 1P → 1D transition becomes the main one. The oscillator strengths are deduced from a fit of the absorption spectra: the results strongly support the $\Delta l = \pm 1$ rule. The oscillator strengths for the allowed transitions given in Table 5.1 are in good agreement with those predicted in tight binding. Note that the values estimated from effective mass wave functions (2.18) are very close to the tight binding ones showing that the envelope function approximation is a good one in that case [140].

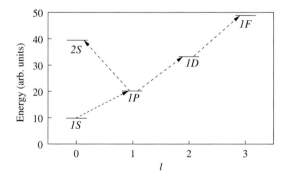

Fig. 5.13. The main optical transitions observed in ZnO nanocrystals by near infrared absorption confirm the $\Delta l = \pm 1$ selection rule [140]

Table 5.1. Dipole oscillator strength for intraband transitions in ZnO nanocrystals summed over all degenerate states [140]. Theoretical values are obtained with a tight binding method. The experimental 1S–1P oscillator strength is set equal to the theoretical value. This allows to compare the oscillator strengths of the 1P–1D, 1D–1F and 1P–2S transitions with theoretical values

Diameter [nm]		1S–1P	1P–1D	1D–1F	1P–2S
3.7	Exp.	6.1	18 ± 2		
	Th.	6.1	18.3	33.8	3.2
4.2	Exp.	6.4	21 ± 2	36 ± 14	4 ± 2
	Th.	6.4	19.2	36.2	3.4
5.2	Exp.	6.6	29 ± 4	54 ± 10	4 ± 3
	Th.	6.6	20.7	40.5	3.7

5.3.3 The Importance of Electron–Phonon Coupling

We consider now the influence of the electron–phonon coupling on the optical transitions in quantum dots. Since quantum dots in the strong confinement regime have an atomic-like energy spectrum, the main source of homogeneous line broadening is expected to be the coupling to vibrations. Thus the knowledge of the electron–phonon or exciton-phonon interaction is critically important, in particular for a number of device applications (e.g., for single photon emitters). In the following sections, we summarize the basic knowledge in this field.

Influence of the Electron–Phonon Coupling on Interband Transitions. In 1987, Schmitt-Rink, Chemla, and Miller [260] argued that the coupling of the exciton to optical phonons through the Fröhlich interaction should vanish in the strong confinement regime. Indeed the coupling arises from the polarization of the lattice by the difference in the electron and hole charge distributions in the exciton, and these distributions are almost identical in strongly confining structures. Thus the total charge density $\rho(\mathbf{r})$ and therefore the coupling term (5.112) vanish in this limit, and the LO phonon features should not be observable. But experiments show that this is not the case, with many reported values of the Huang–Rhys parameter S between 0.1 and 1 [303, 346–352]. On the theoretical side, it has been shown that the coupling is highly sensitive to the form of the electron and hole wave functions. Calculations in the effective mass approximation taking into account the complexity of the valence band predict a non-zero S factor but several orders of magnitude smaller than experiments [329]. Several processes have been invoked to explain the observed values: separation of the electron and hole charge distributions as a result of asymmetric shape [346, 347], the presence of piezo-electric or pyro-electric fields [353, 354] and of additional charges [329, 348, 352]. Concerning the evolution of the coupling strength as a function of nanocrystal size, both experiments and theoretical works produce widely varying results. Thus the problem remains open to further studies and discussions.

Concerning theory, effective mass calculations are probably not sufficiently accurate to predict the correct magnitude of the electron–LO–phonon coupling in the case of intrinsic excitons, since the result depends on fine details of the electron and hole wave functions that also depend on details of the electronic structure. But to our knowledge, there is no microscopic calculations applied to these problems in polar semiconductor quantum dots. Some works have also shown that LO phonon–sidebands are considerably enhanced when one takes into account the non-adiabaticity of the exciton-phonon system, even with relatively weak electron–phonon coupling [333] (see Sect. 5.2.8). Non-adiabatic transitions become important because there are degenerate states or there are exciton states separated by an energy comparable with that of the optical phonons (mainly due to the complex valence band). The non-adiabaticity also enhances the intensity of multi-phonon peaks in nanocrystal Raman spectra [334].

Influence of the Electron–Phonon Coupling on Intraband Transitions. There are much less studies concerning the electron–phonon coupling in intraband transitions. Theoretical works predict that the Fröhlich interaction should be enhanced in small quantum dots of polar semiconductors compared to the bulk and should strongly depend both on the shape and the size distribution [335]. Recent intraband hole burning experiments of CdSe, InP and ZnO colloidal quantum dots of 1.5–2.5 nm radius show an homogeneous line-width for the 1S → 1P transition below 3 meV at 10K [345], probably coming from acoustic phonons. LO-phonon replica are also observed and the corresponding Huang–Rhys factors for CdSe nanocrystals are presented in Fig. 5.14 as a function of size. The measured values are in good agreement with those calculated using the LO-phonon coupling Hamiltonian (5.114) with a charge distribution $\rho(\boldsymbol{r}) = |1P|^2 - |1S|^2$ [345], using spherical Bessel functions (2.18) for the 1S and 1P wave functions.

5.4 Optical Properties of Si and Ge Nanocrystals

Bulk Si and Ge are characterized by a very poor optical radiative efficiency because of their indirect gap. In the past decade, several attempts to improve this efficiency have been proposed, in particular using nanocrystals. One main objective of these studies is to obtain stimulated emission from silicon and to make Si lasers [355]. In this section, we show how the quantum confinement modifies the optical properties of indirect gap semiconductors.

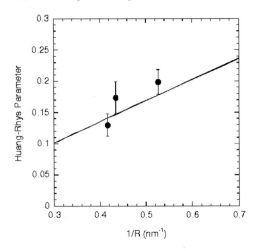

Fig. 5.14. Size dependence of the electron–LO–phonon coupling for intraband transitions in CdSe nanocrystals: experimental results (•) and calculated values using bulk parameters (*straight line*), from [345]

5.4.1 Interband Transitions

No-Phonon Transitions. Because of the indirect gap, band-edge optical transitions in bulk Si or Ge are only possible with the assistance of phonons to supply the momentum (Fig. 5.15). In nanocrystals, the strong confinement of the electron and hole wave functions in real space leads to a spread of the wave functions in momentum space. Thus, radiative recombination or optical absorption can proceed by direct no-phonon transitions (see Sect. 5.2), vertically in $\boldsymbol{k}$ space. To illustrate this effect, we plot in Fig. 5.16 the weight of the lowest electron state Ψ_e and of the highest hole state Ψ_h in momentum space [143] obtained by projecting the eigenstates calculated in tight binding in the basis of the bulk states $u_{n,\boldsymbol{k}}$

$$\Psi_\mathrm{e} = \sum_{n,\boldsymbol{k}} a_{n,\boldsymbol{k}} u_{n,\boldsymbol{k}} ,$$

$$\Psi_\mathrm{h} = \sum_{n,\boldsymbol{k}} b_{n,\boldsymbol{k}} u_{n,\boldsymbol{k}} , \tag{5.178}$$

where n represents the bands. The dipole matrix element is given by

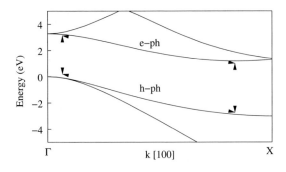

Fig. 5.15. The Si band structure near the gap region along the (100) direction. The processes for phonon-assisted transitions are illustrated

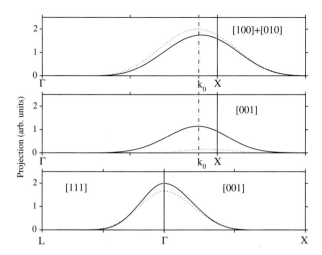

Fig. 5.16. Top: projection ($|a_{n,\boldsymbol{k}}|^2$) of the lowest electron state in a Si ellipsoid on the bulk states $u_{n,\boldsymbol{k}}$ for $\boldsymbol{k}$ along [100] and [010] (sum of the two). Middle: same for $\boldsymbol{k}$ along [001]. Bottom: projection ($|b_{n,\boldsymbol{k}}|^2$) of the highest hole state for $\boldsymbol{k}$ along [001] (right) and [111] (left). Solid lines: ellipsoid long axis of 1.90 nm, short axis of 1.36 nm. Dashed line: long axis of 2.17 nm, short axis of 1.36 nm

$$\langle \Psi_{\mathrm{e}}|\boldsymbol{e}\cdot\boldsymbol{p}|\Psi_{\mathrm{h}}\rangle = \sum_{n,n',\boldsymbol{k}} a_{n,\boldsymbol{k}}^{*} b_{n',\boldsymbol{k}} \langle u_{n,\boldsymbol{k}}|\boldsymbol{e}\cdot\boldsymbol{p}|u_{n',\boldsymbol{k}}\rangle \;, \tag{5.179}$$

where we use the fact that only vertical transitions are allowed. We deduce from Fig. 5.16 that the overlap ($a_{n,\boldsymbol{k}}^{*} b_{n',\boldsymbol{k}}$) between Ψ_{e} and Ψ_{h} in momentum space is small because Ψ_{e} is centered at the conduction band minimum ($\boldsymbol{k} = \boldsymbol{k}_0$) and Ψ_{h} is centered at $\boldsymbol{k} = 0$. Thus we conclude from these general arguments that the efficiency of no-phonon transitions must be small. This is confirmed by tight binding, pseudopotential and ab initio calculations of the recombination rates in Si and Ge nanocrystals [76, 132, 221, 222, 356, 357]. Figure 5.17 shows the radiative lifetime calculated in tight binding for spherical and cubic Ge nanocrystals [132]. Similar results are obtained in Si nanocrystals [143] (Fig. 5.18). There is a huge variation of the lifetime with size, over many decades. Only for very small nanocrystals (< 50 atoms) the lifetime is about 10 ns and becomes comparable to the values in direct gap materials. To understand this behavior in detail, it is interesting to estimate the dipole matrix element in the effective mass approximation [143, 358]. We consider a cubic dot of side L. The envelope function for the ground electron and hole states is (Sect. 2.1.3)

$$\phi(\boldsymbol{r}) = \sqrt{\frac{8}{L^3}} \cos\frac{\pi x}{L} \cos\frac{\pi y}{L} \cos\frac{\pi z}{L} \;,$$
$$\text{with } -L/2 \leq x, y, x \leq L/2 \;. \tag{5.180}$$

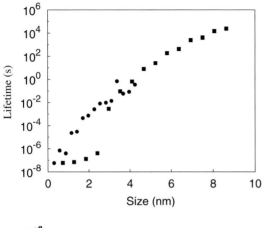

Fig. 5.17. Radiative lifetime calculated for spherical (•) and cubic (■) Ge nanocrystals at 300K. The size is the diameter for spheres and is the diameter of the sphere with the same volume for cubic nanocrystals

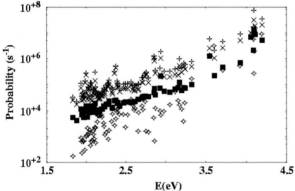

Fig. 5.18. Recombination rate as function of the bandgap of Si nanocrystals: sum of no-phonon and phonon-assisted processes (+), no-phonon transitions (◇), transitions assisted by optical phonons (×) and by transverse acoustic phonons (■)

In the effective mass approximation, the electron and hole wave functions are given by

$$\Psi_{\rm h} = \sum_{\bm{k}} \phi(\bm{k}) u_{{\rm v},\bm{k}} \;,$$

$$\Psi_{\rm e} = \sum_{\bm{k}} \phi(\bm{k} - \bm{k}_{0j}) u_{{\rm c}_j,\bm{k}} \;, \qquad (5.181)$$

where $\phi(\bm{k})$ is the Fourier transform of the envelope function and $\bm{k}_{0j}$ is the wave vector at the conduction band minimum j. We implicitly assumed that the treatment of the degenerate minima can be decoupled. The dipole matrix element (5.179) is given by

$$\langle \Psi_{\rm e} | \bm{e} \cdot \bm{p} | \Psi_{\rm h} \rangle = \sum_{\bm{k}} \phi(\bm{k} - \bm{k}_{0j})^* \phi(\bm{k}) \langle u_{{\rm c}_j,\bm{k}} | \bm{e} \cdot \bm{p} | u_{{\rm v},\bm{k}} \rangle \;. \qquad (5.182)$$

Assuming that the matrix element is independent of $\bm{k}$ (Sect. 5.3.1), we obtain

$$\langle \Psi_\mathrm{e} | \boldsymbol{e} \cdot \boldsymbol{p} | \Psi_\mathrm{h} \rangle \approx M_\mathrm{vc} \sum_{\boldsymbol{k}} \phi(\boldsymbol{k} - \boldsymbol{k}_{0j})^* \phi(\boldsymbol{k}) ,$$

$$= M_\mathrm{vc} \int \exp\left(\mathrm{i} \boldsymbol{k}_{0j} \cdot \boldsymbol{r}\right) |\phi(\boldsymbol{r})|^2 \mathrm{d}\boldsymbol{r} . \tag{5.183}$$

In this expression, we assumed that the optical matrix element is independent on the polarization. The effect of the polarization of the light was examined in [143]. In the case of silicon where the conduction band minima are along [100] directions (e.g. $\boldsymbol{k}_{0x} \approx 0.89(2\pi/a)\boldsymbol{x}$), we have [143, 358] using (5.180)

$$\langle \Psi_\mathrm{e} | \boldsymbol{e} \cdot \boldsymbol{p} | \Psi_\mathrm{h} \rangle \approx -\frac{M_\mathrm{vc}}{\pi k_0} \left(\frac{2\pi}{L}\right)^3 \left[k_0^2 - \left(\frac{2\pi}{L}\right)^2\right]^{-1} \sin\left(\frac{k_0 L}{2}\right) , \tag{5.184}$$

where $k_0 = |\boldsymbol{k}_{0j}| = 0.89(2\pi/a)$. From (5.41), and for L large enough to have $2\pi/L \ll k_0$, we obtain that the spontaneous recombination rate behaves like ω_{21}^3/L^6 where ω_{21} is the transition frequency which depends on the size and the factor L^{-6} comes from the square of the matrix element $\langle \Psi_\mathrm{e} | \boldsymbol{e} \cdot \boldsymbol{p} | \Psi_\mathrm{h} \rangle$. Since ω_{21} remains of the order of $\varepsilon_\mathrm{g}/\hbar$ where ε_g is the bandgap of bulk Si, we deduce that the recombination rate approximately varies like L^{-6} in qualitative agreement with the results of more sophisticated calculations [76, 132, 221, 222, 356, 357] (Fig. 5.18). The optical matrix element (5.184) also contains an oscillating factor $\sin(k_0 L/2)$ which explains the important scattering of the values in Fig. 5.18.

Phonon-Assisted Transitions. We have seen that interband transitions in bulk Si and Ge are only possible with the assistance of phonons because the total momentum must be conserved during an optical transition. In fact, phonon-assisted transitions remain more efficient than no-phonon ones in a wide range of nanocrystal sizes [295, 358] as shown in Fig. 5.18. The values of Fig. 5.18 have been calculated following the method described in Sect. 5.2.4, in particular using (5.82). The calculations are based on a tight binding Hamiltonian for the electron and electron–phonon part together with a valence force-field model for phonons. The heavy part of the work is the evaluation of the coupling coefficients $\boldsymbol{A}_j$ for all phonon modes j of the quantum dot. For each mode, it requires to calculate the wave functions and the optical matrix elements when nuclei are displaced from their equilibrium sites according to the normal modes. The matrix elements of the Hamiltonian are made dependent on the atomic positions following the rules of Harrison [57].

Figure 5.19 shows the recombination rates at 4K calculated for a 2.85 nm diameter nanocrystal with respect to the energy of the phonons involved in the transitions. At this temperature, the phonon absorption process is negligible and the recombination proceeds by phonon emission. We also plot on Fig. 5.19 the photoluminescence spectrum below the no-phonon line calculated assuming that the intensity is directly proportional to the recombination

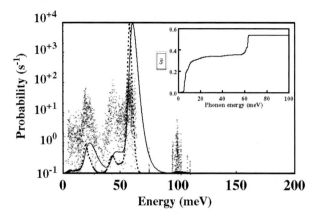

Fig. 5.19. Radiative recombination rate at 4K (small +) with respect to the energy of the phonon involved in the transition for a 2.85 nm hydrogen-passivated quantum dot. Photoluminescence intensity of a single cluster assuming it is directly proportional to the recombination rate, neglecting acoustic phonons (*dashed line*). Photoluminescence intensity including the broadening by multi-phonon process (*straight line*). The insert shows $\bar{S}(\hbar\omega)$ as a function of the phonon energy

rate. It shows that optical modes dominate and that the contribution from transverse acoustic (TA) modes are smaller. However, as discussed in [295], there is a difficulty in this problem due to the fact that the total Huang–Rhys factor $S = \sum_j S_j$ increases rapidly with decreasing size ($\approx 1/R^3$, see Sect. 5.2.7) and can reach values close to unity for bandgaps around 2 eV, meaning that multi-phonon processes become important. From this point of view, it is interesting to analyze the quantity $\bar{S}(\hbar\omega)$ that is the sum of all S_j with $\hbar\omega_j < \hbar\omega$. Figure 5.19 shows that $\bar{S}(\hbar\omega)$ has essentially two contributions: one originating from low-frequency acoustic modes (< 15 meV) and a smaller one from the highest optical modes. The first one corresponds to relaxation effects in the excitonic state, analyzed in Sect. 5.2.7. All this means that for small silicon quantum dots the direct use of (5.82) is no more valid. The way to handle this problem has been discussed in Sect. 5.2.4, using the fact that for low energy acoustic modes the coefficients $\boldsymbol{A}_j$ are negligible. The peaks with an intensity given by (5.82) have just to be broadened by the spectral function describing the coupling to acoustic modes. Since Fig. 5.19 shows that the acoustic modes can safely be considered as degenerate at $\hbar\omega \approx 10$ meV, one can use the spectral function given by (5.72) which was obtained assuming that all phonon frequencies can be approximated by a single one.

The photoluminescence spectrum obtained in this way (Fig. 5.19) is broad, with linewidth of the order of tens of meV for crystallites with a diameter of 3 nm. The broadening arises from two effects. First, the confinement breaks the selection rules and many phonon modes are involved in the transitions. Secondly, the multi-phonon coupling to acoustic modes is substantial. Thus the photoluminescence of single Si nanocrystals cannot be made of very sharp lines as observed in III–V quantum dots. Another important result of the calculations is the relative strength of no-phonon and phonon-assisted transitions. These are plotted in Fig. 5.18 where one sees that phonon-assisted transitions dominate over the whole range of sizes. We also obtain that the ratio of transition rates for one-phonon acoustic W_{ac} and optical processes W_{opt} is $W_{opt}/W_{ac} \approx 10$. Finally, we must note that the rates for no-phonon transitions are extremely sensitive to the presence of defects or disorder [295].

5.4.2 Intraband Transitions

If intraband transitions in direct gap semiconductor quantum dots start to be well understood (Sect. 5.3.2), almost nothing is known experimentally about the intraband transitions in semiconductors characterized by degenerate conduction band minima such as Si or Ge (transitions within the valence band of Si or Ge are basically the same as in III–V or II–VI semiconductors). Here we summarize the results of recent tight binding calculations [144] of intraband transitions in Si nanocrystals doped with one electron. The electron structure is described in detail in Sect. 2.5.2. The quantum confinement gives S and P-like states in each valley but the anisotropy of the effective masses and the inter-valley coupling lift most of the degeneracies of the corresponding levels (Fig. 5.20).

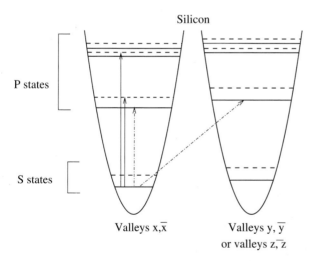

Fig. 5.20. Optical transitions in an anisotropic Si nanocrystal with six valleys in the conduction band. The P levels are split due to the anisotropy of the effective masses and the inter-valley coupling lifts the degeneracies between x and $\bar{x}$ (resp. y and $\bar{y}$, z and $\bar{z}$) valleys (*dashed line*). The optical transitions occur between states in the same valley or in different valleys: no-phonon transitions (*straight arrow*) and one-phonon transitions (*dashed-dotted arrow*)

192   5 Optical Properties

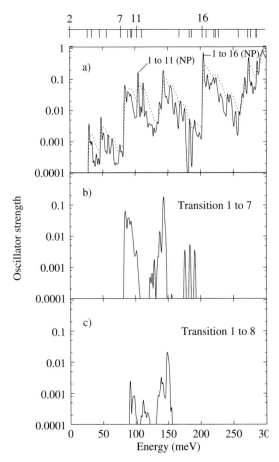

Fig. 5.21. (a) Oscillator strength (*straight line*) calculated for a Si crystallite of dimensions $L_x = 5a$, $L_y = 6a = 3.3$ nm and $L_z = 7a$ where $a = 5.42$Å ($T = 4$K). The calculation includes 25 electron states in the conduction band and 8475 vibrational modes. Oscillator strength including the broadening due to multi-phonon couplings to acoustic phonons (*dashed line*). Some no-phonon transitions are indicated (NP). (b) Contribution of the transition to the seventh level ($\langle 111 \rangle_x^+ \to \langle 211 \rangle_x^-$). (c) Same for the transition to the eighth level ($\langle 111 \rangle_x^+ \to \langle 121 \rangle_y^+$). Top: position of the energy levels with respect ot the ground state, the levels being numbered in order of increasing energy

The optical absorption spectrum at 4K for a crystallite containing 1909 Si atoms is shown in Fig. 5.21. A slightly anisotropic crystallite with $L_x = L-a$, $L_y = L$, $L_z = L+a$ ($L = 3.3$ nm) is considered because the anisotropy lifts all the remaining degeneracies of the levels which helps to identify the nature of the optical transitions. However very close results are obtained for spherical or cubic nanocrystals. At 4 K, the only possible initial state is the ground state ($\mathbf{1} \equiv \langle 111 \rangle_x^+$) and phonon-assisted transitions only proceed by emission of phonons (notations are defined in Sect. 2.5.2). The spectrum consists in series of peaks corresponding to all possible excited states and all phonons j. The energy $h\nu$ of the photon in a no-phonon transition is equal to $\varepsilon_f - \varepsilon_i$ where ε_f (ε_i) is the energy of the excited (initial) state. In the case of a transition with the emission of one phonon of energy $\hbar\omega_j$, $h\nu$ is equal to $\varepsilon_f - \varepsilon_i + \hbar\omega_j$. In contrast to direct gap semiconductors, the spectrum for a single crystallite is broad because there are many excited states derived from the six conduction band minima of Si and because a large number of vibrational modes are involved due to the confinement which leads to a spread of the wave functions in the reciprocal space.

The optical transitions can be classified in three categories. A very interesting case corresponds to transitions within the same valley (here x) since they also occur in semiconductors with a single conduction band minimum. However, the situation is actually more complex due to the fact that there are two coupled equivalent minima at $\boldsymbol{k}$ and $-\boldsymbol{k}$ which for example give rise to two excited states $\langle 211 \rangle_x^+$ and $\langle 211 \rangle_x^-$. Thus the first category of transitions includes those from $\langle 111 \rangle_x^+$ to $\langle 211 \rangle_x^+$ ($\mathbf{1 \to 11}$) or to $\langle 121 \rangle_x^+$ ($\mathbf{1 \to 16}$). They are allowed without phonon, being the analogues of S $\to$ P transitions in direct gap semiconductor nanocrystals, with a similar efficiency (oscillator strengths larger than 0.2). In the second category, transitions like the one from $\langle 111 \rangle_x^+$ to $\langle 211 \rangle_x^-$ ($\mathbf{1 \to 7}$) are totally unusual and original (Fig. 5.21). They are only possible with the assistance of phonons, mainly optical ones at high energy (60–63 meV) and acoustic ones at very low energy (1–10) meV. These modes are mainly derived from phonons at $\boldsymbol{k} \approx 0$ and $\boldsymbol{k} \approx (\pm 2k_0, 0, 0) \equiv (\pm 0.3(2\pi/a), 0, 0)$, as required by the $\boldsymbol{k}$ conservation rule ($x \to x$ and $x \to \bar{x}$). The efficiency of these transitions is high (oscillator strength of the order of 0.5), comparable to direct ones [144].

The third category corresponds to excited states in valleys y and z. The example of the transition $\langle 111 \rangle_x^+ \to \langle 121 \rangle_y^+$ ($\mathbf{1 \to 8}$) is detailed in Fig. 5.21. Because initial and final states are derived from different valleys, the transitions are mainly assisted by phonons, with a main contribution from optical phonons at energy close to 56–60 meV. Once again, these results can be understood in terms of $\boldsymbol{k}$ conservation rules since the transition from valleys x and $\bar{x}$ at $(\pm k_0, 0, 0)$ to valleys y and $\bar{y}$ at $(0, \pm k_0, 0)$ requires phonons with wave vectors close to $(\pm k_0, \pm k_0, 0)$. The efficiency of the transitions is smaller than in the previous categories (oscillator strength of the order of 0.05) but remains substantial.

In conclusion, these calculations show that in Si nanocrystals charged with electrons there are new types of intraband transitions, with no equivalence in direct gap semiconductors. They take place between the confined states in the six valleys of the conduction band. They involve phonons with wave vectors either at the center or at the edge of the Brillouin zone. The efficiency of the main no-phonon and phonon-assisted transitions is comparable to the one in III–V semiconductor quantum dots.

6 Defects and Impurities

The main objective of this chapter is to analyze the influence of the quantum confinement on the electronic levels of point defects and impurities, from quantum wells to quantum dots. In the first two parts, we present the general trends for hydrogenic and deep defects. In the next sections, we consider particular situations: dangling bonds, self-trapped excitons and oxygen related defects at the Si–SiO$_2$ interface.

6.1 Hydrogenic Donors

In this section, we deal with the electronic structure of hydrogenic impurities in quantum confined systems, on the basis of the envelope function approximation or tight binding calculations.

6.1.1 Envelope Function Approximation

The Case of Quantum Wells. The simplest application of the envelope function approximation corresponds to the isolated quantum well for single isotropic band extrema of the same nature in the two materials (Sect. 2.1.1). This occurs for the conduction band of GaAs–Al$_x$Ga$_{1-x}$As systems in which the minimum is of Γ symmetry in both materials when $0 < x < 0.45$. Furthermore, the effective masses are of the same order of magnitude in the two materials. The well and barriers are in the GaAs and GaAlAs parts, respectively. In such a situation the normal boundary conditions should apply and one has thus to solve a simple square well problem, the electron mass being replaced by the effective mass m^*. The super-lattice case is treated as a Krönig–Penney-type model and leads to a similar broadening of the quantum well levels into bands.

The corresponding valence band problem under the same conditions is not as simple. As we have seen in Sect. 2.2.1, it is necessary to solve a set of coupled differential equations which can be more or less simplified after some approximations as discussed in [85, 86].

An interesting application concerns the behavior of hydrogenic impurities in quantum wells. Again the basic case concerns the isotropic band minimum with effective mass m^*. In three dimensions, the binding energy is

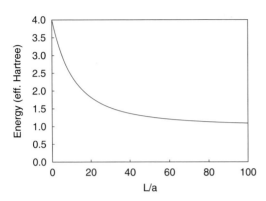

Fig. 6.1. Binding energy of a donor impurity versus the width L of the quantum well expressed in inter-atomic distance a ($= 5$ a.u.), in the case of one single band and of an infinite potential barrier (effective mass $m^* = 0.2 m_0$, dielectric constant $\epsilon_M = 10$) [359]. The energies are given in reduced units E/R^* where R^* is the binding energy of the impurity in the bulk

$R^* = m^* e^4 / (2\hbar^2 \epsilon_M^2)/(4\pi\epsilon_0)^2 \approx 5.8$ meV in GaAs (ϵ_M is the bulk macroscopic dielectric constant). The same problem in two dimensions gives a binding energy equal to $4R^*$. This exact value is of interest in understanding the trends as a function of the quantum well thickness. The first calculation of this problem was performed variationally [359] for a quantum well bounded by two infinite barriers. The variational wave function was written as

$$\phi(\mathbf{r}) = \chi(z) \exp\left[-\alpha\sqrt{\rho^2 + (z - z_i)^2}\right], \tag{6.1}$$

where z is the direction perpendicular to the layer, z_i the impurity position, ρ the in-plane distance from the impurity, and $\chi(z)$ the state of the ground quantum well subband. As expected, the resulting ground state binding energy increases from its three dimensional (3D) value as the thickness decreases (Fig. 6.1). This method runs into difficulties in the small thickness limit where effective mass theory should not apply in the z-direction. In this case, it is more appropriate to perform a strict 2D application of effective mass theory where the variational function is taken as a product of the exact Bloch function at the bottom of the lowest subband times $\exp(-\alpha\rho)$ [360]. This gives the 2D limit exactly by construction and is valid as long as the binding energy is smaller than inter-subband separation. Finally, the case of acceptor impurities is more complex and we do not discuss it here (references can be found in [85, 86]).

The Case of 0D Nanocrystals. It deserves some special consideration. Various effective mass models have been devised [361–368] including finite barrier height effects, dielectric mismatch, etc... which prove to be successful especially for compound semiconductors when the Bohr radius a_B^* is large enough. In the following, we shall consider situations corresponding to $R \ll a_B^*$ when the radius R of the crystallite is small compared to a_B^*. In that case, one can use first order perturbation theory to treat the impurity potential since the splitting between the zeroth order states becomes large compared to the effect of the Coulomb potential. In that situation, the unperturbed effective mass wave function is $\phi \propto \sin(\pi r/R)/r$ and first order perturbation theory gives a shift in energy equal to

$$\Delta E_1 = -\langle \phi | V | \phi \rangle , \tag{6.2}$$

where V is the impurity potential energy. However it is not trivial how ΔE_1 relates to the binding energy.

Indeed, contrary to 1D or 2D systems, for 0D quantum dots the impurity binding energy $E_B(R)$ cannot be calculated as the energy difference between the hydrogenic levels and the continuum as this one does not exist in a cluster which has a discrete energy spectrum. Let us then ionize the hydrogenic impurity by taking the electron (or the hole) from the cluster with the defect into a cluster free of impurity but with the same size and located far away from the first one. In this way, the Coulomb interactions between the cluster electrons and the impurity extra-electron (or hole) are the same before and after impurity ionization and do not contribute to the binding energy. The electron and the impurity nucleus self-energies due to the difference of the dielectric constants are also identical before and after impurity ionization. Thus in such an ideal case, the binding energy is simply the energy difference between the levels of a cluster with and without impurity, i.e. $E_B(R) = -\Delta E_1$ of (6.2). As regards the ionization energy, the situation differs from the bulk where this one is equal to the binding energy since the conduction states form a continuum. In crystallites, this is no more true since the low lying conduction states form a discrete spectrum. In principle, ionization in a perfect crystallite occurs via the continuum of states above the potential barrier which exists at the surface, with an ionization energy $I_0(R)$. For the doped crystallite the ionization energy simply becomes $I_0(R) + E_B(R)$. For an on-center impurity and a dielectric constant ϵ_{out} of the surrounding material equal to unity, we shall see later that the donor (or acceptor) binding energy defined in this way is quite large. It varies from ≈ 1 eV for a cluster diameter close to 3 nm to ≈ 4 eV when the diameter is close to 1 nm. Such an energy range is characteristic of deep levels and one cannot expect impurity ionization even at high temperature. However, it has been shown that the hydrogenic states can remain ionized because their carrier are trapped at deep defects [166]. For example, this could explain why the hydrogenic impurities are not seen in highly porous silicon where the density of dangling bond deep defects is large [369].

6.1.2 Tight Binding Self-Consistent Treatment

We report here the calculations performed in [166, 167]. The first point is that self-consistent screening of the impurity potential is fully calculated using the general technique described in Sects. 1.3.1 and 3.3.2 (in fact [166, 167] make use of a similar but more sophisticated non-orthogonal tight binding method which does not, as we have checked, modify the conclusions). Figure 6.2 gives the results obtained in this way for a donor impurity at the center of a

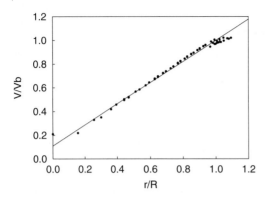

Fig. 6.2. Ratio of the self-consistent values V of the potential to the bare ones V_b as a function of the position in a cluster with 705 Si atoms: tight binding results (•), classical expression (*straight line*)

spherically symmetric silicon nanocrystal passivated by H atoms. The ratio of the screened potential V over the bare potential V_b follows a straight line over practically the whole range of values. This is fully consistent with the finding of Chap. 3 where we have shown that one can still define a local macroscopic dielectric constant $\epsilon_{in} = \epsilon_M$, leading to the classical model of a dielectric sphere embedded in an homogeneous medium of dielectric constants equal to ϵ_{in} and ϵ_{out}, respectively.

According to the considerations detailed in Sect. 3.1.3, this would result in a screened potential energy given by

$$V(r) = -\frac{e^2}{4\pi\epsilon_0}\left[\frac{1}{\epsilon_{in}r} - \frac{1}{R}\left(\frac{1}{\epsilon_{in}} - \frac{1}{\epsilon_{out}}\right)\right] \quad \text{when } r < R. \tag{6.3}$$

We see on Fig. 6.2 that V/V_b is close to the classical expression with $\epsilon_{out} = 1$. Small deviations are due to charge oscillations near the surface of the cluster. When the impurity is not located at the center of the cluster, the screened potential can also be calculated [97, 165] using (3.24) in Chap. 3.

Figure 6.3 gives the binding energy E_B for a donor and an acceptor impurity at the center of a spherical Si nanocrystal versus its radius. One gets pretty high values of E_B even for relatively large clusters. In addition, tight

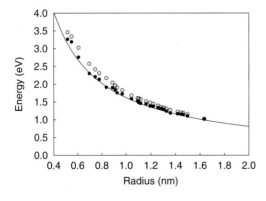

Fig. 6.3. Binding energy of hydrogenic impurities: self-consistent tight binding calculation using $\epsilon_{out} = 1$ for a donor (•) or an acceptor (∘); results deduced from the classical expression (6.4) (*straight line*)

binding calculations [166, 167] show that E_B remains relatively independent on the impurity position (this is only true when $\epsilon_{out} \ll \epsilon_{in}$). The large binding energy and its independence on the impurity location can be easily explained from first-order perturbation theory, the impurity eigenstate being equal to the lowest eigenfunction of the cluster without impurity as given by (6.2). If we take for ϕ the form corresponding to the effective mass approximation, we get for the binding energy of a donor $E_B(R) = -\langle\phi|V|\phi\rangle$. For an on-center impurity, V is given by (6.3) and one gets

$$E_B(R) = \left(\frac{1}{\epsilon_{out}} + \frac{1.44}{\epsilon_{in}}\right) \frac{e^2}{4\pi\epsilon_0 R} . \tag{6.4}$$

Due to the relative values of the dielectric constants, one can see that the main contribution to $E_B(R)$ comes from the ϵ_{out} term when $\epsilon_{out} \ll \epsilon_{in}$. When the impurity is not at the center, V is given by (3.24). Due to the symmetry of the wave function ϕ, only the $n = 0$ term of the sum in (3.24) gives a non-zero contribution to the binding energy. This contribution due to the surface polarization charge does not depend on the impurity location. There is only a slight reduction of the contribution due to the first term in (3.24) but, as we have seen above, this is not the main contribution to the binding energy. Thus the binding energy does not vary very much with the impurity position when $\epsilon_{out} \ll \epsilon_{in}$ and $R \ll a_B^*$.

A final interesting point is that these hydrogenic impurities in nanocrystals remain ionized in the presence of compensating deep defects, like in bulk materials. This is case for example in porous Si where donor and acceptor impurities do not seem to give rise to any electron paramagnetic resonance spectrum whereas there is a huge signal coming from Si dangling bonds at the Si–SiO$_2$ interface [369]. To support this argument, we can consider the reaction

$$D^0 + D_{db}^0 \rightleftarrows D^+ + D_{db}^- , \tag{6.5}$$

where the initial situation consists of a neutral donor and a neutral dangling bond in two different crystallites, the final one resulting from electron transfer between the donor and the dangling bond defect. From (6.4) in the limit $\epsilon_{out} \ll \epsilon_{in}$, electron transfer between the lowest conduction state of the two crystallites costs an energy equal to

$$\Delta E_c^f - \Delta E_c^i + \frac{e^2}{4\pi\epsilon_0\epsilon_{out}}\left(\frac{1}{R_i} - \frac{1}{d}\right) , \tag{6.6}$$

where ΔE_c is the blue shift of the conduction band, R_i is the radius of the crystallite i and d is the distance between crystallites ($i \equiv$ initial, $f \equiv$ final). On the other hand, as discussed in the next section, capture of an electron by the dangling bond implies an energy gain $\Delta E_c^f + 0.3$ eV. Thus the total energy difference ΔE between the final and initial states turns out to be

$$\Delta E = -\left(\Delta E_c^i + 0.3 \text{ eV}\right) + \frac{e^2}{4\pi\epsilon_0\epsilon_{out}}\left(\frac{1}{R_i} - \frac{1}{d}\right) . \tag{6.7}$$

As shown in [166], this formula also holds true within a given crystallite with $d = R_i$. The meaning of (6.7) is obvious, since with $\Delta E_c^i \approx 0.3$ eV, ΔE is likely to be negative in most cases. This means that donor states should remain ionized, the same reasoning holding true for deep defects other than dangling bonds.

6.2 Deep Level Defects in Nanostructures

Deep level defects are characterized by a strongly localized wave function. One then expects that, a few screening lengths away from the boundary, the wave function of the neutral defect will experience the same local potential as in the corresponding bulk material. This means that the neutral deep level itself remains invariant on an absolute scale at the same position as in the bulk material. It will thus not experience a confinement effect as it is the case for the nanostructure bandgap. Such a property has been extensively used to discuss the Stokes shift often observed between luminescence and optical absorption in semiconductor nanocrystals.

However this view is too naive and cannot be directly applied to the so-called occupancy or ionization levels $\varepsilon(n+1, n)$ defined as

$$\varepsilon(n+1, n) = E(n+1) - E(n) , \tag{6.8}$$

where $E(n)$ is the total energy of the system with n electrons on the defect (when the corresponding charge state is stable). These ionization levels are the true observable quantities in capture or emission experiments. To illustrate the situation, we choose the basic example of a non degenerate level for which one can have $n \in \{0, 1, 2\}$ with the neutral state corresponding to $n = 1$. Then one has, if they exist, two ionization levels $\varepsilon(2, 1)$ and $\varepsilon(1, 0)$ which correspond to the addition of an electron or a hole on the defect, respectively (Fig. 6.4). Such quantities are naturally obtained via the resolution of the GW equations of Sect. 1.2.4 which also gives information about the distribution of all other excited quasi-particle states and especially the bandgap

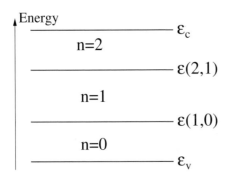

Fig. 6.4. Ionization levels of a defect characterized by a deep level with 0, 1 or 2 electrons. The bandgap limits of the nanostructure are ε_c and ε_v

limits ε_c and ε_v of the nanostructure which are affected by the confinement effects. To get simple but accurate conclusions we proceed as in Chap. 4 and split the single-particle GW equations into a bulk-like contribution and a surface polarization term due to the finite size of the system. The resolution of this problem should then proceed in three steps:

- solve a set of single particle equations, as described in Chaps. 1 and 4, using ab initio or semi-empirical techniques. This will provide us with the single particle energies and wave functions of the system containing the neutral defect: ε_d for the deep level, ε_c and ε_v for the band limits which will include the confinement effect in the presence of the defect. Note that the presence of the defect is not likely to affect seriously the confinement energies since it is an effect of order $1/N$ (N being the number of atoms) while the confinement effect is of order $(N_\mathrm{s}/N)^\nu$ where N_s is the number of surface atoms and the exponent ν is typically between 1 and 2 (the confinement energy in a spherical quantum dot varies like $d^{-\nu}$ where d is the diameter, and $N_\mathrm{s}/N \propto d^{-1}$; see Sect. 2.4)
- add to these single particle energies a bulk contribution $\delta\Sigma_\mathrm{b}$ calculated in the presence of the defect. This problem has not been solved yet. However one might anticipate some elements of solution on the basis of work done in [234]. There, one considers that the local density approximation (LDA) for instance correctly treats the short-range part of the self-energy. $\delta\Sigma_\mathrm{b}$ is then totally determined by the long range part screened by the bulk dielectric constant. This would end up with the bulk $\delta\Sigma_\mathrm{b}$ for the band limits and an intermediate value for the gap state. In any case $\delta\Sigma_\mathrm{b}$, even calculated exactly, would be state dependent. As shown in Chap. 4 for ideal crystallites, this $\delta\Sigma_\mathrm{b}$ should not give rise to appreciable confinement effects
- finally add the surface contributions $\delta\Sigma_\mathrm{surf}$.

Following (4.38) in Chap. 4, the surface contribution $\delta\Sigma_\mathrm{surf}$ takes the general form

$$\delta\Sigma_\mathrm{surf}(\boldsymbol{x},\boldsymbol{x}',\omega) = \frac{1}{\pi}\sum_l u_l(\boldsymbol{x})u_l^*(\boldsymbol{x}') \int_0^\infty \left(\frac{n_l}{\varepsilon_l - \omega - \omega'} - \frac{1-n_l}{\omega - \varepsilon_l - \omega'}\right)$$
$$\times \mathrm{Im}\, W_\mathrm{surf}(\boldsymbol{r},\boldsymbol{r}',\omega')\,\mathrm{d}\omega' , \qquad (6.9)$$

with W_surf being the contribution to (1.89) arising from the surface polarization

$$W_\mathrm{surf}(\boldsymbol{r},\boldsymbol{r}',\omega') = \sum_s \left\{\frac{1}{\omega' + \mathrm{i}\delta - \omega_s} + \frac{1}{\omega' + \mathrm{i}\delta + \omega_s}\right\}[V_s(\boldsymbol{r})V_s^*(\boldsymbol{r}')]_\mathrm{surf} .$$
$$(6.10)$$

Now, for a deep defect, we can obtain the shift $(\delta\varepsilon_\mathrm{d})_\mathrm{surf}$ of the level by perturbation theory under the form

$$(\delta\varepsilon_d)_{\text{surf}} = \langle d|\delta\Sigma_{\text{surf}}(\varepsilon_d)|d\rangle ,$$
$$= \frac{1}{\pi}\int_0^\infty \left(\frac{-n_d}{\omega'} + \frac{1-n_d}{\omega'}\right) \langle u_d u_d|\text{Im}\, W_{\text{surf}}(\omega')|u_d u_d\rangle\, d\omega' , \qquad (6.11)$$

where we have used the fact that W_{surf} is a macroscopic potential which will not mix u_d with other states, i.e. that we can restrict the sum over l in (6.9) to $l = d$ only so that ε_l cancels with $\omega = \varepsilon_d$. Injecting (6.10) into (6.11) then gives

$$(\delta\varepsilon_d)_{\text{surf}} = -(n_d - \frac{1}{2})\langle u_d u_d|W_{\text{surf}}(\omega'=0)|u_d u_d\rangle . \qquad (6.12)$$

To go further, we must notice that there are two deep level states with spin ↑ or ↓. Only one of them is filled or empty. Then we get two possibilities for (6.12)

$$(\delta\varepsilon_d)_{\text{surf}} = \pm\frac{1}{2}\langle u_d u_d|W_{\text{surf}}(\omega'=0)|u_d u_d\rangle . \qquad (6.13)$$

The + sign corresponds to adding an electron in the empty level, i.e. $\delta\varepsilon_{\text{surf}}(2,1)$, while − sign corresponds to adding a hole in the filled level $\delta\varepsilon_{\text{surf}}(1,0)$. Now the expression of the statically screened potential $W_{\text{surf}}(\omega'=0)$ is given by (3.24) so that one gets

$$(\delta\varepsilon_d)_{\text{surf}} = \pm\frac{e^2}{8\pi\epsilon_0}\sum_{n=0}^\infty \frac{(\epsilon_{\text{in}}-\epsilon_{\text{out}})(n+1)\langle u_d u_d|r^n r'^n P_n(\cos\theta)|u_d u_d\rangle}{\epsilon_{\text{in}}[\epsilon_{\text{out}}+n(\epsilon_{\text{in}}+\epsilon_{\text{out}})]R^{2n+1}} . \qquad (6.14)$$

For a strongly localized defect at a distance d from the center large compared to the extension of its wave function, one has $r \approx r' \approx d$ and $P_n(\cos\theta) \approx P_n(1)$ so that

$$(\delta\varepsilon_d)_{\text{surf}} = \pm\frac{e^2}{8\pi\epsilon_0}\sum_{n=0}^\infty \frac{(\epsilon_{\text{in}}-\epsilon_{\text{out}})(n+1)d^{2n}P_n(1)}{\epsilon_{\text{in}}[\epsilon_{\text{out}}+n(\epsilon_{\text{in}}+\epsilon_{\text{out}})]R^{2n+1}} . \qquad (6.15)$$

This is calculated in Appendix C, giving from (C.1) and (C.3)

$$(\delta\varepsilon_d)_{\text{surf}} = \pm\frac{e^2}{8\pi\epsilon_0 R}\frac{\epsilon_{\text{in}}-\epsilon_{\text{out}}}{\epsilon_{\text{in}}[\epsilon_{\text{in}}+\epsilon_{\text{out}}]}\left(\frac{1}{\eta} + J(\eta,x)\right) , \qquad (6.16)$$

where $\eta = \epsilon_{\text{out}}/(\epsilon_{\text{in}}+\epsilon_{\text{out}})$ and $J(\eta,x)$ is given by (C.8) with $x = d^2/R^2$. This self-energy correction differs from that of the highest occupied molecular orbital (HOMO) and lowest unoccupied molecular orbital (LUMO) states which were calculated in Sect. 3.5.1 and which, from (3.66), can be written under the form

$$\delta\varepsilon_c = +\frac{e^2}{8\pi\epsilon_0 R}\frac{\epsilon_{\text{in}}-\epsilon_{\text{out}}}{\epsilon_{\text{in}}[\epsilon_{\text{in}}+\epsilon_{\text{out}}]}\left(\frac{1}{\eta} + \langle J(\eta)\rangle\right) ,$$
$$\delta\varepsilon_v = -\frac{e^2}{8\pi\epsilon_0 R}\frac{\epsilon_{\text{in}}-\epsilon_{\text{out}}}{\epsilon_{\text{in}}[\epsilon_{\text{in}}+\epsilon_{\text{out}}]}\left(\frac{1}{\eta} + \langle J(\eta)\rangle\right) , \qquad (6.17)$$

using the notations defined in Appendix C. $\langle J(\eta) \rangle$ means that one takes the average over $r = r'$ with the wave function of the HOMO and LUMO state. As we have seen before, an excellent approximation is provided by the envelope function approximation in which case $\langle J(\eta) \rangle$ is given by (C.11). Thus the conclusion of this analysis is that the self-energy correction for the defect and the corresponding HOMO or LUMO differs by an amount $J(\eta, x) - \langle J(\eta) \rangle$ which depends on the defect position within the spherical nanocrystal.

One problem arises with $J(\eta, x)$ when the defect is at the surface ($x \to 1$). Indeed, if we use expression (C.9) of Appendix C, i.e.

$$J(\eta, x) = \frac{x}{1-x} - (1-\eta) \ln(1-x), \tag{6.18}$$

we see that it diverges at the surface where $x \to 1$. Obviously this divergence is unphysical and should be removed when averaging over the defect wave function at the interface. To get some feeling about this, let us consider the expansion of $1/(1-x)$ arising from (6.14) or equivalently used in Appendix C. We can rewrite:

$$\frac{1}{R(1-x)} \approx \langle u_d u_d | \sum_{n=0}^{\infty} \frac{(rr')^n}{R^{2n+1}} P_n(\cos \theta) | u_d u_d \rangle. \tag{6.19}$$

Let us, as usual, call $r_>$ and $r_<$ the larger and smaller quantities r and r', and consider the special case where $r_> = R$. Then the sum in (6.19) is given by

$$\sum_{n=0}^{\infty} \frac{r_<^n}{r_>^{n+1}} P_n(\cos \theta) = \frac{1}{|\mathbf{r} - \mathbf{r}'|}. \tag{6.20}$$

This expression is likely to hold true if the defect wave function is strongly localized at the surface. In such a situation, corresponding for instance to a surface dangling bond, we can write

$$\frac{e^2}{4\pi\epsilon_0 R(1-x)} = \langle u_d u_d | \frac{e^2}{4\pi\epsilon_0 |\mathbf{r} - \mathbf{r}'|} | u_d u_d \rangle = U, \tag{6.21}$$

where U is the defect Coulomb term. Writing $U = e^2/(4\pi\epsilon_0 x_0 R)$, the divergence in $1/(1-x)$ can thus be treated simply by the substitution

$$\frac{1}{1-x} \to \frac{1}{1+x_0 - x}. \tag{6.22}$$

Applying this to the corrective term in $\ln(1-x)$, we can rewrite $J(\eta, x)$ for a deep defect under the form

$$J(\eta, x) = \frac{x}{1 + x_0 - x} - (1-\eta) \ln(1 + x_0 - x), \tag{6.23}$$

with $x = d^2/R^2$ and $x_0 = e^2/(4\pi\epsilon_0 U R)$.

Let us then consider in more detail the self-energy correction for a surface defect. From (6.16) and (6.23), it turns out to be given by

$$(\delta\varepsilon_\mathrm{d})_\mathrm{surf} = \pm\frac{1}{2}\frac{\epsilon_\mathrm{in}-\epsilon_\mathrm{out}}{\epsilon_\mathrm{in}[\epsilon_\mathrm{in}+\epsilon_\mathrm{out}]}\left[\frac{e^2}{4\pi\epsilon_0\eta R}+U-\frac{(1-\eta)e^2}{4\pi\epsilon_0 R}\ln\left(\frac{e^2}{4\pi\epsilon_0 RU}\right)\right]. \tag{6.24}$$

This means that, even for a surface defect, the self-energy is size dependent. The first term $e^2/(4\pi\epsilon_0\eta R)$ is also contained in (6.17). At large R, the self-energy contribution is directly proportional to the defect Coulomb term U and is given by

$$\lim_{R\to\infty}(\delta\varepsilon_\mathrm{d})_\mathrm{surf} = \pm\frac{1}{2}\frac{\epsilon_\mathrm{in}-\epsilon_\mathrm{out}}{\epsilon_\mathrm{in}[\epsilon_\mathrm{in}+\epsilon_\mathrm{out}]}U. \tag{6.25}$$

It is of great interest to combine this value to the long range contribution of the bulk defect self-energy which we write $(\delta\varepsilon_\mathrm{d})_\mathrm{b,LR}$. As discussed at the beginning of this section (see also Appendix C and Sect. 4.4), the bulk self-energy can be divided into a short-range metallic like component well described in local density approximation, and a long-range one screened by ϵ_in. One thus directly obtains

$$(\delta\varepsilon_\mathrm{d})_\mathrm{b,LR} \approx \pm\frac{1}{2}\frac{U}{\epsilon_\mathrm{in}}. \tag{6.26}$$

Combining (6.25) and (6.26) gives a total long-range contribution

$$(\delta\varepsilon_\mathrm{d})_\mathrm{LR} \approx \pm\frac{U}{\epsilon_\mathrm{in}+\epsilon_\mathrm{out}}, \tag{6.27}$$

which corresponds, as could be guessed, to the self-energy calculated with the average dielectric constant. Note that exactly the same result is obtained, as it must be, when applying directly similar arguments to the planar interface between the two materials with the use of conventional image charge theory.

The final very important point of this section is to know if similar results can be obtained directly from a self-consistent LDA calculation. For this we use an argument similar to the one used in Sect. 4.4.1 and detailed in [234]. As noted there the ionization level $\varepsilon(n+1,n)$ can be calculated to lowest order (equivalent to linear screening) by using Slater's transition state as $\varepsilon_\mathrm{d}(n+1/2)$, i.e. the defect level calculated with 1/2 excess electron. This corresponds to a bare perturbation $|u_\mathrm{d}(\boldsymbol{r})|^2/2$ which should be screened. Using linear screening theory one directly gets

$$\delta\varepsilon_\mathrm{d} = \frac{1}{2}\langle u_\mathrm{d}u_\mathrm{d}|W(\boldsymbol{r},\boldsymbol{r}')|u_\mathrm{d}u_\mathrm{d}\rangle, \tag{6.28}$$

with

$$W(\boldsymbol{r},\boldsymbol{r}') = \int \epsilon^{-1}(\boldsymbol{r},\boldsymbol{r}'')\frac{e^2}{|\boldsymbol{r}''-\boldsymbol{r}'|}d\boldsymbol{r}'', \tag{6.29}$$

where ϵ^{-1} is the inverse dielectric constant in the limit of static screening. One can, as before, separate the long range component which then exactly leads to (6.27) since both methods use static screening for the long-range component.

6.3 Surface Defects: Si Dangling Bonds

In silicon, dangling bonds correspond to coordination defects in which a silicon atom has only three equivalent covalent bonds (Fig. 6.5). The best known case is the P_b center at the Si(111)–SiO$_2$ interface. Such defects are also expected to occur at the surface of crystallites as was indeed demonstrated by electron paramagnetic resonance studies of porous silicon [369, 370]. In this section, we first review the information on silicon dangling bonds in various situations and then extend these results to describe dangling bonds at the interface of crystallites with their embedding medium like SiO$_2$.

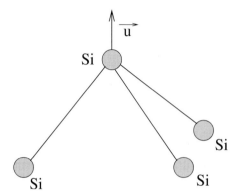

Fig. 6.5. Axial displacement of the tri-coordinated Si atom

6.3.1 Review of the Properties of Si Dangling Bonds

Let us first shortly recall the basic physical properties of dangling bonds. The simplest description comes from a tight binding picture based on an atomic basis consisting of sp^3 hybrid orbitals. The properties of the bulk material are dominated by the coupling between pairs of sp^3 hybrids involved in the same nearest neighbor's bond. This leads to bonding and anti-bonding states which are then broadened by weaker inter-bond interactions to give, respectively, the valence and conduction bands. In the bonding–anti-bonding picture, the rupture of a bond leaves an uncoupled or dangling sp^3 orbital whose energy is midway between the bonding and anti-bonding states. When one allows for inter-bond coupling, this results in a dangling bond state whose energy falls in the gap region and whose wave function is no longer of pure sp^3 character, but is somewhat delocalized along the back-bonds.

Experimentally this isolated dangling bond situation is best realized for the P_b center, i.e. the tri-coordinated Si atom at the Si–SiO$_2$ interface [371–375] but it can also occur in amorphous silicon [376, 377] as well as in grain boundaries or dislocations. It has been identified mainly through electron spin resonance (ESR) [371], deep level transient spectroscopy (DLTS) [373] and capacitance measurements versus frequency and optical experiments [376, 378]. The following picture emerges:

- the isolated dangling bond can exist in three charge states: positive D^+, neutral D^0, and negative D^-. These states correspond to zero, one, and two electrons in the dangling bond level, respectively
- the effective Coulomb term U_{eff}, i.e., the difference in energy between the acceptor and donor levels, ranges from $\approx$ 0.2–0.3 eV in a-Si [376] to about 0.6 eV at the Si–SiO$_2$ interface [373].

The ESR measurements give information on the paramagnetic state D^0 through the g tensor and the hyperfine interaction. Their interpretation indicates that the effective s electron population on the trivalent atom is 7.6% and the p one 59.4 %, which corresponds to a localization of the dangling bond state on this atom amounting to 67% and an s to p ratio of 13% instead of 25% in a pure sp^3 hybrid. This last feature shows a tendency towards a planar sp^2 hybridization.

Several calculations have been devoted to the isolated dangling bond. However, only two of them have dealt with the tri-coordinated silicon atom embedded in an infinite system other than a Bethe lattice. The first one is a self-consistent LDA calculation [379] which concludes that the purely electronic value of the Coulomb term (i.e., in the absence of atomic relaxation) is $U \approx 0.5$ eV. The second one is a tight binding Green's function treatment in which the dangling bond levels are calculated by imposing local neutrality on the tri-coordinated silicon [279]. In this way the donor and acceptor levels are respectively $\varepsilon(1,0) = 0.05$ eV and $\varepsilon(2,1) = 0.7$ eV. Their difference corresponds to $U = 0.65$ eV, which is in good agreement with the LDA results. Both values correspond to a dangling in a bulk system and can be understood simply in the following way: the purely intra-atomic Coulomb term is about 12 eV for a Si atom; it is first reduced by a factor of 2 since the dangling bond state is only localized at 70% on the trivalent atom; finally, dielectric screening reduces it by a further factor of $\epsilon \approx 10$. The final result $6/\epsilon$ gives the order of magnitude 0.6 eV. At the Si–SiO$_2$ interface, however, the situation becomes different because screening is less efficient. A very simple argument leads to the replacement of ϵ by $(\epsilon + 1)/2$ so that the electronic Coulomb term for the P_b center should be twice the previous value, i.e., $U \approx 1.2$ eV. Such a simple estimation is fully confirmed by the detailed analyzes of the previous section.

An extremely important issue is the electron–lattice interaction. There is no reason for this tri-coordinated atom to keep its tetrahedral position (Fig. 6.5). A very simple tight binding model [380] shows that this atom does indeed experience an axial force that depends on the population of the dangling bond state. This is confirmed by more sophisticated calculations [379]. The net result is that, when the dangling bond state is empty (D^+) then the trivalent atom tends to be in the plane of its three neighbors (interbond angle 120°). On the other hand, when it is completely filled (D^-), it moves away to achieve a configuration with bond angles smaller than 109° as for pentavalent atoms. Finally, the situation for D^0 is obviously intermediate with a slight motion towards the plane of its neighbors.

For the three charge states D^+, D^0, and D^-, corresponding to occupation numbers $n = 0, 1$, and 2, respectively, one can then write the total energy in the form

$$E(n,u) = nE_0 + \frac{1}{2}Un^2 - F(n)u + \frac{1}{2}ku^2, \quad (6.30)$$

where u is the outward axial displacement of the tri-coordinated atom (Fig. 6.5), $F(n)$ the occupation dependent force, U the electron–electron interaction, and k the corresponding spring constant which should show little sensitivity to n. We linearize $F(n)$

$$F(n) = F_0 + F_1(n-1), \quad (6.31)$$

and minimize $E(n,u)$ with respect to u to get $E_{\min}(n)$. The first order derivative of $E_{\min}(n)$ at $n = 1/2$ and $n = 3/2$ gives the levels $\varepsilon(1,0)$ and $\varepsilon(2,1)$. The second order derivative gives the effective correlation energy

$$U_{\text{eff}} = U - \frac{F_1^2}{k}. \quad (6.32)$$

Theoretical estimates [379, 380] give $F_1 \approx 1.6$ eV Å^{-1} and $k \approx 4$ eV Å^{-2} [381] so that F_1^2/k is of the order of 0.65 eV. This has strong implications for the dangling bond in a-Si where U_{eff} becomes slightly negative as concluded in [379] but this result should be sensitive to the local environment. On the other hand, with $U \approx 1.2$ eV, the P_b center at the Si–SiO$_2$ interface would correspond to $U_{\text{eff}} \approx 0.6$ eV, in good agreement with experiments.

6.3.2 Si Dangling Bonds at the Surface of Crystallites

From the analysis of the experimental data discussed above, the description which emerges for the P_b center at the Si–SiO$_2$ interface is summarized in Fig. 6.6. The two dangling bond occupancy levels $\varepsilon(2,1)$ and $\varepsilon(1,0)$ are symmetrical with respect to the silicon bandgap. Both ionization energies (to the conduction or valence band) are equal to 0.3 eV. In both cases, the atomic relaxation energies are equal to the same value 0.3 eV.

One can wonder how these properties transfer to the surface dangling bond of a nanocrystal embedded in SiO$_2$. The answer comes from the detailed analyzes performed in Sect. 6.2. One can first express the quantum confinement $\Delta\varepsilon_c$ of the LUMO, the situation for the HOMO being symmetric. We have

$$\Delta\varepsilon_c = \Delta\varepsilon_{c,0} + \frac{e^2}{8\pi\epsilon_0 R} \frac{\epsilon_{\text{in}} - \epsilon_{\text{out}}}{\epsilon_{\text{in}}[\epsilon_{\text{in}} + \epsilon_{\text{out}}]} \left(\frac{1}{\eta} + \langle J(\eta)\rangle\right), \quad (6.33)$$

where $\Delta\varepsilon_{c,0}$ corresponds to the value obtained from a single-particle treatment of the neutral cluster and the last terms represent the surface polarization contribution from (6.17). Similarly the shift $\Delta\varepsilon_d$ of the dangling bond

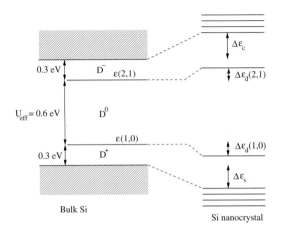

Fig. 6.6. Energy levels of the P_b center at the Si–SiO$_2$ interface and of the dangling bond at the surface of a Si nanocrystal embedded in SiO$_2$

ionization level $\varepsilon(2,1)$ with respect to the planar interface is given by (6.24) to (6.27)

$$\Delta\varepsilon_d = \frac{e^2}{8\pi\epsilon_0 R} \frac{\epsilon_{in} - \epsilon_{out}}{\epsilon_{in}[\epsilon_{in} + \epsilon_{out}]} \left[\frac{1}{\eta} - (1-\eta)\ln\left(\frac{e^2}{4\pi\epsilon_0 RU}\right)\right]. \qquad (6.34)$$

For Si clusters embedded in SiO$_2$ we are in a situation where $\eta = \epsilon_{out}/(\epsilon_{in} + \epsilon_{out}) \to 0$. From (3.67) and (3.68), as well as Appendix C, we get in this limit

$$\Delta\varepsilon_c = \Delta\varepsilon_{c,0} + \frac{e^2}{8\pi\epsilon_0 R}\left\{\frac{1}{\epsilon_{out}} - \frac{1}{\epsilon_{in}} + 0.94 \frac{\epsilon_{in} - \epsilon_{out}}{\epsilon_{in}[\epsilon_{in} + \epsilon_{out}]}\right\}, \qquad (6.35)$$

while for $\Delta\varepsilon_d$ we get

$$\Delta\varepsilon_d = \frac{e^2}{8\pi\epsilon_0 R}\left\{\frac{1}{\epsilon_{out}} - \frac{1}{\epsilon_{in}} - \frac{\epsilon_{in} - \epsilon_{out}}{\epsilon_{in}[\epsilon_{in} + \epsilon_{out}]}\ln\left(\frac{e^2}{4\pi\epsilon_0 RU}\right)\right\}. \qquad (6.36)$$

The evolution of $\Delta\varepsilon_c - \Delta\varepsilon_{c,0}$ and $\Delta\varepsilon_d$ with size is shown in Fig. 6.7. We see that the self-energy shifts tend to be of the same order of magnitude.

Finally, the situation for the HOMO and for the ionization level $\varepsilon(1,0)$ being symmetric to the previous one, the self-energy shifts have the same expression but the opposite sign (Fig. 6.6).

6.3.3 Dangling Bond Defects in III–V and II–VI Semiconductor Nanocrystals

It is believed that the surface of colloidal nanocrystals play an important role in carrier relaxation and recombination processes [382–385] but a complete experimental understanding of the nature and quantitative role of surface effects is not yet available. The importance of deep trap states is clearly evidenced in nanocrystals with unpassivated surface [383–385]. Thus, to improve the yield of the luminescence, various passivation processes have been

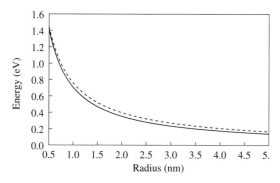

Fig. 6.7. Continuous line: shift $\Delta\varepsilon_c - \Delta\varepsilon_{c,0}$ of the LUMO due to the surface polarization in a Si crystallite with respect to its radius. Dashed line: shift $\Delta\varepsilon_d$ of the dangling bond ionization level $\varepsilon(2,1)$ with respect to the planar interface. The situation for the LUMO and for the ionization level $\varepsilon(1,0)$ being symmetric, the shifts have the opposite sign

explored. For example, CdSe nanocrystals are typically passivated by ligand molecules including TOPO (tri-n-octylphosphine oxide) [386, 387], TOPSE (tri-n-octylphosphine selenide) [386], amines, thiolates, and nitriles [388–391]. The ligands also stabilize the surface and prevent an aggregation of the nanocrystals. Another approach is to overcoat the nanocrystal by an inorganic shell with a higher bandgap, giving rise to so-called core–shell systems [390–395].

In the case of CdSe nanocrystals, the organic ligands bond mainly to the surface Cd atoms whereas the surface Se atoms are often unsaturated [387, 396]. This is confirmed by theoretical works showing that saturation with oxygen ligands (e.g. TOPO) removes Cd dangling bond states from the bandgap region, whereas it leaves Se dangling bond states unpassivated, introducing hole traps in the bandgap [397]. However, it has been also predicted that the surface relaxation removes the Se surface states from the gap region [398]. Indeed, the surface Se atoms are displaced out of the surface, increasing the bond angles and the s character of the surface states. These states shift to lower energy and come in resonance with valence states. Recent experimental results also show that the surface stoichiometry could play an important role in the passivation of surface states [399].

Experimentally, the surface of nanocrystals has been mainly investigated by X-ray photo-electron spectroscopy for CdS [400, 401], ZnS [402], CdSe [387, 403] and InAs [404].

Another system of particular interest concerns the colloidal InP nanocrystals. Indeed, surface states associated with In and P dangling bonds have been suggested by pseudopotential calculations [405]. On the experimental side, optically detected magnetic resonance and electron paramagnetic resonance studies show the existence of surface traps attributed to P surface vacancies [406, 407].

6.4 Surface Defects: Self-Trapped Excitons

A puzzling problem concerning the optical properties of semiconductor nanostructures is that there is sometimes a large difference between luminescence energies and optical absorption energies. For example, in oxidized porous silicon, there is a huge Stokes shift ($\approx$ 1 eV for a crystallite diameter $\approx$ 1.5 nm [408]), much larger than predicted values ($<$ 100 meV, see Sect. 5.2.7). In fact, as shown in [408], optical absorption energy gaps are in agreement with calculated values for crystallites. Only the luminescence energies differ greatly and, for small crystallites, are practically independent of the size. Such behaviors are more consistent with the existence of deep luminescent centers. The problem is that little is known regarding their nature and origin. We discuss here the possibility investigated in [409] of the existence of intrinsic localized states which might behave as luminescent systems. Such states correspond to self-trapped excitons and are stabilized because of the widening of the gap induced by the confinement. This possibility is not restricted to the case of silicon crystallites but is likely to be valid for all types of semiconductor crystallites.

To illustrate the physical basis of such self-trapped excitons let us consider an isolated single covalent bond characterized by a σ bonding state filled with two electrons and an empty σ^* anti-bonding state. The origin of the binding is the gain in energy resulting from having the two electrons in the lower bonding state. Optical absorption in this system leads to the excitation of one electron in the σ^* state. In such a case there is essentially no binding and the repulsive force between the atoms dominates so that the molecule eventually dissociates. If, on the other hand, the molecule is embedded in an elastic medium then it cannot dissociate but one ends up with a large distance between the constituent atoms and a reduced separation between the σ and σ^* states. The resulting luminescence energy is thus much smaller than the optical absorption energy, corresponding to a Stokes shift of the order of the binding energy, i.e. $\approx$ 1 eV.

The applicability of this model to a nanocrystal essentially depends on the possibility of localizing the electron–hole excitation on a particular covalent bond, i.e. of creating a self-trapped exciton. For this, one must be able to draw a configuration coordinate diagram like the one shown in Fig. 6.8 where the configuration coordinate Q corresponds to the stretching of the covalent bond (the notion of configuration coordinate diagram is explained in Sect. 5.2.3). For small Q, the ground and first excited states are delocalized over the crystallite and show a normal parabolic behavior. However, for Q larger than a critical value Q_c, the system localizes the electron–hole pair on one particular single bond, leading to a larger bond length Q_e and a smaller luminescence energy. This self-trapped state can be stable or metastable. An interesting point is that it may exist only for small enough crystallites, in view of the important blue shift as pictured in Fig. 6.8. Such a self-trapped

6.4 Surface Defects: Self-Trapped Excitons

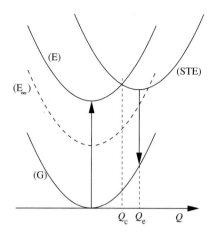

Fig. 6.8. Schematic configuration coordinate diagram showing the energies of the ground state (G), the free exciton state (E) and the self-trapped exciton state (STE). The curve (E_∞) corresponds to a very large crystallite with no blue shift, showing that the STE state might not exist for very large crystallites

exciton is likely to be favored at surfaces of crystallites where the elastic response of the environment is weaker than in the bulk.

In [409], two different techniques have been used for the calculations. The first one is a total energy semi-empirical tight binding technique which allows the treatment of relatively large crystallites (≈ 180 atoms). The second one is based on an ab initio technique in the local density approximation (LDA) which has already been applied to silicon clusters [128] (Sect. 2.2.3). Because of computation limits, the clusters studied in LDA are restricted to a maximum of ≈ 30 Si atoms which is not a severe restriction since we are interested in localized surface states. With the two techniques, the total energy is minimized with respect to all the atom positions to get the stable atomic configuration for the ground and first excited states. Only spherical crystallites centered on a silicon atom with the dangling bonds saturated by hydrogen atoms are considered. When needed, one surface dimer is created by removing the two closest hydrogen atoms of the second neighbor silicon atoms at the surface (see schematic side views in Fig. 6.9). We present here the results for two crystallites: one with 29 silicon and 36 saturating hydrogen atoms (diameter = 1 nm, tight binding energy gap = 3.4 eV, LDA gap = 3.5 eV) where tight binding and LDA techniques predict similar behavior. Then one can use with confidence tight binding for a much bigger crystallites (123 silicon atoms, 1.7 nm, tight binding gap = 2.63 eV).

If one first minimizes the total energy of excited crystallites starting from the atomic positions corresponding to the ground state situation, one obtains for very small crystallites (< 100 atoms) that the system in its excited state relaxes in highly distorted configurations with low symmetry. But, for large enough crystallites, the exciton remains delocalized, and there is a small lattice relaxation as discussed in Sect. 5.2.7. The situation gets different when considering the case of Si-H surface bonds. One finds that it is possible to trap an exciton when these are sufficiently stretched. Then the minimum of

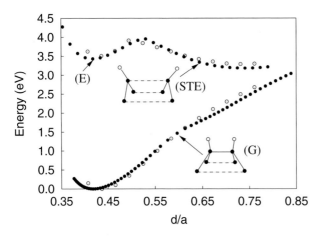

Fig. 6.9. Total energy (• = tight binding, ○ = LDA) of a spherical crystallite with 29 Si atoms in the ground state and in the excitonic state as a function of the dimer inter-atomic distance d ($a = 0.54$ nm). Schematic side views of the cluster surface dimer in the ground (G) and in the self-trapped state (STE) are also shown

energy corresponds to the broken bond, i.e. to hydrogen desorption. In the same spirit, one can get SiH$_3$ desorption by breaking the Si–Si back-bond in a process similar to polysilanes [410].

A more interesting situation is obtained when stretching the Si–Si bond of a surface dimer. Then the stable atomic configuration for the excited state corresponds to the surface Si atoms returned to their original lattice sites (Fig. 6.9). The electron and the hole are localized on the weakly interacting Si dangling bonds (second nearest neighbors) which form bonding and anti-bonding states separated by 0.72 eV in the tight binding calculation for the 1 nm crystallite (0.80 eV in LDA). Figure 6.9 fully corresponds to the general schematic picture of Fig. 6.8. As expected for a localized state, the self-trapped exciton bandgap only slightly depends on the crystallite size with a value of 0.52 eV for the 1.7 nm crystallite. We see in Fig. 6.10 that for this larger crystallite the self-trapped exciton becomes metastable because the free exciton bandgap has decreased in energy. Figure 6.10 also gives the radiative lifetime in the excited state. In the free exciton state (E), the lifetime is long because of the indirect nature of the silicon bandgap (Sect. 5.4). Increasing the dimer bond length, the lifetime in the self-trapped state first decreases because the localization of the exciton on one bond relaxes the selection rules. Finally, the lifetime increases because the optical matrix element between the two silicon atoms of the dimer decreases with the bond length. From this, one can conclude that light emission is possible in the self-trapped state. At high temperature, the recombination could be at some intermediate coordinate Q with a smaller lifetime and a larger emission energy. More details concerning the calculation can be found in [409].

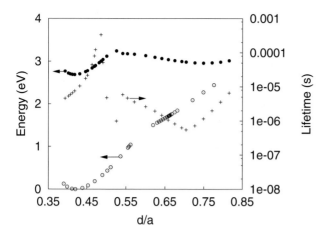

Fig. 6.10. Total energy of a spherical crystallite with 123 silicon atoms (diameter = 1.67 nm) in the ground state (○) and in the excitonic state (●) as a function of the dimer inter-atomic distance d ($a = 0.54$ nm). The crosses represent the radiative lifetime in the excitonic state

General statements about the conditions favoring the existence of such self-trapped states for a given bond are the following:

- the elastic response of the environment must be as weak as possible, which is best realized near surfaces
- the size of the nanocrystal must be small, favoring a large blue shift and the stabilization of locally distorted excited states
- the capture of the exciton must allow the release of local stresses. This is the case of the Si–Si dimer where the stresses correspond to the bending of the back-bonds in the free exciton state. Such self-trapped states are likely to be metastable in most cases. The question then arises if and how they can be excited. One answer is provided by the well documented example of the EL2 defect in GaAlAs which can be optically excited with a long lifetime [411].

In conclusion of this section, total energy calculations demonstrate the existence of self-trapped excitons at some surface bonds of Si crystallites. These give a luminescence energy almost independent of size and can explain the Stokes shift observed for small crystallites. On the experimental side, self-trapped excitons have been invoked in Si–SiO_2 multi-layers [412] and in small Si particles [413, 414]. Such self-trapped excitons are not specific to Si nanostructures but should also manifest themselves in crystallites obtained from other semiconductors.

6.5 Oxygen Related Defects at Si–SiO$_2$ Interfaces

We present here another possible interpretation of the huge Stokes shift in porous Si discussed in the previous section. This comes from a combined experimental and theoretical study [248] concluding about the important role of Si=O bonds which act as surface deep defects. We briefly summarize here the essential results of this work.

Let us first examine the experimental data. Figure 6.11a shows the photoluminescence (PL) spectra of 5 types of oxygen-free porous silicon samples with different porosities. Red, orange, yellow, green and blue spectra, in increasing order of porosity, are obtained and measured for samples kept under Ar environment. The PL intensity increases by several orders of magnitude as the PL wavelength changes from the red to the yellow, consistent with the quantum confinement model. Figure 6.11b shows how the spectra are modified after the samples have been exposed to air for 24 hours. Two major trends are observed. The PL from the samples emitting in the blue to orange region is red shifted, and the PL intensities decrease. The magnitude of the red shift increases with increasing porosities, ranging from 60 meV for the orange sample up to 1 eV for the blue sample. The PL peak saturates near 2.1 eV for the green and blue samples. Investigation of the PL in different gas environments shows that a large red shift is observed as soon as the samples are transferred from Ar to a pure oxygen atmosphere. In contrast, no red shift at all is detected when the samples are kept in pure hydrogen or in vacuum. Thus a natural hypothesis is that the large red shift is related to surface passivation, and probably the presence of oxygen.

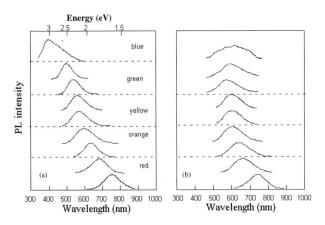

Fig. 6.11. Room temperature photoluminescence spectra from porous silicon samples with different porosities, kept under Ar atmosphere (**a**) and after exposure to air (**b**) (from [248])

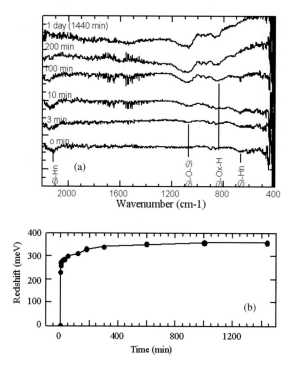

Fig. 6.12. Evolution of FTIR transmission spectra (**a**) and photoluminescence red shift (**b**), of a blue–green sample as a function of time exposed to air (from [248])

To test this hypothesis, the evolution of the chemical coverage of an Ar-stored sample as it was exposed to air was studied by Fourier transform infrared spectroscopy (FTIR). Figure 6.12a shows transmission spectra of a blue–green sample before ($t = 0$) and after exposure to air ($t > 0$). The spectrum of the fresh sample shows strong absorption bands near 2100 cm^{-1} and 664 cm^{-1}, associated with the stretching and deformation of SiH$_x$ ($x = 1 - 3$), and no sign of an oxygen peak, which confirms that the samples stored in Ar were well passivated by hydrogen and free of oxygen. As fast as 3 minutes after exposure to air, a Si–O–Si feature at 1070 cm^{-1} appears and gradually becomes dominant. In addition, after 100 minutes, a new peak is observed at 850 cm^{-1} related Si–O–H. The SiH$_x$ peaks at 2100 cm^{-1} decrease progressively with time and disappear after 24 hours. No significant change in the Si–O–Si and Si–O–H peaks are observed, indicating stabilization of the chemical coverage. As the surface passivation is gradually changing, the PL is red shifted. Figure 6.12b shows the progressive red shift of the PL with time. Most of it is obtained in the first few minutes of exposure, and stabilization is achieved after aging for 200 minutes. This result correlates well with the change of the surface passivation.

All these results suggest that the electron–hole recombination in oxidized samples occurs via carriers trapped in oxygen-related localized states that are stabilized by the widening of the gap induced by quantum confinement. This is confirmed by electronic structure calculations performed for various situations involving oxygen atoms at the surface of Si clusters [248]. As expected in view of the large offset between bulk SiO_2 and Si (≈ 4 eV), normal Si–O–Si bonds do not produce any localized gap state. Similar results are obtained for Si–O–H bonds. But, when nanocrystalline Si is oxidized and a Si–O–Si layer is formed on the surface, the Si–Si or Si–O–Si bonds are likely to weaken or break in many places because of the large lattice mismatch at the Si–SiO_2 interface [415]. Several mechanisms can act to passivate the dangling bonds [416] but a Si=O double bond is likely to be formed and stabilize the interface, since it requires neither a large deformation energy nor an excess element. It would also terminate two dangling bonds. Such bonds have been suggested at the Si–SiO_2 interface [416]. The electronic structure of Si clusters with one Si=O bond (the other dangling bonds being saturated by hydrogen atoms) has thus been calculated as a function of the cluster size using a self-consistent tight binding method (for details see [248]).

The calculated electronic states in oxidized Si nanocrystals are presented in Fig. 6.13 which shows that the recombination mechanism depends on both

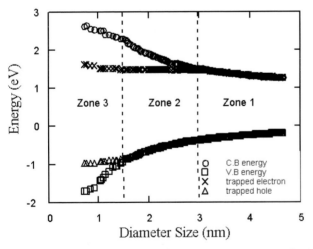

Fig. 6.13. Electronic states in Si nanocrystals as a function of size and surface passivation. The electron state is a p-state localized on the Si atom of the Si=O bond and the hole state is a p-state on the oxygen atom (from [248])

surface passivation and nanocrystal size. The model suggests that when a Si cluster is passivated by hydrogen, recombination is via free exciton states for all sizes. The PL energy is equal to the free exciton bandgap and follows the quantum confinement model. However, if the Si nanocrystal is passivated by oxygen, an electronic state (trapped exciton) stabilizes on the Si=O covalent bond. The electron state is a p-state, localized on the Si atom, and the hole state is a p-state localized on the oxygen atom. For oxygen-passivated clusters, three different recombination mechanisms are suggested, depending on the size of the cluster. Each zone in Fig. 6.13 corresponds to a different mechanism. In zone I, recombination is via free excitons. As the cluster size decreases, the PL energy increases, exactly as predicted by quantum confinement. There is no red shift whether the surface termination is hydrogen or oxygen, since the bandgap is not wide enough to stabilize the Si=O surface state. In zone II recombination involves a trapped electron and a free hole. As the size decreases, the PL emission energy stays constant, and there is a large PL red shift when the nanocrystal surface is oxidized.

In order to compare quantitatively the calculations with experimental results, it is necessary to evaluate the nanocrystal size. In ultra high porosity samples, the crystallites are very small (≤ 2 nm) and there is no obvious way to measure their size reliably. If one accepts that the PL in porous silicon stored under an Ar atmosphere is due to recombination via free excitonic states, the PL energy itself can be used to deduce the average size. Therefore, one can equate the calculated excitonic bandgap and the peak PL energy to obtain the size of the nanocrystals in each porous silicon sample. Figure 6.14

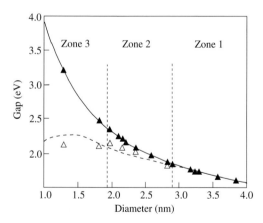

Fig. 6.14. Comparison between experimental and theoretical results as a function of crystallite sizes. The upper line is the calculated free exciton bandgap and the lower line the calculated lowest transition energy in the presence of a Si=O bond. The full and open dots are the peak PL energies obtained from Fig. 6.11a and Fig. 6.11b, respectively (▲: in Ar; △: in air). In zone I, the PL peak energies are identical whether the samples have been exposed to oxygen or not (adapted from [248])

presents the experimental PL (measured in Ar and air), and the calculated PL (free exciton energy and lowest transition energy for a nanocrystal with a Si=O bond) as a function of nanocrystal sizes. The agreement between experiments and theory is good, despite the simplicity of the model. The magnitude of the measured red shift is as calculated in the model. In addition, the experimental and theoretical PL decay lifetimes have the same order of magnitude and show similar trends. Therefore the model proposed for the electronic states and the luminescence of porous silicon quantum dots explains the experimental data. Note finally that these calculations have been basically confirmed by density functional and quantum Monte Carlo calculations [417].

7 Non-radiative and Relaxation Processes

This chapter deals with the importance of non-radiative processes which can severely limit the luminescence properties of nanocrystals. We give two detailed examples of such processes: multi-phonon capture at surface point defects and Auger recombination of electron–hole pairs. Both are known to play a central role not only for silicon but also III–V and II–VI semiconductor nanocrystals embedded in different types of matrices. As a typical example of surface point defect, we choose the dangling bond for silicon crystallites in a SiO_2 matrix. The reason is that the properties of such defects at the planar Si–SiO_2 interface are well-known. Extrapolation of these results shows that one dangling bond is enough to kill the luminescence of the crystallites. In the second part, we describe a calculation of a phonon assisted Auger recombination process. This turns out to be efficient, in the nanosecond to 10 picosecond range for small crystallites, which is shown to explain several experimental observations on nanostructures. Finally, we concentrate on hot carrier relaxation processes. We first discuss the predicted existence of a phonon bottleneck for small crystallites which is an intrinsic effect limiting their optical properties. We end up this section by reviewing different processes which can overcome this limitation, again based essentially on Auger processes or capture on point defects followed by re-emission.

7.1 Multi-phonon Capture at Point Defects

As for bulk semiconductors, carrier capture at deep level defects can occur via a multi-phonon mechanism. One of the best documented cases is the P_b center, or dangling bond, known to occur at the planar Si–SiO_2 interface [371–375]. For obvious reasons, this coordination defect is also likely to occur at the surface of silicon crystallites embedded in a SiO_2 matrix. This has been demonstrated experimentally in [369, 370] and we thus choose this case as a typical example. The derivation of this section follows closely the work detailed in [222].

The aim here is to estimate the probability per unit time that an electron–hole pair created in a silicon crystallite recombines on one dangling bond at the surface of the crystallite. This probability is related to the probabilities W that an electron and a hole in delocalized states are captured in the localized defect states. W is related to the capture coefficient c by $c = \Omega W$ [418] where Ω is the nanocrystal volume. The validity of the relation $c = \Omega W$ is discussed in the appendix of [222]. A theoretical estimate of c is a difficult task [419]. However, one can reasonably suppose that the physics of the capture in a crystallite is not very different from the capture at the planar Si–SiO$_2$ interface provided that the crystallite is not too small. For example, electron paramagnetic resonance experiments show that dangling bond states in porous silicon are very close to the (111) surface dangling bonds P$_b$ [369, 370]. In the following, instead of using c, we follow the common procedure and introduce the capture cross section defined as $\sigma = c/v_{\text{th}}$, where v_{th} is the average thermal velocity approximately equal to $\sqrt{8kT/(\pi m^*)}$, m^* being the effective mass of the trapped carrier [418].

The situation we discuss now corresponds to a crystallite with an electron–hole pair and a neutral dangling bond at the surface. The electron–hole recombination on the dangling bond can be seen as a two step process: first a carrier is captured by the neutral dangling bond and then the second carrier is captured by the charged dangling bond. Cross sections corresponding to the capture of an electron or a hole by a neutral silicon dangling bond at the planar Si–SiO$_2$ interface are measured in the 10^{-14}–10^{-15} cm^2 range at 170K [419]. Cross sections for a capture by a charged dangling bond are not known experimentally. Therefore we first concentrate on the capture of a carrier on a neutral dangling bond. We start by considering a simple classical model corresponding to the configuration coordinate diagram of Fig. 7.1 where the electron–lattice coupling is dominated by one local lattice coordinate Q (for details on configuration coordinate diagrams, see Sect. 5.2.3). In this case, capture of the carrier from the initial state i' into the final state f occurs with thermal activation over the barrier E_b [420] leading to

$$\sigma \propto \exp\left(-\frac{E_b}{kT}\right), \qquad (7.1)$$

the barrier height being given by

$$E_b = \frac{(E_0' - d_{\text{FC}})^2}{4 d_{\text{FC}}}, \qquad (7.2)$$

where E_0' is the ionization energy of the defect and d_{FC} is the Franck–Condon shift equal to the energy gain due to lattice relaxation after capture. As defined in Sect. 5.2.3, d_{FC} is related to the phonon energy $\hbar\omega$ by $d_{\text{FC}} = S\hbar\omega$ where S is the Huang–Rhys factor. It can be shown (in the following and also see Sect. 5.2.4) that (7.2) is valid only under restrictive conditions which are fulfilled in the case of the dangling bond at the planar Si–SiO$_2$ interface [418]: strong electron–phonon coupling ($S \gg 1$), high temperature and $E_0 \approx d_{\text{FC}}$. The fact that the Franck–Condon shift is close to the ionization energy at the planar Si–SiO$_2$ interface ($E_b \approx 0$, i.e. a negligible barrier) explains why the cross section is weakly thermally activated [419]. This situation is summarized on the configuration coordinate diagram of Fig. 7.1 which is qualitatively valid

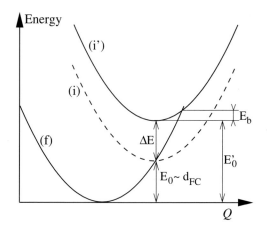

Fig. 7.1. Configuration coordinate diagram representing the variation of the total energy versus the atomic displacement for two charge states of the defect (initial (i), final (f)). Two initial states are indicated, one (i) at the planar Si–SiO$_2$ interface (ionization energy E_0) and the other (i') in a silicon crystallite (ionization energy $E'_0 = E_0 + \Delta E$). The situation in bulk silicon corresponds to a negligible barrier for the capture. In crystallites, the increase in ionization energy creates a barrier E_b for the recombination

both for the capture of a hole or an electron by a dangling bond (energies are quite similar for the two processes [419]).

For reasons discussed in detail in Sect. 6.3, the excitation energy for dangling bonds in silicon crystallites embedded in SiO$_2$ differs from its value E_0 at the planar Si–SiO$_2$ interface. As shown in this section, the difference is dominated by the confinement energy ΔE, i.e. the excitation energy in crystallites (difference between the minima of (i') and (f) in Fig. 7.1) becomes $E'_0 = E_0 + \Delta E$. The shift is not the same for the hole and for the electron. On the contrary, the Franck–Condon shift is unaffected by the confinement because it only depends on the local atomic relaxation.

From (7.1) we thus expect that the cross section will behave as $\sigma \approx \sigma_0 \exp(-\Delta E/kT)$ and will exhibit a strong decrease with the confinement. But to estimate this change, (7.1) is no longer valid because the condition $E'_0 \approx d_{\text{FC}}$ is not verified anymore when ΔE is important (since $E_0 \approx d_{\text{FC}}$). We are also interested in the dependence of the cross section over a wide range of temperatures for which (7.1) is not accurate enough. To improve on this, one can make use of an analytic formulation of the capture coefficient which remains valid over the whole temperature range, any ionization energy and any strength of the coupling between the lattice and the defect [418].

The full treatment proceeds in the same manner as in Sects. 5.2.3 and 5.2.4 for the determination of the phonon line-shape of optical transitions. There is however a major difference due to the fact that the transition matrix element is not purely electronic as for optical transitions but is now due to the electron–phonon coupling. The capture coefficient is again given by the Fermi golden rule

$$c = \Omega W = \frac{2\pi\Omega}{\hbar} \sum_{i',n'} p(i',n') \left[\sum_{f,n} |\langle \Psi_{i',n'} | h_{\text{e-ph}} | \Psi_{f,n} \rangle|^2 \delta(E_{i',n'} - E_{f,n}) \right], \tag{7.3}$$

with the same notations as in Sects. 5.2.3 and 5.2.4, i.e. $\Psi_{i',n'}$ and $\Psi_{f,n}$ are Born–Oppenheimer products corresponding to the initial (i') and final (f) states of Fig. 7.1 with respective phonon states n' and n. One can notice that W corresponds to the optical line-shape at zero frequency except that the optical matrix element is replaced by the element of the electron–phonon interaction $h_{\text{e-ph}}$ which is linear in the phonon coordinates. Apart from this, one can then follow the same derivation as in Sect. 5.2.4 or the formulation of [418] in terms of the Fourier transform of the line-shape function. The capture coefficient c can thus be written as $c_0 R$ where c_0 is a coefficient whose dependence on T and E'_0 is weak and R is a dimensionless function in which the dependence on the same parameters is important. R is given by [418]

$$R = \frac{1}{\sqrt{2\pi}} (p^2 + z^2)^{-1/4} \left(\frac{z}{p + \sqrt{p^2 + z^2}} \right)^p$$
$$\times \exp \left\{ -S \coth \left(\frac{\hbar\omega}{2kT} \right) + \frac{E'_0}{2kT} + \sqrt{p^2 + z^2} \right\}, \tag{7.4}$$

which corresponds to (5.72) with a continuous approximation to the Bessel function I_p given by (5.73) with

$$p = \frac{E'_0}{\hbar\omega} \quad \text{and} \quad z = \frac{S}{\sinh\left(\frac{\hbar\omega}{2kT}\right)}. \tag{7.5}$$

Note that (7.1) corresponds to another limit (5.74) of the Bessel function.

On Fig. 7.2, the capture cross section of a single dangling bond is plotted versus the shift ΔE. The cross section is taken equal to 10^{-15} cm^2 at 170K which is a lower limit of the measured values [419]. We also take $\hbar\omega = 20$ meV and $S = 15$. The dependence of σ on ΔE is very strong, over several decades. In the classical approximation (7.1), this is due to the increase of the energy barrier for carrier capture with ΔE. Figure 7.2 shows that the dependence of σ on temperature is weak when ΔE is small since the energy barrier is negligible. But when the ionization energy increases, the dependence becomes important as expected.

Now we can compare the non radiative capture due to a single dangling bond in a crystallite with the intrinsic radiative recombination. The non radiative capture rate W is estimated with (7.4) using the confinement energies ΔE for the electron and the hole given in Table 2.2 of Sect. 2.4 [64] (note that

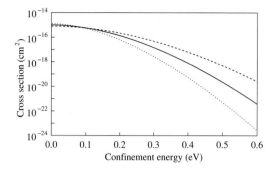

Fig. 7.2. Dependence of the capture cross section σ of an isolated dangling bond with respect to the confinement energy ΔE at 300K (*dashed line*), 170K (*straight line*) and 20K (*dotted line*)

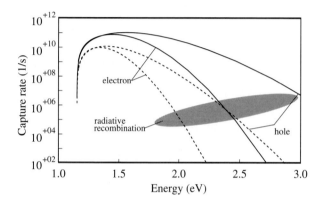

Fig. 7.3. Probability of capture W of an electron or a hole on a neutral dangling bond with respect to the gap energy of the crystallites at 10K (*dashed line*) and 300K (*continuous line*). The gray area denotes the radiative recombination rate for an electron–hole pair including phonon-assisted processes, according to the results of Fig. 5.18, Sect. 5.4

the present values of W differ from those of [222] at large band-gap energies because we use here more accurate confinement energies [64]). On Fig. 7.3, W is plotted with respect to the electron–hole energy in the crystallites. For comparison, we also show the calculated recombination rates. W decreases at high energy because of the increase in the barrier. It decreases faster at 10K than at 300K because the process is strongly thermally activated when the energy barrier becomes important. At energies close to the bulk band gap, W also decreases very quickly because the volume of the corresponding crystallite tends to infinity and the probability to be captured by a single dangling bond vanishes. For electron–hole energies corresponding to the visible luminescence of silicon crystallites (Sect. 5.4), the non radiative capture is much faster than the radiative recombination. This means that the presence of one silicon dangling bond at the surface of the crystallite kills its luminescence above 1.1 eV, in agreement with experiments [370, 421]. One can wonder if the capture on the dangling bond could be radiative. In [64], the radiative capture rate of the electron or the hole on the neutral dangling bond was calculated and was found to be always much smaller than the non radiative capture rate.

From the above discussion, the presence of a neutral dangling bond leads to non radiative capture of the electron or the hole (most probably the hole). In any case, this leaves a free carrier in the conduction band or the valence band which will then be captured by the charged dangling bond to complete the recombination process. The situation is summarized on the configuration coordinate diagram of Fig. 7.4. The lower energy curve corresponds to the ground state of the crystallite with one neutral dangling bond at the surface. The higher curve describes the system after optical excitation of an electron in the conduction band. The intermediate curve is the total energy after capture of the first carrier on the dangling bond. E_{b1} is the energy barrier (in a classical point of view) for capture of the first carrier which has been analyzed previously and is given by (7.2). E_{b2} is the energy barrier for the capture of the second carrier. Compared to the first capture, it involves much larger energies, because the sum of the thermal ionization energies is equal to the band-gap energy. From Fig. 7.4 and using (7.2), we obtain

$$E_{b2} = \frac{(E_{g0} + \Delta E - 2d_{FC})^2}{4d_{FC}}, \tag{7.6}$$

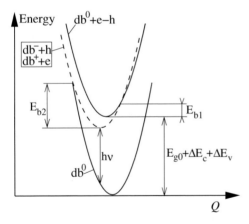

Fig. 7.4. Configuration coordinate diagram representing the variation of the total energy of a crystallite with one dangling bond at the surface. The ground state (lower curve) corresponds to filled valence states, empty conduction states and the neutral dangling bond (db^0). The higher curve represents the same system with one electron–hole pair. The intermediate curve (equivalent to curve (f) of Fig. 7.1) represents the system after the capture of one hole or one electron, the dangling bond becoming charged

where E_{g0} is the bulk silicon band-gap energy, ΔE is the confinement energy of the band corresponding to the carrier involved in the second capture (ΔE_c for the electron, ΔE_v for the hole). For a confinement energy of 0.3 eV, E_{b2} is equal to 0.53 eV which is a very large barrier for a multi-phonon capture. Therefore the capture cross section for the second capture should be strongly reduced compared to the first one (injecting the appropriate values in (7.4) gives a reduction factor of 3×10^{-7} at 300K and 5×10^{-11} at 10K). But this does not take into account the fact that the dangling bond is charged and that the capture must be enhanced by the Coulomb interaction. The numerical estimation of this enhancement is a difficult task and will not be done here. Anyway, due to the large barrier E_{b2}, we can conclude that the second capture become a radiative process, at least at low temperature. The energy of the emitted photon ($h\nu$ on Fig. 7.4) should be equal to $E_{g0} + \Delta E - 2d_{FC}$ which is about 0.8 eV for $\Delta E = 0.3$ eV. Note that an infrared emission from porous silicon has been reported and interpreted as due to the radiative recombination on silicon dangling bonds [422]. The above results support this interpretation.

7.2 Auger Recombination

Non radiative Auger recombination is known to be fast, in the nanosecond range, for bulk semiconductors [423]. It is thus of interest to examine whether this process remains efficient for the corresponding nanocrystals. We shall present in this section a theoretical calculation of this effect with comparison to the bulk values. In a second part, we review some experimental evidences that the Auger mechanism is efficient in semiconductor nanocrystals, confirming the calculated values.

7.2.1 Theoretical Calculation

Auger recombination is a three particle mechanism schematically illustrated on Fig. 7.5. In the eeh mechanism of Fig. 7.5a, for instance, an electron–hole (e–h) pair has been created (e.g. by optical absorption) in the presence of an extra electron. As shown on the right of Fig. 7.5a, recombination of the e–h pair can occur if the energy gained in this process is transferred to the extra electron. This one can either leave the nanocrystal if its energy is high enough to fall in the continuum of propagating states (Auger auto-ionization) or relax to the lowest unoccupied state (LUMO) in the opposite case. The symmetrical situation holds for the ehh process.

Let us review briefly the situation for bulk semiconductors. Then the probability per unit time for the occurrence of an Auger event is related to each carrier density and can be written $Apn^2 + Bnp^2$ where p, n are the hole and electron concentrations, respectively. Usually, A and B are not known accurately. For instance, values are reported between 10^{-30} and 10^{-32} cm^6/s

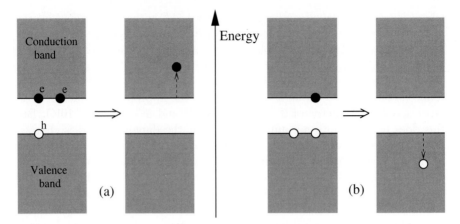

Fig. 7.5. eeh (**a**) or ehh (**b**) Auger recombination mechanisms

[423, 424] for silicon. It is tempting, although unjustified, to extrapolate these data to nanocrystals taking the concentration n and p to correspond to one carrier confined in a spherical volume $4\pi R^3/3$. The result, for silicon, is given in Fig. 7.6 showing that such extrapolated Auger lifetimes τ lie between 0.1 and 100 nanoseconds for crystallite radius $R < 2.5$ nm, which is several

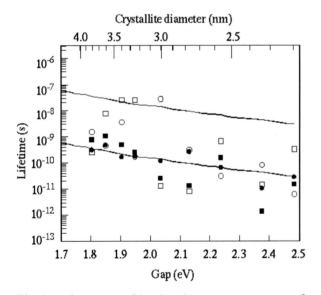

Fig. 7.6. Auger recombination time versus energy gap for the eeh (*circles*) and ehh (*squares*) processes in spherical silicon crystallites. The empty symbols correspond to a level broadening induced by the electron–lattice coupling and full symbols to a level broadening of 0.1 eV. Upper and lower bounds of an extrapolation from the bulk values (*straight line*)

orders of magnitude faster that the radiative lifetime (see Sect. 5.4). If true, recombination in silicon crystallites should occur by an Auger process instead of light emission.

However, this extrapolation of the Auger coefficients A and B from bulk to nanocrystals is questionable. If one excludes the case of Auger auto-ionization where the electron in the final state is propagating and belongs to a continuum, the energy quantization (level spacing ≈ 10 meV) in crystallites makes it in general impossible to find a final state with energy matching exactly that of the initial state. Furthermore, any irreversible decay requires a transition between an initial state and a continuum of final states. This means that we need a source of broadening corresponding to a coupling to the environment. For a crystallite embedded in a matrix, e.g. Si in SiO_2, a likely source is the coupling to phonons. We have seen previously (Sect. 5.2.7) that the Franck–Condon shift d_{FC}, taken as the relaxation energy of the crystallite following electron–hole excitation, is likely to be in the range of a few tens of meV for diameters below 4 nm. With such electron–lattice coupling, we can calculate the probability per unit time for the Auger transitions defined in Fig. 7.5, between an initial state i and a final state f. Using again the Fermi golden rule, we have

$$W = \frac{2\pi}{\hbar} \sum_{n,n'} p(n)|\langle \Psi_{i,n}|V|\Psi_{f,n'}\rangle|^2 \delta(E_{f,n'} - E_{i,n}) , \qquad (7.7)$$

where, as in Sects. 5.2.3 and 5.2.4, the states $\Psi_{i,n}$ and $\Psi_{f,n'}$ are Born–Oppenheimer products

$$|\Psi_{i,n}\rangle = |\phi_i\rangle|\chi_{i,n}\rangle ,$$
$$|\Psi_{f,n'}\rangle = |\phi_f\rangle|\chi_{f,n'}\rangle , \qquad (7.8)$$

$|\phi_i\rangle, |\phi_f\rangle$ corresponding to the electronic parts, $|\chi_{i,n}\rangle$ and $|\chi_{f,n'}\rangle$ being products of harmonic oscillators centered on the initial and final stable atomic configurations, respectively, and $p(n)$ is the thermal equilibrium occupation of $|\chi_{i,n}\rangle$. If we anticipate that V is a purely electronic operator (to be shown in the following), then we recover the following expression

$$W = \frac{2\pi}{\hbar}|\langle \phi_i|V|\phi_f\rangle|^2 \sum_{n,n'} p(n)|\langle \chi_{i,n}|\chi_{f,n'}\rangle|^2 \delta(E_{f,n'} - E_{i,n}) , \qquad (7.9)$$

which is completely similar to the optical broadening function of Sect. 5.2.4, except for the difference in the electronic matrix element and for the fact that it is calculated at zero frequency.

We now come to the most difficult part which concerns the definition of the electronic matrix element $\langle \phi_i|V|\phi_f\rangle$. It requires a physically meaningful definition of the initial and final state and of their coupling V. First of all, we make use of the notion of quasi-particles: extra electron, extra hole, electron–hole pair. We have shown in Chap. 1 how these notions are based essentially on the use of Slater determinants with optimized single-particle orbitals. They provide good descriptions not only for the ground state but also for the one-particle excitations (GW method). In the same chapter, we have also demonstrated that it is possible to extend the Hartree–Fock picture to low-lying excited states but at the condition of renormalizing the electron–electron interactions which become screened by a frequency dependent dielectric function via the higher energy plasmon-like excitations. Once renormalized, electron–electron interactions have a much weaker effect. This conclusion, valid for the bulk, holds even more true for crystallites with radius smaller than the exciton Bohr radius since correlation effects become smaller than typical single-particle level splittings.

We thus start with such Slater determinants as our zeroth-order states. In this picture, the ground state of the neutral crystallites corresponds to a Slater determinant with all valence states occupied by electrons. For the Auger eeh mechanism, the ground state is a Slater determinant with all valence states occupied plus the extra electron in the LUMO. Now the initial state is obtained from this, for instance by optical excitation. This can have in principle two effects:

1. excite an electron–hole pair across the gap with no change in energy for the excess electron,
2. excite the extra electron without creating an electron–hole pair.

However, it is well established both for the bulk [425] and nanocrystals that any excited carrier relaxes very fast (in times of the order of 10^{-14} to 10^{-12} sec) in its own band to reach the equilibrium distribution within the band. For case 2, this leads directly to the original ground state distribution. Only case 1 corresponds to the correct initial situation of an excited e–h pair in the presence of the extra electron. The statistical distribution of these initial states for the eeh situation will be fixed by the combined equilibrium distributions of two electrons in the conduction states and one hole in the valence state.

The final state in this Auger process obviously corresponds to a Slater determinant in which the valence states are again filled and the extra electron is excited by an amount almost equal to the e–h recombination energy, the rest of the energy being supplied by phonons. Once excited, the extra electron relaxes again very fast to the ground state to achieve the Auger recombination process.

From this discussion, the initial and final states $|\phi_i\rangle$ and $|\phi_f\rangle$ are Slater determinants corresponding to the physical situation pictured in Fig. 7.5. Such determinants can obviously be coupled by the electron–electron interactions $\sum_{i>j} e^2/r_{ij}$. However, one can extend qualitatively the arguments of Chap. 1 (Sect. 1.2) to conclude that one should use interactions screened by a frequency dependent dielectric function. As for the excitons (Sect. 1.2.5) for low-lying excited states, a good approximation would be to use the static dielectric constant so that $V = \sum_{i>j} e^2/(\epsilon r_{ij})$. This is the commonly used approximation in Auger calculations [423].

It is of some interest to discuss further the nature of the initial and final state. Instead of the Slater determinants $|\phi_i\rangle$ and $|\phi_f\rangle$, one could think that a better idea would be to work with combinations of them which are true eigenstates of the electron Hamiltonian of the crystallite including the electron–electron interactions V. However, such mixed states are not of interest here for two reasons:

- the matrix elements of V are much smaller than the average splitting between the Slater determinants
- even if such mixed state exists, e.g. $\alpha|\phi_i\rangle+\beta|\phi_f\rangle$, it has a very short lifetime since $|\phi_f\rangle$ relaxes very fast so that one is left only with $|\phi_i\rangle$.

From this, it seems clear that the limiting rate for the Auger transitions is provided by the matrix elements of $\sum_{i>j} e^2/(\epsilon r_{ij})$ between two Slater determinants corresponding to Fig. 7.5.

The Auger recombination rate in Si nanocrystals has been calculated in [426, 427] by using expressions (7.7) and (7.8). The initial and final state Slater determinants are built from one-electron spin orbitals obtained from a tight binding calculation [171]. The matrix elements of the screened electron–electron interaction V are calculated from the rules given in Chap. 1, in the manner also described in [171].

The calculated Auger lifetimes are plotted in Fig. 7.6 and lie mostly in the range 0.1–1 nsec. The scattering is relatively large because the e–h energy can be more or less close to the possible excitation energies of the third carrier. The results are close to those obtained from the extrapolation of the bulk results discussed above because the broadening induced by the electron–phonon coupling is sufficiently large to smooth the effect of the quantization.

Another calculation of the Auger lifetime has been reported in [428] for the case of Auger auto-ionization. This corresponds to the situation of Fig. 7.7, i.e. a quantum dot embedded in a semiconductor matrix, for which the e–

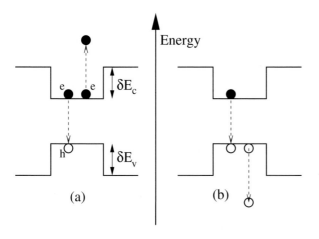

Fig. 7.7. eeh (**a**) or ehh (**b**) Auger auto-ionization mechanisms

h recombination energy E_g becomes larger than the band offsets δE_c, δE_v (or the two). In such cases, the final electron (or hole) state belongs to the continuum of propagating states of the embedding matrix. There is thus no need of the phonon continuum to match the initial and final state as was the case previously. The calculation of [428] has been performed using the envelope function approximation. The results exhibit strong oscillatory character with size with a typical value of 1 nsec for clusters with radius $R = 1.4$ nm and a size dependence $R^{-\nu}$ with $\nu \approx 5$ to 7. Such results are quite comparable to those reported in Fig. 7.6 for silicon clusters with localized final states. This means that one-phonon emission or capture is a very efficient process.

7.2.2 Experimental Evidence for Auger Recombination

There is a growing evidence that non radiative Auger recombination plays a central role in the properties of nanocrystals and quantum dots. We summarize here some observations and their interpretation, first for silicon nanostructures, second for nanocrystals of III–V and II–VI semiconductors in glasses or in colloidal suspension.

Silicon Nanocrystals. We report here results detailed in [426, 427] consisting of three distinct observations made on porous silicon:

– saturation of the luminescence at high excitation power
– voltage quenching of the luminescence
– voltage tunable electro-luminescence.

Saturation of the Luminescence at High Power. Time evolution of the photoluminescence typically shows two components [216]: a fast one with a lifetime smaller than 20 nsec associated with a defect either in the oxide or at the interface with silicon [429]; a slow one which is often attributed to recombination in confined silicon structures [216, 430]. Typical results are plotted in Fig. 7.8. The intensity of the fast component and the photo-acoustic signal vary linearly with excitation intensity showing that the absorbed light is proportional to the incident one. In contrast, the intensity of the slow component saturates at high flux. The Auger effect gives a simple and natural explanation of this saturation if one assumes that the slow band comes from the radiative recombination in luminescent crystallites. As long as the excitation remains weak, there is only one e–h pair per crystallite and the luminescence follows linearly the excitation power. At higher flux, when two e–h pairs are created in the same crystallite, the first one quickly recombines non radiatively by the eeh or hhe Auger effect.

Voltage Quenching of the Photo-luminescence. The photoluminescence is observed [431] on n-type porous silicon samples cathodically polarized in an aqueous solution of sulfuric acid. The measured spectra are very

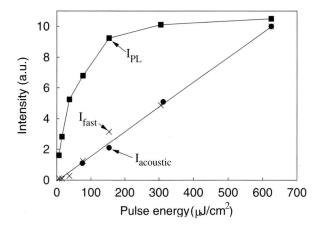

Fig. 7.8. Comparison of the photoluminescence intensity (for the slow and fast components) and the photoacoustic signal as a function of the excitation intensity for 1μm 65% anodic oxidized sample [426, 427]

close to those taken in air for a polarization between 0 and -1 V, but as the potential increases, the red part of the spectrum is gradually quenched with an energy cut-off which increases linearly with the potential (complete quenching is achieved at 1.5 V). The explanation uses the injection level $\varepsilon(1/0)$ for the filling of the lowest conduction state by one electron (equivalent to the quasi-particle level defined in Sect. 4.1) which increases in energy when the radius R of the crystallite decreases (Fig. 7.9). We also define an electron injection level $\varepsilon_F(V)$ (the effective Fermi level) which depends linearly on the applied voltage. All the crystallites with $\varepsilon(1/0)$ below $\varepsilon_F(V)$ have at least one electron injected in the conduction band. When excited, they are non luminescent because of the fast Auger recombination mediated by the injected

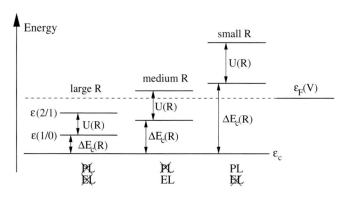

Fig. 7.9. Injection levels in Si crystallites of different sizes R corresponding to the filling of the lowest conduction state by one electron $\varepsilon(1/0)$ or two electrons $\varepsilon(2/1)$. The difference $\Delta E_c(R)$ between $\varepsilon(1/0)$ and the bottom of the bulk conduction band (ε_c) is due to the confinement. In each case, the possibility of photoluminescence (PL) and electro-luminescence (EL) is indicated (from [427])

electron. They are the larger crystallites characterized by a small confinement energy $\Delta E_c(R)$. Therefore the photoluminescence from the crystallites with a small bandgap (red part) is quenched by the Auger effect.

Voltage Tunable Electro-luminescence. The Auger effect in addition to Coulomb charging effects also explains the peculiar spectral width ($\approx$ 0.25 eV) of the electro-luminescence of n-type porous silicon cathodically polarized in a persulfate aqueous solution [432]. Under electron injection, because of the charging effects, the energy of the lowest conduction band level depends on its population (Chap. 4). If we define $\varepsilon(2/1)$ the ionization level for the filling by a second electron, the difference $U(R) = \varepsilon(2/1) - \varepsilon(1/0)$ is the average Coulomb electron–electron interaction. For a crystallite in an aqueous medium characterized by a large dielectric constant (≈ 80), $U(R)$ given by (3.72) is in the range of 0.1 eV. Due to the Auger process, only the crystallites with only one electron can be luminescent when a hole is injected. Thus, the electro-luminescence is only possible in crystallites for which $\varepsilon_F(V)$ lies between $\varepsilon(2/1)$ and $\varepsilon(1/0)$. In that case, we see in Fig. 7.9 that $\Delta E_c(R)$ must be restricted to an energy window defined by $U(R)$. With $\Delta E_g(R) \approx 2\Delta E_c(R)$, one obtains that the width of the electro-luminescence is approximately equal to $2 \times U(R) \approx 0.2$ eV in excellent agreement with experiments [432].

Other evidences of the Auger effect in Si nanocrystals are reviewed in [433].

III–V and II–VI Nanocrystals. Another indication of the importance of Auger recombination is provided by the observation of random intermittency of the photoluminescence intensity in spectroscopy of single nanocrystals of CdSe [344, 434–436], CdS [437, 438], CdTe [439, 440], InP [441], InAlAs [442], InGaN [443] and Si [444]. This means that the photoluminescence intensity exhibits a sequence of on and off periods under constant excitation conditions in a way analogous to the random telegraph signal. This effect was discussed in [428, 445] and can be interpreted on the basis of the Auger recombination. Indeed, the normal process is that low intensity optical excitation creates an e–h pair which should recombine radiatively and emit light (on period) except if, occasionally, and for some characteristic time τ, a non radiative process is induced which suppresses the emission of light (off period). In [445], such a process is induced either by Auger auto-ionization or thermal ionization followed by capture of the ionized carrier on a nearby trap state. This leaves the nanocrystal in a charged state for which any luminescence would be quenched by a non radiative Auger process as discussed before. This corresponds to the off period, τ being the time for the trapped carrier to be released to the quantum dot. On this basis, Monte Carlo simulations of the time dependence of the photoluminescence intensity were performed in [445] resulting in a satisfactory description of the sequence of on and off periods for reasonable values of the different time constants.

The Auger effect was also invoked to explain the photo-darkening effect, i.e. the decrease of the photoluminescence intensity with time under steady state excitation conditions [428, 446, 447].

7.3 Hot Carrier Relaxation: Existence of a Phonon Bottleneck

As discussed before, one usually assumes that excited carriers (electrons or holes) almost instantaneously relax to the lowest states of their respective band (or set of discrete states for nanostructures) in order to reach an equilibrium within this band (thermalization). In bulk systems, this is known to occur via a cascade of one-phonon processes which is an extremely efficient and fast process [425], of the order of 0.1 eV psec^{-1}. As shown in Fig. 7.10, the difference between the initial and final state energies of the carrier $\varepsilon_i - \varepsilon_f$ is matched by a phonon energy $\hbar\omega$. In the bulk pseudo-continuum, this can be achieved without problem. However, quantum dots are characterized by discrete levels. When they become small, the energy level spacing can become comparable to or even larger than typical phonon energies. If so, this prevents one-phonon processes to occur. As multi-phonon processes are much less efficient, this leads to a phonon bottleneck [448, 449] corresponding to a suppression of carrier relaxation rates. As argued in [449], carriers will remain trapped in excited states from which radiative recombination is less efficient. This is thus an intrinsic mechanism which should reduce the luminescence of 0D nanostructrures compared to 1D or 2D systems.

On the experimental side, it has been quite difficult to produce evidence of the phonon bottleneck effect. In [450] for instance, direct injection of carriers in the nanocrystal excited state leads to a fast relaxation consistent with intra-dot electron–hole scattering. This effect was also invoked in [451] but other authors also conclude that the phonon bottleneck effect can also be circumvented by various Auger [451, 452] and multi-phonon processes [453, 454]. In fact, the observation of a phonon bottleneck seems to require special care [455], under conditions of low carrier injection. This provides quantum dots either with one carrier or with one electron–hole pair. Time resolved differential transmission measurements then show that the first case leads to a reduced relaxation rate.

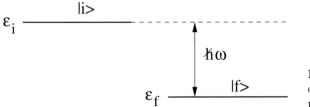

Fig. 7.10. Relaxation of a carrier by a one-phonon process

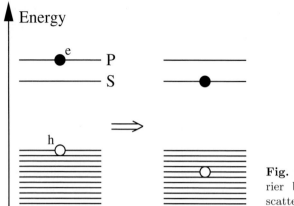

Fig. 7.11. Relaxation of a carrier by electron–hole inelastic scattering

Let us now examine briefly different effects which can mask the phonon bottleneck and lead to fast relaxation effects. The first one is the Auger-like thermalization process [456, 457] of Fig. 7.11 which occurs for e–h pairs and in which the electron transfers its energy to the hole which can then relax to its ground state in cases where the hole state splittings are smaller than the phonon energies. This Auger thermalization also called electron–hole inelastic scattering can be very fast, of the order of 2 psec [457]. This type of process has been confirmed experimentally, in quantum dots [458–460] and in quantum wires [461].

Among other possibilities, let us mention the relaxation of a carrier in a quantum dot after capture of a second carrier from the wetting layer. Here again, one of the carriers relaxes while the second one is excited into the continuum, i.e. is re-emitted. The calculation of [462] performed for self-organized InAs/GaAs dots shows that both capture and Auger relaxation are fast with characteristic time of the order of 1 psec. Another mechanism which has been proposed in [463] is the multi-phonon capture and re-emission of the quantum dot carrier by a nearby defect, the relaxation energy corresponding to the number of phonons emitted in the process. In principle, this is possible but unlikely in view of the constrains to be imposed on the model parameters to reach the psec range for the process.

In conclusion of this section, we have seen that several mechanisms could be operative to get fast relaxation in small quantum dots. However, the situation has yet to be cleared up both with the help of more quantitative calculations and of experiments showing clearly under what conditions the phonon bottleneck effect is observable.

8 Transport

The research on transport properties of nanoelectronic devices has become a worldwide effort due to the possibility to fabricate structures at the nanometer scale. Metal–Oxyde–Semiconductor transistors with channel lengths as small as 10 nm are now being actively studied both theoretically and experimentally [464]. Remarkable experiments have been performed to measure the current I through single-quantum systems, such as molecules [465–472] or semiconductor quantum dots [249, 473–478]. In these experiments, the molecules or the quantum dots are connected to metallic electrodes under bias φ using scanning tunneling microscopy tips [249, 465, 468, 476], nanometer-size electrodes [469, 477] or break junctions [470, 472]. Measurements display features arising from the quantum states of the system and from Coulombic effects (see Chap. 4). Peaks in the conductance $dI/d\varphi$ characteristics are attributed to resonant tunneling through discrete levels. Also, semiconductor nanocrystals can be assembled to form artificial materials with interesting transport properties [479–481].

From the theoretical point of view, the simulation of such experiments is a serious challenge. The aim of this chapter is to show how the electronic structure calculations can be adapted to these problems. We present different approaches to compute the conduction properties of nanostructures.

In the first section, we give a basic description of systems consisting of single nanostructures connected to two electrodes and of their boundary conditions. In the next section, we deal with the limit of weak coupling between the electrodes and the nanostructure where perturbation theory can be used. In the third section, we go beyond perturbation theory using the scattering formalism. In the fourth part, we present simplified methods to include electron–electron interactions in a mean-field theory through self-consistent calculations and we discuss open issues concerning the treatment of electronic correlations. In the final section, we consider the transport in networks of nanostructures.

8.1 Description of the Systems and of the Boundary Conditions

The main problem that we shall address in this chapter is that of a nano-device (e.g. the channel of a nano-transistor, molecules or quantum dots) connected to two electrodes with electrochemical potentials μ_L (left) and μ_R (right). When μ_L and μ_R are not equal due to an external bias ($\mu_L - \mu_R = e\varphi$), the nano-device is in a non-equilibrium state and there is a net electron flow through the system (Fig. 8.1). The two electrodes are macroscopic conducting leads (electron reservoirs) which can be simulated as semi-infinite metals or semiconductors. The reservoirs are large enough that the bulk μ_L and μ_R are not perturbed by the current I. We assume that the leads can be described by a one-particle Hamiltonian, and thus the electrons are viewed as non-interacting except for an overall mean-field potential. In contrast, electron–electron interactions usually play an important role in the nano-device because electrons are confined in a small region: their treatment represents a major challenge that we will address in Sect. 8.4. Another issue is that the electronic transport is extremely sensitive to bonding and surface chemistry at the contacts between the nano-device and the leads. For example, molecules can be attached to metallic contacts through weak van der Waals bonds or through strong covalent bonds. In consequence, we will see that there is not yet a unified description of the conduction in all these systems in spite of recent progresses in ab initio computational techniques [482–487].

Standard ab initio and semi-empirical methods used in the electronic structure calculations (Chap. 1) are not directly applicable to transport problems because

- they usually apply to closed systems either periodic or finite
- the electronic system must be in equilibrium, whereas electronic conduction through nanostructures involves open systems (infinite and non-periodic) in non-equilibrium.

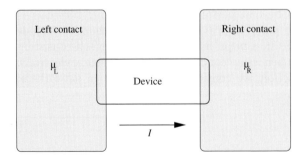

Fig. 8.1. Nano-device coupled to semi-infinite left (L) and right (R) electrodes with different Fermi levels μ_L and μ_R

The usual way to deal with open systems is to partition them into three regions, the device and the two contacts (Fig. 8.1), and to perform the calculations in three steps. The first one is to calculate the electronic structure of the contacts. It must be done only one time, for example at zero bias, because the contacts are defined in such a manner that a change in the applied bias just corresponds to a rigid energy shift of their electronic levels. This step requires computational methods which were mostly developed to study surfaces of metals or semiconductors. The second step is the resolution of the Schrödinger equation in the device region using an Hamiltonian which is renormalized to take into account the effect of the contacts on the device. This renormalization can be achieved by adding self-energy terms in the Hamiltonian. The resolution may be done iteratively when the calculation of the potential and of the eigenstates in the nano-device is performed self-consistently. The third step is the calculation of the current which leads to define a non-equilibrium density operator (or matrix) with the constraint that deep in the electrodes the electronic levels are filled according to their Fermi levels μ_L and μ_R.

This approach also works when the nano-device is connected to microscopic leads provided that these leads are coupled to macroscopic conductors acting as particle reservoirs. In that case it is often necessary to include a part of the leads into the device region (Fig. 8.1). In the general case, the device region includes the parts of the leads where the electron density differs importantly from the bulk or the free surface situation. In the case of metallic leads, these regions are small due to the strong screening of the electric fields.

The partition of the system requires to solve the quantum mechanical problem in an adapted representation. One possibility is to work directly in real space using a discrete computational grid. Another one is to use a tight binding representation in which each region is represented by a finite set of atomic orbitals whose the overlaps are neglected. These two representations lead to similar matrix formulations which we consider in the following. Extensions to non-orthogonal tight binding models may be found in [487–490].

8.2 Weak Coupling Limit

In this section, we consider the case where the nano-device is only weakly coupled to the two electrodes which allows to use perturbation theory. In many situations this approximation is well justified and is the basis of important theoretical developments to describe the conduction through small metallic or semiconducting islands.

8.2.1 Perturbation Theory

We treat here the transfer of electrons between two electrodes in perturbation theory (Fig. 8.2). We write the total Hamiltonian of the system as $H = H_0 + V$

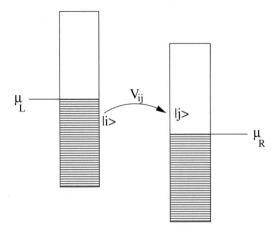

Fig. 8.2. An electron has a small probability to be transferred from a filled state $|i\rangle$ in the left electrode to an empty state $|j\rangle$ in the right electrode due to the coupling $V_{ij} = \langle i|V|j\rangle$

where H_0 is the Hamiltonian of the free electrodes with their corresponding bias and V is their coupling which takes into account the presence of the nano-device. We assume that the right and left electrodes are characterized by quasi-continuum of states $|j\rangle$ and $|i\rangle$, respectively. The Fermi golden rule provides the transfer rate (probablity per unit time) of an electron between these states

$$W_{ij} = \frac{2\pi}{\hbar}|V_{ij}|^2 \delta(\varepsilon_i - \varepsilon_j) , \qquad (8.1)$$

where $V_{ij} = \langle i|V|j\rangle$ and ε_i (ε_j) is the energy of the state $|i\rangle$ ($|j\rangle$). The current I is given by the net difference between the electron flow from the left to the right and the flow from the right to the left. Summing over all states and taking into account the occupation of the levels, we obtain

$$I = (-e) \sum_{i \in L,\ j \in R} W_{ij}\, f(\varepsilon_i - \mu_L)\, [1 - f(\varepsilon_j - \mu_R)]$$
$$-(-e) \sum_{i \in L,\ j \in R} W_{ij}\, f(\varepsilon_j - \mu_R)\, [1 - f(\varepsilon_i - \mu_L)] , \qquad (8.2)$$

where f is the Fermi-Dirac distribution function

$$f(\varepsilon) = \left[1 + \exp\left(\frac{\varepsilon}{kT}\right)\right]^{-1} . \qquad (8.3)$$

Equation (8.2) simplifies to

$$I = \frac{2\pi e}{\hbar} \sum_{i \in L,\ j \in R} |V_{ij}|^2 \{f(\varepsilon_j - \mu_R) - f(\varepsilon_i - \mu_L)\}\, \delta(\varepsilon_i - \varepsilon_j) . \qquad (8.4)$$

In the usual case of spin degeneracy, a factor 2 can be factorized. Introducing the density of states in the right and left electrodes as

$$n_R(\varepsilon) = \sum_{j \in R} \delta(\varepsilon - \varepsilon_j) , \quad n_L(\varepsilon) = \sum_{i \in L} \delta(\varepsilon - \varepsilon_i) , \qquad (8.5)$$

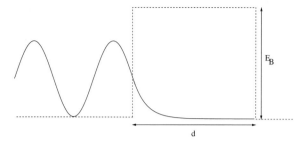

Fig. 8.3. Energy barrier in a metal-insulator-metal junction

we derive a well-known formula [491] for the current

$$I = \frac{2\pi e}{\hbar} \int |V(\varepsilon)|^2 \left\{ f(\varepsilon - \mu_\mathrm{R}) - f(\varepsilon - \mu_\mathrm{L}) \right\} n_\mathrm{L}(\varepsilon)\, n_\mathrm{R}(\varepsilon) \mathrm{d}\varepsilon \;, \tag{8.6}$$

where $V(\varepsilon)$ is the coupling of the states at energy ε (assuming that they are not degenerate).

These expressions have been used to calculate the current in metal–insulator–metal junctions [492] and in tunneling microscopy [493, 494]. In particular Bardeen has established general rules to calculate the coupling matrix elements V_{ij} when the central region is an insulating barrier [492]. In the limit of a thick barrier, the Wentzel–Kramers–Brillouin approximation gives

$$V_{ij} \propto \exp\left(-d\, \frac{\sqrt{2m_0 E_\mathrm{B}}}{\hbar}\right) \;, \tag{8.7}$$

where E_B is the barrier height and d is the barrier thickness (Fig. 8.3).

The formalism of Bardeen has been used to describe the current through Langmuir–Blodgett films of porphyrin [495]. But cases where a nano-device can be reduced to an insulating barrier for the electrical behavior are scarce. In particular, there are many situations where electronic levels of the nano-device lie between the Fermi levels μ_L and μ_R. We describe these situations in the next section.

8.2.2 Orthodox Theory of Tunneling

This theory describes nano-systems coupled to electrodes by tunnel junctions in the limit where the coupling can be treated to first order in perturbation theory (golden rule). It deals with incoherent transport through the device, when electrons tunnel sequentially. It was developed for small metallic islands and was referred to as the orthodox theory of tunneling. Then it was adapted to semiconductor quantum dots and molecules characterized by discrete energy levels [250, 251, 496, 497]. The orthodox theory assumes that the coupling is so weak that the spectral density of the nano-device is not influenced and the current can be calculated using classical master equations. To illustrate the basic physics included in the theory, we start with the simplest case of a very small device with just one level. Then we will consider the general situation.

The Case of an Island with a Single Level ε_0. We assume that ε_0 is between μ_L and μ_R. We define the transfer rates W_L and W_R through the left and right junctions, respectively (Fig.8.4). Following the Fermi golden rule, we have

$$W_L = \frac{2\pi}{\hbar} \sum_{i \in L} |V_{i0}|^2 \delta(\varepsilon_i - \varepsilon_0) = \frac{\Gamma^L(\varepsilon_0)}{\hbar} ,$$

$$W_R = \frac{2\pi}{\hbar} \sum_{j \in R} |V_{0j}|^2 \delta(\varepsilon_j - \varepsilon_0) = \frac{\Gamma^R(\varepsilon_0)}{\hbar} , \quad (8.8)$$

where Γ^L and Γ^R have the dimension of an energy and describe the coupling of the level ε_0 with the electrodes. The net current through the left and right junctions is given by

$$I_L = (-e)[f(\varepsilon_0 - \mu_L) - f_0] W_L ,$$
$$I_R = (-e)[f_0 - f(\varepsilon_0 - \mu_R)] W_R , \quad (8.9)$$

where f_0 is the mean occupation of the island state ε_0. In a permanent regime, we have $I = I_L = I_R$ and we deduce that

$$f_0 = \frac{W_L f(\varepsilon_0 - \mu_L) + W_R f(\varepsilon_0 - \mu_R)}{W_L + W_R} . \quad (8.10)$$

Injecting (8.10) into (8.9) and using (8.8), we obtain

$$I = \frac{e}{\hbar} [f(\varepsilon_0 - \mu_R) - f(\varepsilon_0 - \mu_L)] \frac{\Gamma^L(\varepsilon_0) \, \Gamma^R(\varepsilon_0)}{\Gamma^L(\varepsilon_0) + \Gamma^R(\varepsilon_0)} . \quad (8.11)$$

This simple expression and (8.10) show that the difference between the chemical potentials in the two reservoirs creates a continuous flow of electrons

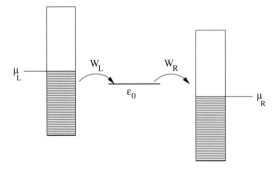

Fig. 8.4. Transfer of electrons between two electrodes through an island characterized by a single level ε_0

through the island level for which the occupation f_0 is intermediate between $f(\varepsilon_0 - \mu_R)$ and $f(\varepsilon_0 - \mu_L)$ [498, 499].

The General Situation. We consider electrons tunneling at an energy ε through the left and right junctions with the respective rates $\Gamma^L(\varepsilon)/\hbar$ and $\Gamma^R(\varepsilon)/\hbar$. The orthodox theory assumes that $\Gamma^L(\varepsilon), \Gamma^R(\varepsilon) \ll kT \ll U$ where U is the average Coulomb charging energy of the nanostructure (see Chaps. 3 and 4). Thus, at each instant, the total charge q of the island is well defined and must be an integer in unit of the electron charge. At a given q, the nanostructure can be in different electronic configurations (of index n) characterized by a total energy $E_n(q, \varphi)$. The knowledge of these energies and of the tunneling rates completely determines the $I(\varphi)$ curve, the main features of which have been discussed by several authors [250, 251, 496, 497].

The Transition Levels. The current I is the resultant of several tunneling processes. For example, an electron can tunnel from the left electrode to the island, which goes from a configuration of energy $E_n(q, \varphi)$ to a configuration of energy $E_m(q-1, \varphi)$ (for simplicity, the charge q is defined in atomic units). At $T \to 0K$, this process is possible only if

$$\mu_L > \varepsilon_{nm}(q|q-1, \varphi) , \tag{8.12}$$

where

$$\varepsilon_{nm}(q|q-1, \varphi) = E_m(q-1, \varphi) - E_n(q, \varphi) , \tag{8.13}$$

which we define as the transition levels (corresponding to ε_0 in the previous section). The position of these levels with respect to μ_L and μ_R determines which tunneling processes are possible at a given bias φ. Therefore the $I(\varphi)$ curve looks like a staircase, exhibiting a step each time a Fermi level crosses a transition level.

The transition levels are obtained by calculating the total energies $E_n(q, \varphi)$, or by using empirical expressions of these energies. For example, we have seen in Sect. 4.6.1 that the total energy of a semiconductor quantum dot charged with n electrons and p holes can be approximated by [138, 250, 251, 496, 497]

$$E(\{n_i\}, \{p_i\}) = \sum_i n_i \varepsilon_i^e - \sum_i p_i \varepsilon_i^h + \eta e \varphi q + \frac{1}{2} U q^2 , \tag{8.14}$$

where ε_i^e and ε_i^h are the electron and hole energy levels, n_i and p_i are the electron and hole occupation numbers ($n = \sum_i n_i$, $p = \sum_i p_i$ and $q = p - n$).

Calculation of the Distribution Functions. The amplitude of the steps in the $I(\varphi)$ characteristic depends on the transfer rates $\Gamma^L(\varepsilon)/\hbar$ and $\Gamma^R(\varepsilon)/\hbar$ and on the probabilities to find the nanostructure in the different electronic configurations. We illustrate now the calculation of these probabilities in the case of semiconductor quantum dots. We assume an efficient relaxation of the carriers so that the electrons and holes remain in equilibrium in their respective energy levels subsets $\{\varepsilon_i^{\rm e}\}$ and $\{\varepsilon_i^{\rm h}\}$. The recombination between electrons and holes will be introduced in the master equations. The single-particle distribution function $g_i^{\rm e}(n)$ for n electrons in the nanostructure is [250]

$$g_i^{\rm e}(n) = Z_{\rm e}^{-1}(n) \sum_{\{n_j\}_n / n_i = 1} \exp\left(-\beta \sum_j n_j \varepsilon_j^{\rm e}\right), \qquad (8.15)$$

where

$$Z_{\rm e}(n) = \sum_{\{n_j\}_n} \exp\left(-\beta \sum_j n_j \varepsilon_j^{\rm e}\right). \qquad (8.16)$$

$\{n_j\}_n$ stands for any configuration with n occupied energy levels $\varepsilon_j^{\rm e}$ and $\beta = 1/kT$. A similar expression holds for the single-particle distribution function $g_i^{\rm h}(p)$ for p holes in the system. The total rates $w_\pm^{\rm e\alpha}(n,p)$ for the tunneling of electrons through the junction α (=L,R) into (+) or out of (-) the system charged with n electrons and p holes can be written [250] as

$$w_+^{\rm e\alpha}(n,p) = \sum_i \frac{\Gamma^\alpha}{\hbar} f(\varepsilon_i^{\rm e}(q|q-1,\varphi) - \varepsilon_f^\alpha)[1 - g_i^{\rm e}(n)],$$

$$w_-^{\rm e\alpha}(n,p) = \sum_i \frac{\Gamma^\alpha}{\hbar}[1 - f(\varepsilon_i^{\rm e}(q+1|q,\varphi) - \varepsilon_f^\alpha)]g_i^{\rm e}(n). \qquad (8.17)$$

The total rates $w_\pm^{\rm h\alpha}(n,p)$ for the tunneling of holes can be written in the same way. The probability $\sigma_{n,p}$ to find n electrons and p holes is solution of master equations

$$\frac{\rm d}{{\rm d}t}\sigma_{n,p} = R(n+1, p+1)\,\sigma_{n+1,p+1} - R(n,p)\,\sigma_{n,p}$$
$$+ w_+^{\rm e}(n-1, p)\,\sigma_{n-1,p} + w_-^{\rm e}(n+1, p)\,\sigma_{n+1,p}$$
$$+ w_+^{\rm h}(n, p-1)\,\sigma_{n,p-1} + w_-^{\rm h}(n, p+1)\,\sigma_{n,p+1}$$
$$- [w_+^{\rm e}(n,p) + w_-^{\rm e}(n,p) + w_+^{\rm h}(n,p) + w_-^{\rm h}(n,p)]\,\sigma_{n,p}, \qquad (8.18)$$

where

$$w_\pm^{\rm e}(n,p) = w_\pm^{\rm eL}(n,p) + w_\pm^{\rm eR}(n,p),$$
$$w_\pm^{\rm h}(n,p) = w_\pm^{\rm hL}(n,p) + w_\pm^{\rm hR}(n,p). \qquad (8.19)$$

$R(n,p)$ is the recombination rate from the charge state (n,p) to the charge state $(n-1, p-1)$. In a permanent regime (${\rm d}\sigma_{n,p}/{\rm d}t = 0$), the current I is the same through the left and right junctions and is given by the sum of the electron and hole contributions

$$I = I^{\rm eL} + I^{\rm hL} = I^{\rm eR} + I^{\rm hR}, \qquad (8.20)$$

where for example

$$I^{eR} = -e \sum_{n,p} [\omega_+^{eR}(n,p) - \omega_-^{eR}(n,p)] \sigma_{n,p} ,$$

$$I^{hR} = e \sum_{n,p} [\omega_+^{hR}(n,p) - \omega_-^{hR}(n,p)] \sigma_{n,p} . \tag{8.21}$$

The stationary solution of (8.18) must be obtained under the constraint $\sum_{n,p} \sigma_{n,p} = 1$.

An application of the orthodox theory to InAs nanocrystals probed by a scanning tunneling microscope was presented in Sect. 4.6.1 where the transition levels are obtained by a self-consistent tight binding calculation. The theory provides a detailed interpretation of the experimental results [138, 173], showing that in certain conditions the injection of both electrons and holes into a nanostructure is possible.

Addition and Excitation Spectra. Figure 8.5 shows two situations for semiconductor nanostructures where the interpretation of the $I(\varphi)$ characteristic can be straightforward. We consider the specific case where the electrons flow from the left to the right. When $\Gamma^L \gg \Gamma^R$, the nanostructure remains close to the equilibrium with the left electrode (see (8.10)). If the applied voltage increases, new channels open for the tunneling. However, as the evacuation through the junction on the right side is not fast enough, the nanostructure remains charged with the maximum number of electrons, on average. Thus, the most visible steps in the $I(\varphi)$ curve correspond to the

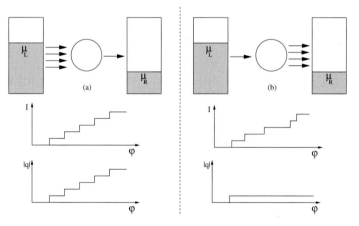

Fig. 8.5. Electrons tunneling from the left electrode to the right one through a semiconductor nanostructure. (**a**) Addition spectrum ($\Gamma^L \gg \Gamma^R$): each step of the current corresponds to the addition of one electron in the nanostructure. (**b**) excitation spectrum ($\Gamma^L \ll \Gamma^R$): each step corresponds to the tunneling of one electron through excited states of the nanostructure

opening of new charge states (addition steps, also called shell-filling spectroscopy [141]). In the opposite situation, where $\Gamma^{\rm L} \ll \Gamma^{\rm R}$, the evacuation is so fast that the nanostructure remains neutral, on average, and the steps in the current correspond to different excited configurations of one electron (excitation steps, or shell-tunneling spectroscopy [141]).

Metallic Islands and Coulomb Blockade Effect. The application of the orthodox theory to metallic islands has been extensively described in the literature [500, 501]. If the island is not too small, quantum confinement effects can be neglected and the transport is dominated by Coulomb blockade effects. Then the transition levels are simply given by

$$\varepsilon(q|q-1,\varphi) = \left(-q+\frac{1}{2}\right)U - \eta e\varphi , \qquad (8.22)$$

and the separation between the steps in the $I(\varphi)$ curve is constant and proportional to the charging energy U.

A particularly important application of these studies is the single-electron transistor shown in Fig. 8.6. The potential in the island is controlled capacitively by a gate voltage. This leads to an additional term $-e\varphi_{\rm G}$ in (8.22) where $\varphi_{\rm G}$ is a linear function of the gate voltage and of the coupling capacitances [500]. Thus by varying the gate voltage, the ionization levels can be tuned and the current shows Coulomb oscillations, i.e. a periodic dependence of the conductance on $\varphi_{\rm G}$ (Fig. 8.6c).

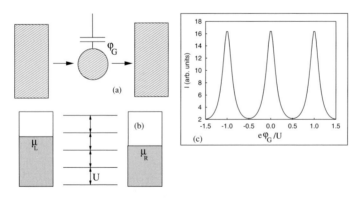

Fig. 8.6. (a) A single electron transistor with a metallic island. (b) The ionization levels of the island are equally spaced by the charging energy U. (c) Shifting the ionization levels of a quantity $-e\varphi_{\rm G}$ using the gate voltage, the current presents peaks each time an ionization level crosses the window between the Fermi levels

8.3 Beyond Perturbation Theory

In the previous sections, we assumed that the coupling between the contacts and the nano-device could be treated in perturbation. For metallic islands, this is valid when $\Gamma^\mathrm{L}, \Gamma^\mathrm{R} \ll U$ where U is the average Coulomb charging energy. For molecules or semiconductor nanostructures, there is a further requirement that $\Gamma^\mathrm{L}, \Gamma^\mathrm{R} \ll \Delta\varepsilon$ where $\Delta\varepsilon$ is the average splitting between quantum-confined states. But in many situations the coupling parameters Γ^L and Γ^R are of the same order of magnitude or larger than U and $\Delta\varepsilon$. Thus there is a need for a computational theory valid for any coupling strength. In addition, one must be able to treat correlation effects up to a certain degree, in particular to describe charging effects when $U \approx \Gamma^\mathrm{L}, \Gamma^\mathrm{R}$. Such a theory does not exist yet even if there is a general formalism based on non-equilibrium Green's functions [498, 502] which, in principle, is able to incorporate all these effects. This formalism (referred to as the Keldysh [503] or the Kadanoff–Baym [504] formalism) has been applied for the first time to tunneling processes by Caroli et al. [505, 506], and is now used in combination with density functional theory to calculate the current through very small molecules [485–487, 499]. It also provides a conceptual framework to take into account electron–electron and electron–phonon interactions in a nano-device [506–508].

In this section, we only consider the case of non-interacting electrons making use of the fact that most of the electronic structure calculations resolve single-particle equations. In this limit, the non-equilibrium Green's function formalism leads to simple expressions for the current [507, 508] which can be established by other means [487, 498, 509], in particular using the more transparent formalism of the elastic scattering [488, 498, 510–513] that we present hereafter.

8.3.1 Elastic Scattering Formalism

We consider once again the system of Fig. 8.1 divided into three regions. We neglect inelastic scattering processes within the islands and at the contacts, which turns out to be a good approximation for nano-devices in which the transport is often coherent. The elastic scattering formalism allows to calculate the current through the structure using the eigenstates of the total Hamiltonian $H = H_0 + V$ where H_0 is the Hamiltonian of the three uncoupled regions and V is their coupling. Among the eigenstates of H, we consider in the following two groups of states: those ($|i_+\rangle$) of energy ε_i incident from the left lead, partially reflected back, and partially transmitted into the right lead ; the symmetric states ($|j_-\rangle$) of energy ε_j incident from the right lead (Fig. 8.7). Since the transport is coherent, the states $|i_+\rangle$ are in equilibrium with the left reservoir, i.e. are occupied by electrons according to the Fermi function $f(\varepsilon_i - \mu_\mathrm{L})$. Symmetrically, the states $|j_-\rangle$ are filled up according to

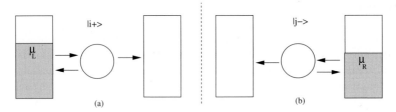

Fig. 8.7. In a non-equilibrium situation, the eigenstates of the total Hamiltonian are divided into two groups: **(a)** those incident from the left lead are filled up to the Fermi level μ_L and **(b)** those incident from the right lead are filled up to the Fermi level μ_R

the Fermi function $f(\varepsilon_j - \mu_R)$. The states $|i_+\rangle$ are solutions of the Schrödinger equation

$$(\varepsilon_i - H)|i_+\rangle = V|i_+\rangle \ . \tag{8.23}$$

If $V = 0$, the solutions correspond to the eigenstates $|i\rangle$ of H_0 in the left lead:

$$(\varepsilon_i - H_0)|i\rangle = 0 \ . \tag{8.24}$$

Thus the formal solutions of (8.23) are

$$|i_+\rangle = |i\rangle + (\varepsilon_i - H_0)^{-1} V |i_+\rangle \ . \tag{8.25}$$

Now we define the (retarded) Green's functions [516] which will be particularly useful in the following:

$$G_0(\varepsilon) = \lim_{\eta \to 0^+} (\varepsilon - H_0 + i\eta)^{-1} \ ,$$

$$G(\varepsilon) = \lim_{\eta \to 0^+} (\varepsilon - H + i\eta)^{-1} \ . \tag{8.26}$$

The small imaginary part is introduced to avoid problems of division by zero when ε is equal to an eigenvalue of H (or H_0). A more explicit form of G (similarly G_0) is

$$G(\varepsilon) = \lim_{\eta \to 0^+} \sum_k \frac{|k\rangle\langle k|}{\varepsilon - \varepsilon_k + i\eta} \ , \tag{8.27}$$

where the vectors $|k\rangle$ are the eigenstates of H.

The injection of G_0 from (8.26) into (8.25) leads to the Lippmann–Schwinger equation

$$|i_+\rangle = |i\rangle + G_0(\varepsilon_i) V |i_+\rangle \ . \tag{8.28}$$

The symmetrical equation to (8.28) expressing $|i\rangle$ in terms of $|i_+\rangle$ is obtained by reversing V and by changing G_0 into G:

$$|i\rangle = |i_+\rangle - G(\varepsilon_i) V |i\rangle \ , \tag{8.29}$$

or,

$$|i_+\rangle = |i\rangle + G(\varepsilon_i)V|i\rangle .\tag{8.30}$$

The Green's functions G and G_0 are connected by the Dyson's equation [516]:

$$G = G_0 + G_0 V G .\tag{8.31}$$

Let us consider now the total current I. It can be calculated using a generalization [514] of the Ehrenfest theorem [515]. In the following, we use an equivalent approach based on the density matrix formalism. We write

$$I = \left(\frac{dQ_R}{dt}\right)_0 ,\tag{8.32}$$

where Q_R is the total charge in the right region which is given by

$$Q_R = \sum_{j \in R} (-e)\langle j|\rho|j\rangle ,\tag{8.33}$$

where ρ is the non-equilibrium density operator which we will define hereafter and the kets $|j\rangle$ are the eigenstates of H_0 in the right lead. The subscript 0 in (8.32) means that one must consider only the charges circulating across the interface between the right lead and the nano-device because, in a permanent regime, this contribution to dQ_R/dt is fully compensated by the charges transferred through the circuitry since the total charge in the right lead remains constant over the time. Writing the Liouville equation which describes the time evolution of ρ,

$$\frac{d\rho}{dt} = \frac{1}{i\hbar}[H_0, \rho] + \frac{1}{i\hbar}[V, \rho] ,\tag{8.34}$$

this means that only the second term arising from the coupling V must be considered because, if $V = 0$, there is no current circulating though the nano-device.

Let us write now the total current as $I = I^+ + I^-$, separating the contribution from each group of states. The current I^+ comes from electrons in states $|i_+\rangle$, injected from the left side and scattered by the potential V. Thus, from (8.32), (8.33) and (8.34), we have

$$I^+ = \frac{(-e)}{i\hbar} \sum_{j \in R} \langle j|[V, \rho_+]|j\rangle ,\tag{8.35}$$

where ρ_+ is the contribution of the states $|i_+\rangle$ in ρ:

$$\rho_+ = \sum_{i \in L} |i_+\rangle f(\varepsilon_i - \mu_L)\langle i_+| .\tag{8.36}$$

Injecting (8.36) into (8.35), we obtain

$$I^+ = \frac{(-e)}{i\hbar} \sum_{j \in R,\, i \in L} \{\langle j|V|i_+\rangle\langle i_+|j\rangle - \text{c.c.}\} f(\varepsilon_i - \mu_L) ,\tag{8.37}$$

where c.c. denotes the complex conjugate of the previous term. Using (8.30) and the fact that $\langle j|i\rangle = 0$, we get

$$I^+ = \frac{(-e)}{i\hbar} \sum_{j\in R,\, i\in L} \left\{\langle j|t(\varepsilon_i)|i\rangle\langle i|VG^+(\varepsilon_i)|j\rangle - \text{c.c.}\right\} f(\varepsilon_i - \mu_L)\,, \quad (8.38)$$

where G^+ is the adjoint operator of G and t is the scattering operator given by

$$t(\varepsilon) = V + VG(\varepsilon)V\,. \quad (8.39)$$

Using (8.31), we replace G^+ in (8.38) by $G_0^+ + G^+VG_0^+$. Since the vectors $|j\rangle$ are eigenstates of H_0 and thus of G_0, we have

$$G_0^+(\varepsilon_i)|j\rangle = \frac{|j\rangle}{\varepsilon_i - \varepsilon_j - i\eta}\,, \quad (8.40)$$

and we deduce that

$$\langle i|VG^+(\varepsilon_i)|j\rangle = \frac{\langle i|t^+(\varepsilon_i)|j\rangle}{\varepsilon_i - \varepsilon_j - i\eta}\,. \quad (8.41)$$

Injecting (8.41) into (8.38), we obtain

$$I^+ = \frac{(-e)}{i\hbar} \sum_{j\in R,\, i\in L} \left\{\frac{|\langle i|t(\varepsilon_i)|j\rangle|^2}{\varepsilon_i - \varepsilon_j - i\eta} - \text{c.c.}\right\} f(\varepsilon_i - \mu_L)\,, \quad (8.42)$$

which, in the limit $\eta \to 0^+$, becomes:

$$I^+ = \frac{2\pi(-e)}{\hbar} \sum_{i\in L,\, j\in R} |\langle j|t(\varepsilon_i)|i\rangle|^2 f(\varepsilon_i - \mu_L)\delta(\varepsilon_i - \varepsilon_j)\,. \quad (8.43)$$

Using a similar derivation for I^-, one obtains the total current

$$I = \frac{2\pi e}{\hbar} \sum_{i\in L,\, j\in R} |\langle j|t(\varepsilon_i)|i\rangle|^2 \left\{f(\varepsilon_j - \mu_R) - f(\varepsilon_i - \mu_L)\right\} \delta(\varepsilon_i - \varepsilon_j)\,. \quad (8.44)$$

We recover the expression (8.4) obtained in perturbation theory except that V has been replaced by the scattering operator $t(\varepsilon)$. In fact, V is the first-order term in the development of $t(\varepsilon)$ in powers of V, as shown by injecting the Dyson's equation (8.31) into (8.39)

$$t = V + VG_0V + VG_0VG_0V + \ldots \quad (8.45)$$

Therefore (8.4) corresponds to the Born approximation of the scattering theory.

Let us now consider the common situation when the two leads are only coupled through the nanostructure ($\langle i|V|j\rangle = 0$). We have

$$t_{ij} = \langle i|t|j\rangle = \sum_{n,m} V_{in} G_{nm} V_{mj}\,, \quad (8.46)$$

where n and m denote eigenstates ($|n\rangle$ and $|m\rangle$) of H_0 in the decoupled nano-device (the energy dependence of G and t is implicit). Making use of the fact that $G_{nm}^* = G_{mn}^+$, we have

$$|t_{ij}|^2 = \sum_{n,m,n',m'} V_{in} G_{nm} V_{mj} V_{n'i} G_{m'n'}^+ V_{jm'} . \tag{8.47}$$

In analogy with (8.8), let us define the coupling matrices $\Gamma^{\mathrm{L}}(\varepsilon)$ and $\Gamma^{\mathrm{R}}(\varepsilon)$ by

$$\Gamma_{nm}^{\mathrm{L}}(\varepsilon) = 2\pi \sum_{i \in \mathrm{L}} V_{ni} V_{im} \delta(\varepsilon - \varepsilon_i) ,$$

$$\Gamma_{nm}^{\mathrm{R}}(\varepsilon) = 2\pi \sum_{j \in \mathrm{R}} V_{nj} V_{jm} \delta(\varepsilon - \varepsilon_j) . \tag{8.48}$$

The introduction of (8.47) and (8.48) into (8.44) leads to a compact expression for the current

$$I = \frac{e}{h} \int \sum_{n,n',m,m'} \Gamma_{n'n}^{\mathrm{L}}(\varepsilon) \Gamma_{mm'}^{\mathrm{R}}(\varepsilon) G_{nm}(\varepsilon) G_{m'n'}^+(\varepsilon) \{f(\varepsilon - \mu_{\mathrm{R}}) - f(\varepsilon - \mu_{\mathrm{L}})\} \, \mathrm{d}\varepsilon,$$

$$= \frac{e}{h} \int \mathrm{Tr} \left[\Gamma^{\mathrm{L}} G \Gamma^{\mathrm{R}} G^+ \right] \{f(\varepsilon - \mu_{\mathrm{R}}) - f(\varepsilon - \mu_{\mathrm{L}})\} \, \mathrm{d}\varepsilon , \tag{8.49}$$

where the trace (Tr) of the matrix is taken over all the basis states within the decoupled nano-device subspace. This expression can be rewritten as

$$I = \frac{e}{h} \int T(\varepsilon) \{f(\varepsilon - \mu_{\mathrm{R}}) - f(\varepsilon - \mu_{\mathrm{L}})\} \, \mathrm{d}\varepsilon , \tag{8.50}$$

which is a generalization of the Landauer formula that relates the current to the transmission coefficient $T(\varepsilon) = \mathrm{Tr} \left[\Gamma^{\mathrm{L}} G \Gamma^{\mathrm{R}} G^+ \right]$ across the nano-device region [517–520] (a factor 2 is usually factorized to account for the spin degeneracy).

8.3.2 Calculation of the Green's Functions

In the previous section, we have introduced the Green's functions in the framework of the scattering formalism. Here we discuss how these functions can be calculated in a practical way. We will see that the Green's functions formalism is particularly interesting to describe open systems, here the nano-device connected to semi-infinite contacts.

Self-Energy. The notion of self-energy has been introduced in Sect. 1.2.4 to describe electronic correlations. In that case, the coupling of single-particle excitations to plasmons leads to a renormalization of electron–electron interactions. The same concept can be used to account for the interaction of the nano-device with the contacts (Fig. 8.8). We will see that the Hamiltonian of the nano-device is renormalized due to the coupling with the leads.

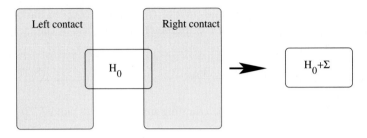

Fig. 8.8. The interaction of the nano-device with the contacts leads to a renormalization of its Hamiltonian by a self-energy Σ

The expression (8.49) of the current shows that the Green's functions must be determined only in the nano-device region. These Green's functions are represented by a matrix which may be of quite small size for a nano-device. According to the partition of the system, the matrix of the Hamiltonian $H = H_0 + V$ has the following form in terms of block matrices

$$\begin{bmatrix} [H_0]_{00} & [V]_{0L} & [V]_{0R} \\ [V]_{L0} & [H_0]_{LL} & 0 \\ [V]_{R0} & 0 & [H_0]_{RR} \end{bmatrix}, \tag{8.51}$$

where the labels 0, L, R refer to the nano-device, the left and right reservoirs, respectively. Using (8.26), we have ($\eta \to 0^+$)

$$[G] = \begin{bmatrix} (\varepsilon + i\eta)[I]_{00} - [H_0]_{00} & -[V]_{0L} & -[V]_{0R} \\ -[V]_{L0} & (\varepsilon + i\eta)[I]_{LL} - [H_0]_{LL} & 0 \\ -[V]_{R0} & 0 & (\varepsilon + i\eta)[I]_{RR} - [H_0]_{RR} \end{bmatrix}^{-1}, \tag{8.52}$$

where $[I]$ is the unit matrix. We only need the nano-device part $[G]_{00}$ which, after straightforward algebra based on the Dyson's equation (8.31), is given by

$$[G]_{00} = \{(\varepsilon + i\eta)[I]_{00} - [H_0]_{00} - [\Sigma]\}^{-1}, \tag{8.53}$$

where

$$\begin{aligned}[] [\Sigma] &= [V]_{0R}\,[G_0]_{RR}\,[V]_{R0} + [V]_{0L}\,[G_0]_{LL}\,[V]_{L0}, \\ &= [\Sigma^R] + [\Sigma^L], \end{aligned} \tag{8.54}$$

and

$$\begin{aligned}[] [G_0]_{RR} &= \{(\varepsilon + i\eta)[I]_{RR} - [H_0]_{RR}\}^{-1}, \\ [G_0]_{LL} &= \{(\varepsilon + i\eta)[I]_{LL} - [H_0]_{LL}\}^{-1}. \end{aligned} \tag{8.55}$$

8.3 Beyond Perturbation Theory

Thus in (8.53) the coupling to the contacts is represented by a self-energy Σ that renormalizes the Hamiltonian H_0 of the nano-device (Fig. 8.8) and replaces its discrete spectrum by a continuous one. The expression of Σ greatly simplifies when the matrices $[G_0]_{RR}$ and $[G_0]_{LL}$ are written in the basis of the eigenstates for the free electrodes R and L. From (8.27), we see that these matrices are diagonal and thus, from (8.54), we calculate the matrix elements of Σ

$$\Sigma_{nm} = \Sigma^R_{nm} + \Sigma^L_{nm},$$
$$= \sum_{i \in L} \frac{V_{ni} V_{im}}{\varepsilon - \varepsilon_i + i\eta} + \sum_{j \in R} \frac{V_{nj} V_{jm}}{\varepsilon - \varepsilon_j + i\eta}, \quad (8.56)$$

where again the indexes n, m refer to states within the nano-device. The self-energy is a complex, non-Hermitian, and energy dependent operator. Its imaginary part is related to the coupling matrices (8.48) by

$$\Gamma^L_{nm}(\varepsilon) = -2\mathrm{Im}\left(\Sigma^L_{nm}\right),$$
$$\Gamma^R_{nm}(\varepsilon) = -2\mathrm{Im}\left(\Sigma^R_{nm}\right). \quad (8.57)$$

A Nano-Device with a Single Level. In order to discuss the physical meaning of the self-energy, we consider once again the case of a nano-device represented by a discrete and non-degenerate level of energy ε_0. The current is given by (8.49) with $n, m, n', m' = 0$ and

$$G_{00}(\varepsilon) = \frac{1}{\varepsilon - \varepsilon_0 - \Lambda(\varepsilon) + i[\Gamma(\varepsilon)/2]}, \quad (8.58)$$

where ($\Gamma^L_{00} \equiv \Gamma^L$, $\Gamma^R_{00} \equiv \Gamma^R$)

$$\Gamma(\varepsilon) = \Gamma^L(\varepsilon) + \Gamma^R(\varepsilon), \quad \Lambda(\varepsilon) = \mathrm{Re}(\Sigma_{00}). \quad (8.59)$$

Now we assume for simplicity that $\Gamma^L(\varepsilon)$, $\Gamma^R(\varepsilon)$, $\Gamma(\varepsilon)$ and $\Lambda(\varepsilon)$ are constant in the energy range where $f(\varepsilon - \mu_R) - f(\varepsilon - \mu_L)$ is significant. Thus the current is given by the Landauer formula (8.50) with

$$T(\varepsilon) = \frac{\Gamma^L \Gamma^R}{\Gamma^L + \Gamma^R} A(\varepsilon), \quad (8.60)$$

where the line-shape is given by the Breit–Wigner formula

$$A(\varepsilon) = \frac{\Gamma}{(\varepsilon - \varepsilon_0 - \Lambda)^2 + (\Gamma/2)^2}. \quad (8.61)$$

When $kT \ll \Gamma$, (8.50) leads to an analytic expression for the current through the single level

$$I = \frac{e}{h} \frac{\Gamma^L \Gamma^R}{\Gamma^L + \Gamma^R} \int_{\mu_L}^{\mu_R} A(\varepsilon) d\varepsilon,$$
$$= \frac{2e}{h} \frac{\Gamma^L \Gamma^R}{\Gamma^L + \Gamma^R} \left[\arctan\left(\frac{\mu_R - \varepsilon_0 - \Lambda}{\Gamma/2}\right) - \arctan\left(\frac{\mu_L - \varepsilon_0 - \Lambda}{\Gamma/2}\right) \right].$$
$$(8.62)$$

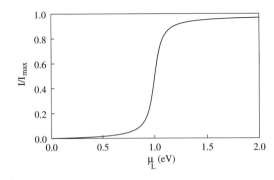

Fig. 8.9. Variation of I/I_{max} as a function of the Fermi level μ_L for a nano-device with a single non-degenerate level ($\Gamma^L = \Gamma^R = 50$ meV, $\varepsilon_0 - \Lambda = 1$ eV, $\mu_R = 0$ eV)

We plot in Fig. 8.9 the current as a function of μ_L for $\mu_R = 0$ eV and $\varepsilon_0 - \Lambda = 1$ eV. It presents a step when μ_L comes in resonance with $\varepsilon_0 - \Lambda$ and the broadening of the step is determined by Γ. When $\mu_L - (\varepsilon_0 - \Lambda) \gg \Gamma$ and $(\varepsilon_0 - \Lambda) - \mu_R \gg \Gamma$, the intensity of the current is maximum and is given by

$$I_{max} = \frac{-e}{\hbar} \frac{\Gamma^L \Gamma^R}{\Gamma^L + \Gamma^R}, \tag{8.63}$$

which corresponds to the value (8.11) obtained in perturbation theory.

This simple example illustrates that the real part of $\Sigma(\Lambda)$ describes the shift of the resonance with respect to the energy ε_0. The imaginary part expresses the broadening of the level due to the coupling to the leads. The perturbation theory corresponds to the limit where the shift and the broadening are vanishingly small.

Note that $kT \ll \Gamma$ corresponds to the resonant tunneling regime. In the opposite limit $kT \gg \Gamma$, the width of the conductance peak is determined by the thermal broadening kT. This is the regime of sequential tunneling.

Green's Functions of the Leads. In principle, the evaluation of the Green's functions (8.55) of the semi-infinite contacts requires to invert infinite matrices. However, if one writes them in the real space or in a tight binding representation, only the matrix elements at the surface of the leads and in the vicinity of the nano-device are needed because the coupling V describes short-range interactions. The surface Green's functions can be obtained by treating a finite slab of the corresponding metal or semiconductor. In that case, one often takes advantage to make it periodic which allows to use the Bloch theorem. Another possibility is to consider real semi-infinite contacts. Then the surface Green's functions can be obtained with the recursion or decimation methods [521, 522].

8.4 Electron–Electron Interactions Beyond the Orthodox Theory

Electron-electron interactions have a major impact on the transport properties of nano-devices. On one hand, we have seen in Sect. 8.2.2 that it is possible to treat Coulomb blockade effects quite accurately when the coupling to the leads is weak ($\Gamma^\mathrm{L}, \Gamma^\mathrm{R} \ll U$), provided that one is capable to calculate the total energy of the nanostructure versus charge state and occupation numbers. On the other hand, scattering approaches treat independent particles. In this section, we try to go beyond these limits which requires to consider approximate descriptions of electron–electron interactions. In a first part, we deal with mean-field approaches and we show how to implement them in non-equilibrium problems. In the second part, we discuss the limits of these methods and we consider possible ways of improvements. Note that we will not investigate complex problems such as the Kondo effect which arises from the interaction of the electrons in the leads with the spin of an unpaired electron stored in a quantum dot [523–526] or in a molecule [527, 528] (when $kT \ll \Gamma$). We will restrict our discussion to the treatment of Coulomb blockade effects using standard computational methods of the electronic structure.

8.4.1 Self-Consistent Mean-Field Calculations

We assume that the system can be described by a single-electron equation in which the potential depends on the electron density $n(\boldsymbol{r})$. This is the case of methods based on the Hartree approximation or on the density functional theory (see Chap. 1). Schrödinger–Poisson solvers used in device simulation also work on the same basis. When the nano-device is connected to the leads with different Fermi levels, there are some variations in the electron density and in the potential that must be calculated self-consistently (Fig. 8.10).

The self-consistent part of the potential consists of two contributions: the Hartree potential and, in density functional theory, the exchange–correlation potential. The Hartree potential can be determined from $n(\boldsymbol{r})$ with the Poisson's equation exactly like for the system in equilibrium. Concerning

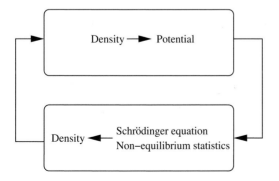

Fig. 8.10. The electron density and the potential must be calculated self-consistently taking into account the non-equilibrium occupation of the levels

the exchange–correlation potential, one always assumes that the commonly used exchange–correlation functionals are able to describe electrons in non-equilibrium situations [485–487, 499].

We must evaluate the electron density $n(\boldsymbol{r})$ which, by hypothesis, differs from the equilibrium situation only in the nano-device region (see Sect. 8.1). Equivalently, in a tight binding representation, we must calculate the quantities $\langle m|\rho|m\rangle$ where ρ is the density operator and the kets $|m\rangle$ represent atomic orbitals within the nano-device. As shown previously, the density operator is built according to the fact that the states $|i_+\rangle$ and $|j_-\rangle$ are in equilibrium with the left and right reservoirs, respectively. It is given by

$$\rho = \rho_+ + \rho_-,$$
$$= \sum_{i \in L} |i_+\rangle\langle i_+| \, f(\varepsilon_i - \mu_L) + \sum_{j \in R} |j_-\rangle\langle j_-| \, f(\varepsilon_j - \mu_R) \, . \tag{8.64}$$

Using the Lippmann-Schwinger equation (8.30), we obtain

$$\rho_+ = \int f(\varepsilon - \mu_L)(I + GV) \left[\sum_{i \in L} |i\rangle\langle i| \delta(\varepsilon - \varepsilon_i) \right] (I + VG^+) \mathrm{d}\varepsilon \, . \tag{8.65}$$

Writing the delta function in terms of the Green's functions, i.e.

$$\delta(\varepsilon - \varepsilon_i) = \frac{i}{2\pi} \langle i| G_0 - G_0^+ |i\rangle \, , \tag{8.66}$$

and neglecting the overlaps between the atomic orbitals, we obtain the matrix of ρ_+ within the nano-device region

$$[\rho_+]_{00} = \frac{i}{2\pi} \int f(\varepsilon - \mu_L) [G]_{00} [V]_{0L} \left([G_0]_{LL} - [G_0^+]_{LL}\right) [V]_{L0} [G^+]_{00} \mathrm{d}\varepsilon \, , \tag{8.67}$$

where we assume that there is no direct coupling between the contacts. With (8.54) and (8.57), we deduce

$$[\rho_+]_{00} = \frac{1}{2\pi} \int f(\varepsilon - \mu_L) [G]_{00} [\Gamma^L] [G^+]_{00} \mathrm{d}\varepsilon \, . \tag{8.68}$$

With similar equations for ρ_-, we obtain finally

$$[\rho]_{00} = \frac{1}{2\pi} \int f(\varepsilon - \mu_L) [G]_{00} [\Gamma^L] [G^+]_{00} + f(\varepsilon - \mu_R) [G]_{00} [\Gamma^R] [G^+]_{00} \mathrm{d}\varepsilon \, . \tag{8.69}$$

The Population of a Nano-Device with a Single Level. Injecting (8.58) into (8.69) and using (8.60), we obtain the population n_0 on the orbital within the nano-device

$$n_0 = \frac{1}{2\pi} \int \left\{ f(\varepsilon - \mu_L) \frac{\Gamma^L}{\Gamma^L + \Gamma^R} + f(\varepsilon - \mu_R) \frac{\Gamma^R}{\Gamma^L + \Gamma^R} \right\} A(\varepsilon) \mathrm{d}\varepsilon \, . \tag{8.70}$$

8.4 Electron–Electron Interactions Beyond the Orthodox Theory

In the limit of vanishing coupling where $A(\varepsilon) \to 2\pi\delta(\varepsilon-\varepsilon_0)$, we recognize that n_0 is equal to f_0 given in (8.10) which was obtained using first-order perturbation theory. Evaluating (8.70) in the limit $kT \ll \Gamma^{\rm L}, \Gamma^{\rm R}$, we obtain

$$n_0 = \frac{\Gamma^{\rm L}}{\Gamma^{\rm L} + \Gamma^{\rm R}} \left[\frac{1}{\pi} \arctan\left(\frac{\mu_{\rm L} - \varepsilon_0 - \Lambda}{\Gamma/2}\right) + \frac{1}{2} \right]$$
$$+ \frac{\Gamma^{\rm R}}{\Gamma^{\rm L} + \Gamma^{\rm R}} \left[\frac{1}{\pi} \arctan\left(\frac{\mu_{\rm R} - \varepsilon_0 - \Lambda}{\Gamma/2}\right) + \frac{1}{2} \right]. \tag{8.71}$$

8.4.2 The Self-Consistent Potential Profile

The one-electron potential (Hartree, exchange–correlation) within the nano-device can be calculated self-consistently using (8.69). From this, the evolution of the potential profile within the system can be studied as a function of the applied bias. In conventional wires in which the transport is diffusive, the potential varies linearly along the wire according to the Ohm's law. In nano-devices or molecular wires, the electrostatic potential may have a more complex profile, as shown recently in a large number of works [485, 487, 499, 529–532].

In order to illustrate these problems, we present the results of recent calculations [531] on two molecular wires, a carbon chain (C_{15}) and a C_7-Si-C_7 chain bonded to two metal electrodes. The Si atom can be considered as an impurity in the wire. We show in Fig. 8.11 the change in the electrostatic potential in the biased molecular devices (3V) with respect to the zero bias situation. In the case of C_{15}, the voltage drop mainly occurs near the two metal-wire interfaces because the resistance of the junction arises from the scattering at these interfaces. The potential remains almost constant (disregarding the oscillations) inside the wire because there is no scattering. In contrast, in the case of C_7-Si-C_7, the Si atom introduces scattering at the center of the wire and the potential appears to drop almost linearly. Note that quite similar results have been obtained in the case of silicon nano-transistors [499].

In the case of ballistic systems with small barriers at the interface between the nano-device and the contacts, the resistance of the central region is small, and therefore a substantial part of the bias is dropped inside the leads, and not inside the nano-device [498, 499, 518]. To understand this, let us consider the eigenstates $|i_+\rangle$ and $|j_-\rangle$ of H at a given energy $\varepsilon = \varepsilon_i = \varepsilon_j$. Inside the leads, these states are given by combinations of Bloch waves. Near the contact, $|i_+\rangle$ is the sum of incident and reflected components in the left lead, and there is a transmitted component in the right lead (Fig. 8.12), the situation for $|j_-\rangle$ being symmetric. When the system is biased with $\mu_{\rm L} > \varepsilon > \mu_{\rm R}$, the state $|i_+\rangle$ is occupied by electrons whereas $|j_-\rangle$ is empty (at $T \to 0{\rm K}$) and therefore only $|i_+\rangle$ contributes to the electron density $n(\boldsymbol{r})$. Thus, in the right lead, near the contact, if the incident wave contributes to $n(\boldsymbol{r})$ with some weight set arbitrarily to 1, then the reflected one contributes with a

256   8 Transport

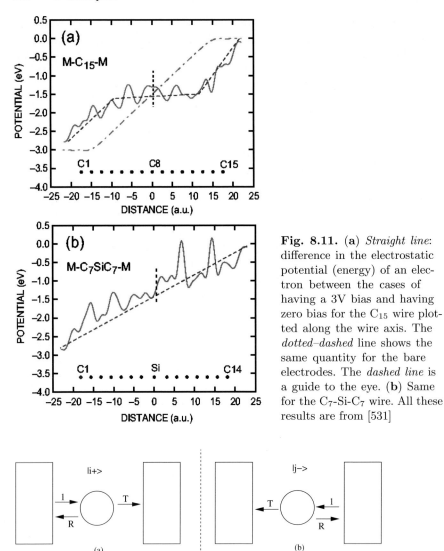

Fig. 8.11. (a) *Straight line*: difference in the electrostatic potential (energy) of an electron between the cases of having a 3V bias and having zero bias for the C_{15} wire plotted along the wire axis. The *dotted–dashed* line shows the same quantity for the bare electrodes. The *dashed line* is a guide to the eye. (b) Same for the C_7-Si-C_7 wire. All these results are from [531]

Fig. 8.12. (a) The states coming from the left side are partially reflected back with a reflection coefficient R and partially transmitted to the right with a transmission coefficient T ($R + T = 1$). (b) Same for the states coming from the right side

weight R, where R is the reflection coefficient (Fig. 8.12). The situation is different deep inside the left contact because the electronic waves are scattered by phonons, defects or impurities, and both states traveling to the right or to the left contribute statistically with a weight 1, like in the bulk material characterized by a chemical potential μ_L. Thus, the electron density is not the same deep inside the lead and close to the contact. As a consequence, the self-consistency induces a potential drop inside the leads, at the origin of the so-called contact resistance. If R is close to unity, we recover near the contact a situation close to the bulk one, and the voltage drop is small.

For all these reasons, the potential at each end of the island cannot be imposed *a priori*. Thus, in microscopic calculations, since it is not possible to describe the entire leads (including scattering events), it was proposed to impose zero-field conditions at the boundaries of the central region and to let the potential float to whatever it chooses to [499].

Another important factor which determines the potential profile is the electrostatic screening inside the wire. For example, we show in Fig. 8.13 the potential distribution in a Langmuir–Blodgett mono-layer of γ–hexadecylquinolinium tricyanoquinodimethanide ($C_{16}H_{33}Q$–3CNQ) sandwiched between Al electrodes, as calculated in [533]. Q–3CNQ is characterized by a large internal dipole because it is a D–π–A molecule, where D and A are, respectively, an electron donor and an electron donor, and π is a pi-bridge. The dipole layer gives rise to a built-in potential (Fig. 8.13). In the presence of the electrodes and at zero bias, opposite charges appear on the electrodes, so that the induced potential drop exactly cancels the one created by the dipole layer. When a bias is applied to the system, a large part of the potential drop takes place in the alkyl chains ($C_{16}H_{33}$), which have a small polarizability due to their large gap. This asymmetry in the potential profile explains the current–voltage characteristics of these layers which show rectifying behavior [534, 535].

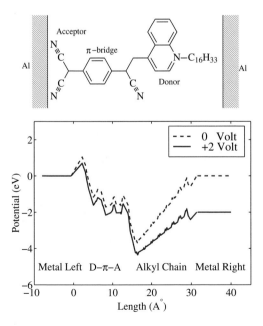

Fig. 8.13. Electrostatic potential in the metal | $C_{16}H_{33}Q$–3CNQ film | metal system at zero bias (*dashed line*) and at +2V (*straight line*) [533]. The potential is defined as an average value in a lattice unit cell

8.4.3 The Coulomb Blockade Effect

The application of a mean-field theory to the transport through nanostructures is better justified when the coupling is strong ($\Gamma^L, \Gamma^R \gg U$). However, as already mentioned, many experiments actually belong to an intermediate regime where the coupling coefficients and the charging energy have similar magnitudes. These situations are usually investigated using model Hamiltonians [500, 536, 537]. Therefore our aim in this section is to point out the main deficiencies of computational methods of the electronic structure to describe the Coulomb blockade effect which is the more obvious consequence of electron–electron interactions in nanostructures. We shall consider the limit $\Gamma^L, \Gamma^R \ll U$ in order to compare with the predictions of the orthodox theory. It will give us the opportunity to judge the mean-field approaches in the worst situation.

We consider the simplest model of a nano-device characterized by a single level twofold degenerate due to the spin. We assume that this level is empty in the neutral state and that the total energy of the system can be written as

$$E(n) = n\varepsilon_0 + \frac{1}{2}Un^2 , \tag{8.72}$$

where n is the number of electron in the island ($= 0, 1, 2$) and U is the Coulomb charging energy which takes into account the dielectric environment of the nanostructure. We suppose for simplicity that the triplet and singlet states for $n = 2$ have the same energy. We also neglect the dependence of the energy levels with the applied bias. Thus, following (8.13), we define two ionization energies

$$\varepsilon(0|-1) = \varepsilon_0 + U/2 ,$$
$$\varepsilon(-1|-2) = \varepsilon_0 + 3U/2 , \tag{8.73}$$

and the current is determined by the position of the chemical potentials with respect to these levels.

Orthodox Model. We consider the situation of Fig. 8.14b where a bias voltage shifts the chemical potential μ_L of the left electrode. When $\mu_L > \varepsilon(0|-1) > \mu_R$, the current flows through the nanostructure which is in the charge state 0 or -1 with the respective probabilities σ_0 and σ_1. Using the master equations given in Sect. 8.2.2, we obtain

$$\sigma_0 = \frac{\Gamma^R}{\Gamma^R + 2\Gamma^L} , \quad \sigma_1 = \frac{2\Gamma^L}{\Gamma^R + 2\Gamma^L} , \tag{8.74}$$

where for simplicity we assume that the coupling coefficients Γ^R and Γ^L are independent of the energy. In these conditions, the current is given by

$$I = \frac{-e}{\hbar} \frac{2\Gamma^L \Gamma^R}{\Gamma^R + 2\Gamma^L} . \tag{8.75}$$

8.4 Electron–Electron Interactions Beyond the Orthodox Theory

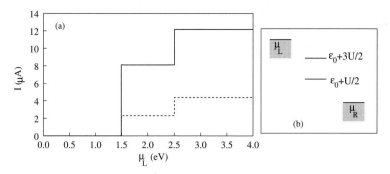

Fig. 8.14. (a) Current through the nano-device as a function of the Fermi level μ_L in the orthodox theory ($\varepsilon_0 = 1$ eV, $\mu_R = 0$ eV, $U = 1$ eV): $\Gamma^L = \Gamma^R = 50$ meV (*straight line*) and $\Gamma^L = 90$ meV, $\Gamma^R = 10$ meV (*dashed line*). (b) Schematic representation of the energy levels

When $\mu_L > \varepsilon(-1| - 2)$ and $\varepsilon(0| - 1) > \mu_R$, the three charge states 0, -1 and -2 are possible with the respective probabilities

$$\sigma_0 = \left(\frac{\Gamma^R}{\Gamma^R + \Gamma^L}\right)^2 , \quad \sigma_1 = \frac{2\Gamma^R \Gamma^L}{(\Gamma^R + \Gamma^L)^2} , \quad \sigma_2 = \left(\frac{\Gamma^L}{\Gamma^R + \Gamma^L}\right)^2 , \quad (8.76)$$

and the current becomes

$$I = \frac{-e}{\hbar} \frac{2\Gamma^L \Gamma^R}{\Gamma^R + \Gamma^L} . \quad (8.77)$$

The current–voltage characteristic in the limit $T \to 0$K is shown in Fig. 8.14 for two sets of values for Γ^R and Γ^L. It is a staircase function due to the use of a perturbation theory. The true characteristic should be more like the function (8.62) with a broadening of the steps of the order of Γ^R and Γ^L (neglecting Kondo effect).

Self-Consistent Approach. If now the same problem is investigated using a self-consistent approach, for example using the density functional theory in the local density approximation (LDA), the potential in the nano-device is a function of the mean occupation $\bar{n}$ of the energy level. From (8.62), the current is given by

$$I = \frac{4e}{h} \frac{\Gamma^L \Gamma^R}{\Gamma^L + \Gamma^R} \left[\arctan\left(\frac{\mu_R - \varepsilon_s}{\Gamma/2}\right) - \arctan\left(\frac{\mu_L - \varepsilon_s}{\Gamma/2}\right)\right] , \quad (8.78)$$

where ε_s is the self-consistent energy which can be approximated by

$$\varepsilon_s \approx \varepsilon_0' + U'\bar{n} , \quad (8.79)$$

where ε_0' is the energy of the empty state (the Kohn–Sham eigenvalue in LDA) and U' is an effective charging energy. Taking into account the spin degeneracy, $\bar{n}$ is equal to $2n_0$ where n_0 is given in (8.71). We have argued in Sects. 4.4 and 4.6 that the charging energy U' obtained in LDA (or in Hartree–

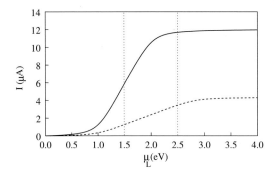

Fig. 8.15. Current through the nano-device as a function of the Fermi level μ_L. It is calculated using a self-consistent approach with $\varepsilon_0 = 1$ eV, $\mu_R = 0$ eV, $U = 1$ eV: $\Gamma^L = \Gamma^R = 50$ meV (*straight line*) and $\Gamma^L = 90$ meV, $\Gamma^R = 10$ meV (*dashed line*). The position of the steps predicted in the orthodox theory are indicated (*dotted line*)

like approaches) must be close to the true value U in semiconductor quantum dots. Thus we set $U' = U$ and, accordingly, we assume for the moment that ε'_0 is equal to ε_0. The current–voltage characteristic obtained in these conditions is shown in Fig. 8.15. Compared to Fig. 8.14, the current does not present the two sharp steps but a very broad transition. The threshold of the current is largely below the first step corresponding to $\varepsilon(0|-1)$. The broadening is due to the fact the occupancy $\bar{n}$ of the level varies continuously from ≈ 0 to $\approx 2\Gamma^L/(\Gamma^L + \Gamma^R)$ whereas in the true system the island can be occupied only by an integer number of electrons ($n = 0$, 1 or 2). When $\Gamma^L = \Gamma^R$, the current saturates well before the second step predicted in the orthodox theory because the maximum value of $\bar{n}$ is 1 whereas the ionization level $\varepsilon(-1|-2)$ corresponds to an occupancy of 1.5 electrons (corresponding to the Slater's transition state [236]).

Another difficulty with calculations based on the LDA is that the gap is underestimated. Consequently, the assumption $\varepsilon'_0 \approx \varepsilon_0$ is not justified because ε'_0 is probably well below ε_0. As discussed in Chaps. 1 and 4, the issue is that the exchange–correlation potential varies when one electron is added to the nanostructure, and this change is not described in LDA. Therefore there are good reasons to believe that the threshold of the current predicted in LDA is incorrect, at least in the limit of weak coupling.

GW Approach. We have seen in Sects. 1.2.4 and 4.4.2 that the GW approximation provides a prescription to calculate the quasi-particle spectrum starting from the one-particle LDA spectrum. Applying GW self-energy corrections in our model system, the lowest unoccupied level shifts from the Kohn–Sham eigenvalue ε'_0 to a quasi-particle energy ε^{qp} corresponding to the lowest energy at which an electron can be injected into the nano-device.

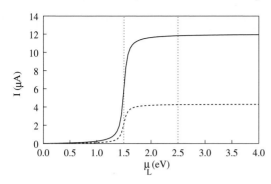

Fig. 8.16. Same as Fig. 8.15 but the current is calculated using a non self-consistent approach with an energy level $\varepsilon_0 + U/2$

Thus we can write

$$\varepsilon^{\text{qp}} \approx \varepsilon_0 + \frac{U}{2} = \varepsilon(0| - 1) , \qquad (8.80)$$

as shown by GW calculations in the case of semiconductor nanocrystals (Chap. 4).

We deduce that the GW approximation is probably a good approach to predict the threshold of the current as proposed in [538]. However, GW has not been adapted to non-equilibrium problems. One possible way to circumvent this difficulty is to calculate the self-energy corrections for the system at equilibrium and to assume that these corrections remain constant when the system is biased [533]. Doing this, one must take care that the current must be calculated using the non self-consistent Green's functions, i.e. without calculating the change in the Hartree and exchange–correlation potentials, because the charging energy is already included in the self-energy corrections.

According to this procedure, we present in Fig. 8.16 the current obtained when ε_s is replaced by ε^{qp} in (8.78). Comparing with Fig. 8.14, the threshold of the current is obviously improved, but there is only one step since the lowest state of energy ε^{qp} is twofold degenerate. One consequence of this discrepancy is that the amplitude of the current is overestimated just above the threshold.

Unrestricted Approach. One possible way to improve the results is to use an unrestricted approach [539], which means that the energies of the spin up and spin down electronic levels are allowed to be different. The motivation is that unrestricted Hartree–Fock calculations provide a good description of two-electron systems in the strongly correlated limit [77]. In this approximation, the current is given by

$$I = \frac{2e}{h} \frac{\Gamma^{\text{L}} \Gamma^{\text{R}}}{\Gamma^{\text{L}} + \Gamma^{\text{R}}} \left[\arctan\left(\frac{\mu_{\text{R}} - \varepsilon_\uparrow}{\Gamma/2}\right) - \arctan\left(\frac{\mu_{\text{L}} - \varepsilon_\uparrow}{\Gamma/2}\right) \right.$$
$$\left. + \arctan\left(\frac{\mu_{\text{R}} - \varepsilon_\downarrow}{\Gamma/2}\right) - \arctan\left(\frac{\mu_{\text{L}} - \varepsilon_\downarrow}{\Gamma/2}\right) \right] , \qquad (8.81)$$

where $\varepsilon_\uparrow$ and $\varepsilon_\downarrow$ are the self-consistent energies of the spin up and spin down states, respectively. Let us assume that these energies can take the following form

$$\varepsilon_\uparrow = \varepsilon_0 + \frac{U}{2} + U\bar{n}_\downarrow,$$
$$\varepsilon_\downarrow = \varepsilon_0 + \frac{U}{2} + U\bar{n}_\uparrow, \quad (8.82)$$

where $\bar{n}_\uparrow$ and $\bar{n}_\downarrow$ are the mean occupancies of spin up and spin down levels, respectively. $\bar{n}_\uparrow$ and $\bar{n}_\downarrow$ are given by (8.71) in which ε_0 is replaced by $\varepsilon_\uparrow$ and $\varepsilon_\downarrow$, respectively. In (8.82), the self-interaction terms are removed (Sect. 1.1.3). The equations (8.82) and (8.71) are solved self-consistently, and the result is shown in Fig. 8.17. When μ_L increases, one of the spin levels has a lower energy and is progressively occupied while the other has a higher energy and is empty [539]. The higher spin level starts to be occupied only when the lowest one is filled up to its maximum ($\approx \Gamma^L/(\Gamma^L + \Gamma^R)$). In consequence, the current–voltage characteristic presents two steps as it must be. However, in contrast to Fig. 8.14, the energy splitting between the two steps depends on the coupling coefficients Γ^L and Γ^R. In spite of this discrepancy, Fig. 8.17 shows that unrestricted calculations could lead to substantial improvements.

Finally, the expressions (8.82) can be obtained as follows. First, the electronic structure of the system at equilibrium is calculated in LDA. Second, GW self-energy corrections are applied, shifting the lowest unoccupied state to $\approx \varepsilon_0 + U/2$. Third, the potential variations are calculated self-consistently following an unrestricted prescription and applying self-interaction corrections.

It is clear that this approach is quite empirical and thus there is a need for more elaborate schemes. A simpler approach could be to perform self-consistent unrestricted calculations in the local spin density approximation, but the threshold of the current is likely to be underestimated like in LDA.

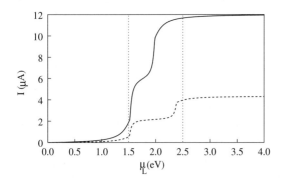

Fig. 8.17. Same as Fig. 8.15 but the current is calculated using a self-consistent and unrestricted approach

8.5 Transport in Networks of Nanostructures

In this section, we deal with the problem of electron (or hole) transport in networks of nanostructures weakly coupled by tunnel junctions. First, we consider the tunneling between two neighboring sites and, second, we present a method to calculate the conductivity of the network.

8.5.1 Tunneling Between Nanostructures

In the case of metallic islands in which the density of states is high, Coulomb blockade effects dominate the transport properties of the networks [481, 500, 540]. In the case of semiconductor nanocrystals, the conductivity is also determined by the discretization of the energy levels induced by the quantum confinement (Chap. 2). In both cases, the disorder arising for example from the dispersion in size and shape of the nanostructures plays an essential role. Therefore, elastic tunneling between neighbor nanostructures is rather unlikely (Fig. 8.18) and one must consider inelastic tunneling between non resonant states, which requires to take into account the electron–phonon coupling. We assume that an injected electron only couples to phonons localized in the nanostructure where it resides, which is a reasonable assumption for weakly coupled nanostructures. Atomic vibrations in the barrier may lead to a modulation of the barrier height [541, 542] but we do not consider this effect here. When an electron is transferred from one site to another, there is an emission or an absorption of phonons as required by the conservation of the total energy.

The coupling mechanism has been basically described in Sect. 5.2. When an extra carrier is introduced into a nanostructure, there is a relaxation of the atoms toward a new equilibrium situation. The relaxation energy is defined as the Franck–Condon shift (e.g. $d_{\rm FC}^{(1)}$ for site 1). In order to simplify the

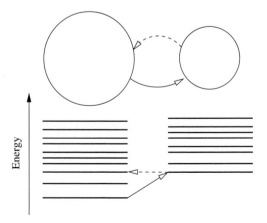

Fig. 8.18. Tunneling of an electron between two nanostructures. The arrows indicate the most efficient tunnel processes

problem, we assume as usual that the total energy is a quadratic function of $3N_1 + 3N_2$ configuration coordinates where N_1 and N_2 are the numbers of atoms in the nanostructures 1 and 2, respectively (see Sect. 5.2.3). The probability per unit time for the transfer of an electron from a site 1 to a site 2 is obtained from the Fermi golden rule

$$W_{1\to 2} =$$
$$\frac{2\pi}{\hbar} \sum_{i,n_1,n_2} p(i,n_1,n_2) \left[\sum_{f,n'_1,n'_2} |\langle f,n'_1,n'_2|V|i,n_1,n_2\rangle|^2 \delta(E_{f,n'_1,n'_2} - E_{i,n_1,n_2}) \right], \quad (8.83)$$

where $|i,n_1,n_2\rangle$ and $|f,n'_1,n'_2\rangle$ denote the initial and final states of energy E_{i,n_1,n_2} and E_{f,n'_1,n'_2}, respectively, and $p(i,n_1,n_2)$ is the probability to find the system in the state $|i,n_1,n_2\rangle$. The integers n_1, n_2, n'_1, n'_2 label the vibronic configurations on each site. Since we assume that the vibrations of the two sites are uncoupled, we write

$$|i,n_1,n_2\rangle = |\phi_i^{(1)}\rangle \, |\chi_{n_1}(Q_1 - Q_1^0)\rangle \, |\chi_{n_2}(Q_2)\rangle \,,$$
$$|f,n'_1,n'_2\rangle = |\phi_f^{(2)}\rangle \, |\chi_{n'_1}(Q_1)\rangle \, |\chi_{n'_2}(Q_2 - Q_2^0)\rangle \,, \quad (8.84)$$

where Q_1^0 denotes the equilibrium configuration of the site 1 with one extra electron, $Q_1 = 0$ being the equilibrium configuration for the neutral nanostructure. $|\phi_i^{(1)}\rangle$ and $|\phi_f^{(2)}\rangle$ are the electronic states. The vibronic states χ_{n_1} and χ_{n_2} are given by the product of $3N_1$ and $3N_2$ harmonic oscillators like in (5.62). We suppose that the matrix element of the tunneling operator V between the electronic states does not depend on the phonon quantum numbers. Thus we can factorize in (8.83) the terms $\langle \chi_{n'_1}(Q_1)|\chi_{n_1}(Q_1 - Q_1^0)\rangle$ and $\langle \chi_{n'_2}(Q_2 - Q_2^0)|\chi_{n_2}(Q_2)\rangle$ which are given by products of overlaps between displaced harmonic oscillators. Since the coupling to anyone mode is of the order of $1/N_1$ or $1/N_2$, we have seen in Sect. 5.2.4 that one can keep only first-order terms corresponding to change in phonon quantum numbers equal to 0, +1 and -1, all the other terms being of higher order. In that case, the probability of the transition can be obtained exactly if all the electron–phonon coupling coefficients are calculated.

Simpler expressions can be obtained when all phonons frequencies in a nanostructure can be approximated by a single one (here ω_1 and ω_2). Then one can sum the intensity of all possible transitions corresponding to the same difference in total energies between the final and initial states, for example differing by p_1 phonons of energy $\hbar\omega_1$ and p_2 phonons of energy $\hbar\omega_2$. Then we have

$$E_{f,n'_1,n'_2} - E_{i,n_1,n_2} = \varepsilon_f - d_{FC}^{(2)} - \varepsilon_i + d_{FC}^{(1)} + p_1\hbar\omega_1 + p_2\hbar\omega_2 \,, \quad (8.85)$$

where the Franck–Condon shifts correspond to the relaxation energy when the nanostructure is occupied by one electron. The total intensity of these transitions is equal to $W_{p_1} \times W_{p_2}$ where W_p is given by (5.72) in Sect. 5.2.4. Injecting this in (8.83), we obtain

$$W_{1\to 2} = \frac{2\pi}{\hbar} \sum_{i,f} p(i)|\langle \phi_f^{(2)}|V|\phi_i^{(1)}\rangle|^2$$
$$\times \left[\sum_{p_1,p_2} W_{p_1} W_{p_2} \delta(\varepsilon_f - d_{FC}^{(2)} - \varepsilon_i + d_{FC}^{(1)} + p_1\hbar\omega_1 + p_2\hbar\omega_2) \right], \quad (8.86)$$

which can be written as a convolution of two phonon line-shapes (see (5.79)).

In the case of strong electron–phonon coupling ($S_1 = d_{FC}^{(1)}/\hbar\omega_1 \gg 1$ and $S_2 = d_{FC}^{(2)}/\hbar\omega_2 \gg 1$), we can use (5.74) to derive a simpler expression. Considering p_1 and p_2 as continuous variables (i.e. $\sum_{p_1,p_2} \to \int dp_1 dp_2$), the probability per unit time becomes

$$W_{1\to 2} = \frac{2\pi}{\hbar} \sum_{i,f} p(i)|\langle \phi_f^{(2)}|V|\phi_i^{(1)}\rangle|^2 \frac{1}{\pi\mu} \exp\left[-\frac{(\Delta - d_{FC}^{(1)} - d_{FC}^{(2)})^2}{\mu^2}\right], (8.87)$$

where

$$\mu = \sqrt{2S_1(\hbar\omega_1)^2 \coth\left(\frac{\hbar\omega_1}{2kT}\right) + 2S_2(\hbar\omega_2)^2 \coth\left(\frac{\hbar\omega_2}{2kT}\right)}, \quad (8.88)$$

and $\Delta = \varepsilon_f - d_{FC}^{(2)} - \varepsilon_i + d_{FC}^{(1)}$ is the energy required to transfer the electron from the site 1 to the site 2. An example of variation of $W_{1\to 2}$ with respect to Δ is presented in Fig. 8.19.

As a final remark, we point out that in the limit of strong electron–phonon coupling, it is not necessary to assume a single phonon frequency in each nanostructure. It can be shown, using the method of moments [288, 289], that the expression (8.87) can be recovered in the general case using the first and second moments of the phonon line-shape, at the condition to write

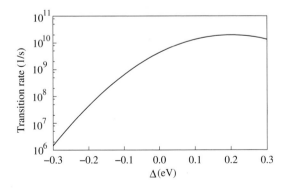

Fig. 8.19. Probability per unit time for the transition between two nanostructures as a function of the energy Δ ($S_1 = S_2 = 2$, $\hbar\omega_1 = \hbar\omega_2 = 50$ meV, $\langle \phi_f^{(2)}|V|\phi_i^{(1)}\rangle = 1$ meV, $T = 300$K)

$$d_{\text{FC}}^{(1)} + d_{\text{FC}}^{(2)} = \sum_i S_i^{(1)} \hbar \omega_i^{(1)} + \sum_j S_j^{(2)} \hbar \omega_j^{(2)} \,, \tag{8.89}$$

where the sums run over all the phonon modes of energy $\hbar\omega_i^{(1)}$ and $\hbar\omega_j^{(2)}$ in nanostructures 1 and 2, respectively, and at the condition to replace (8.88) by

$$\mu = \sqrt{M^{(1)} + M^{(2)}} \,, \tag{8.90}$$

where

$$M^{(1)} = (d_{\text{FC}}^{(1)})^2 + \sum_i (2\bar{n}_i^{(1)} + 1) S_i^1 (\hbar\omega_i^{(1)})^2 \tag{8.91}$$

with $\bar{n}_i^{(1)}$ given by (5.69) for the phonon frequency $\omega_i^{(1)}$ (a similar expression holds for $M^{(2)}$).

8.5.2 Hopping Conductivity

We consider now a network of nanostructures. Our aim is to present a computational method to calculate the conductivity. We consider systems where the degree of randomness is sufficiently large that the transport of carriers takes place by hopping between neighboring nanostructures. The disorder may arise under different forms such as the topological or cellular disorder [543, 544]. It was shown that a fixed array of sites can serve as a useful model for topologically disordered systems [543, 544]. Therefore, in the following, we assume a fixed array of sites in which the activated hopping between neighboring sites i and j is defined by the probability per unit time $W_{i \to j}$ that was calculated in the previous section and from which we want to determine the dynamic conductivity $\sigma(\omega)$.

The combination of the disorder and of a particular topology of the array may dramatically influence the electrical transport, in particular when the system is close to the percolation limit. Due to these constraints, the diffusion of the electron is anomalous at the microscopic scale [545], in the sense that the diffusion coefficient D depends on time (note that the same theory obviously applies to the holes). However, at the mesoscopic scale, when the mean displacement of the electron becomes larger than the correlation length which characterizes the system, a constant diffusion coefficient can be defined, and the diffusion becomes normal. These problems have been extensively discussed in [148, 543, 544, 546–550]. Here we give a simplified presentation of this theory.

Diffusion Coefficient. The diffusion of electrons of density $n(\boldsymbol{r}, t)$ at a position $\boldsymbol{r}$ and at time t is given by the Fick law and the charge conservation equation

$$\frac{\partial n(\boldsymbol{r},t)}{\partial t} = \frac{1}{e} \nabla \cdot \boldsymbol{J} \,,$$
$$\boldsymbol{J} = eD \, \nabla n(\boldsymbol{r},t) \,, \tag{8.92}$$

8.5 Transport in Networks of Nanostructures

where $\boldsymbol{J}$ is the current density. To solve these equations, we define the following Laplace transform of a function $f(t)$ as

$$F(\omega) = \mathcal{L}\left[f(t)\right] = \int_0^\infty e^{-(\alpha+i\omega)t} f(t) \mathrm{d}t \,, \quad \alpha \to 0^+. \tag{8.93}$$

Applying this to (8.92) in the case of a 1D system, we obtain

$$i\omega N(x,\omega) = D \frac{\partial^2 N(x,\omega)}{\partial x^2} \,, \tag{8.94}$$

which leads to

$$N(x,\omega) = \frac{1+i}{2} \sqrt{\frac{\omega}{2D}} \frac{n_0}{i\omega} \exp\left(-\sqrt{\frac{\omega}{2D}} |x| (1+i)\right) . \tag{8.95}$$

The coefficient in front of the exponential has been obtained by the Laplace transform of the normalization condition for n_0 electrons in the system

$$\int_{-\infty}^{+\infty} n(x,t) \mathrm{d}x = n_0 \Rightarrow \int_{-\infty}^{+\infty} N(x,\omega) \mathrm{d}x = \frac{n_0}{i\omega} \,. \tag{8.96}$$

In order to characterize the diffusion of the electrons, we calculate the mean square displacement of the electrons $\overline{x^2(t)}$ at time t defined by

$$\overline{x^2(t)} = \frac{1}{n_0} \int_{-\infty}^{+\infty} x^2 n(x,t) \mathrm{d}x \,. \tag{8.97}$$

Using (8.95), we obtain that the Laplace transform of $\overline{x^2(t)}$ is related to the generalized diffusion coefficient $D(\omega)$ by

$$\mathcal{L}\left[\overline{x^2(t)}\right] = \frac{1}{n_0} \int_{-\infty}^{+\infty} x^2 N(x,\omega) \mathrm{d}x = \frac{2D(\omega)}{(i\omega)^2} \,, \tag{8.98}$$

which can be generalized to a system of dimension d ($=1,2,3$) as

$$D(\omega) = -\frac{\omega^2}{2d} \int_0^{+\infty} e^{-i\omega t} \overline{r^2(t)} \mathrm{d}t \,. \tag{8.99}$$

In the case of a normal diffusion, D is a constant and we recover the linear dependence of $\overline{r^2(t)}$ with time t

$$\overline{r^2(t)} = 2dDt \,. \tag{8.100}$$

The expression (8.99) is particularly interesting in the case of the hopping transport on an array of localized sites defined by vectors $\boldsymbol{s}$. Let us define the probability $p(\boldsymbol{s},t|\boldsymbol{s}_0)$ to find an electron on the site $\boldsymbol{s}$ at time t whereas it was on the site $\boldsymbol{s}_0$ at time $t=0$. Then we can write

$$\overline{r^2(t)} = \sum_{\boldsymbol{s}} (\boldsymbol{s}-\boldsymbol{s}_0)^2 \langle p(\boldsymbol{s},t|\boldsymbol{s}_0)\rangle \,, \tag{8.101}$$

where the brackets denote the average on the sites s_0. Note that this average is required when the system is disordered. From (8.99), we deduce the diffusion coefficient

$$D(\omega) = -\frac{\omega^2}{2d} \sum_{s} (s - s_0)^2 \langle P(s,\omega|s_0) \rangle \ . \tag{8.102}$$

The conductivity $\sigma(\omega)$ is related to the diffusion coefficient by the Einstein relation

$$\sigma(\omega) = \frac{ne^2}{kT} D(\omega) \ , \tag{8.103}$$

where n is the carrier density. We deduce that

$$\sigma(\omega) = -\frac{ne^2}{kT}\frac{\omega^2}{2d} \sum_{s} (s - s_0)^2 \langle P(s,\omega|s_0) \rangle \ . \tag{8.104}$$

One can wonder if the Einstein relation still holds for hopping transport. Actually, (8.104) can be obtained from the Kubo formula [148, 543, 546].

8.5.3 Coherent Potential Approximation

It remains to calculate the terms $\langle P(s,\omega|s_0) \rangle$. The probability $p(s,t|s_0)$ is solution of a master equation

$$\frac{\partial p(s,t|s_0)}{\partial t} = -\left[\sum_{s' \neq s} W_{s \to s'}\right] p(s,t|s_0) + \sum_{s' \neq s} W_{s' \to s}\, p(s',t|s_0) \ , \tag{8.105}$$

where the term with a minus sign describes the outgoing flux of particle from the site s to neighbor sites s', the last term corresponds to the ingoing flux, and $W_{s \to s'}$ is the transfer rate from s to s' which can be calculated as shown in Sect. 8.5.1. In the following, we will neglect all $W_{s \to s'}$ beyond first nearest neighbors. Thus we will not take into account the variable range hopping [551, 552] which may be important at low temperature.

Approximations are required to solve (8.105). The simplest one is to model the real system by an effective one where the transfer rate on each bond is replaced by its average value ($W_m = <W_{s \to s'}>$). In the case of a 1D lattice of parameter a, (8.105) becomes

$$\frac{\partial p(x,t|x_0)}{\partial t} = -W_m\left[p(x+a,t|x_0) + p(x-a,t|x_0) - 2p(x,t|x_0)\right] \ , \tag{8.106}$$

which after Laplace transform gives

$$i\omega P(x,\omega|x_0) - p(x,t=0|x_0)$$
$$= W_m\left[P(x+a,\omega|x_0) + P(x-a,\omega|x_0) - 2P(x,\omega|x_0)\right] \ . \tag{8.107}$$

Multiplying each term by $(na)^2$, and summing over n from $-\infty$ to $+\infty$, we obtain

$$i\omega \sum_{n=-\infty}^{+\infty} (na)^2 P(na+x_0,\omega|x_0) = \frac{2aW_{\mathrm{m}}}{i\omega}. \tag{8.108}$$

Injecting this result into (8.102), the generalized diffusivity simply becomes

$$D(\omega) = a^2 W_{\mathrm{m}}, \tag{8.109}$$

which can be easily generalized to 2D and 3D lattices. Therefore this model predicts a conductivity which is independent of the frequency. It does not describe correctly the anomalous diffusion, for example in percolating systems.

A better description is given by the Coherent Potential Approximation (CPA) which has been extensively used to study the electronic structure of disordered solids [148] and has been adapted to the problem of transport in percolating lattices [548, 549]. The idea is to build an effective medium where the fluctuating hopping rates $W_{s \to s'}$ are replaced by a uniform frequency-dependent rate $W_{\mathrm{c}}(\omega)$. At each frequency, $W_{\mathrm{c}}(\omega)$ is calculated in a self-consistent way from the statistical distribution of the $W_{s \to s'}$. To do that, we take the Laplace transform of (8.105)

$$\left(i\omega + \sum_{s' \neq s} W_{s \to s'}\right) P(s,\omega|s_0) - \sum_{s' \neq s} W_{s' \to s} P(s',\omega|s_0) = \delta(s,s_0), \tag{8.110}$$

which can be rewritten in matrix form as

$$(i\omega[I] - [H])[P(\omega)] = [I], \tag{8.111}$$

where the diagonal terms of $[H]$ are equal to $-\sum_{s' \neq s} W_{s \to s'}$ and the non-diagonal ones are equal to $W_{s' \to s}$. We recognize in (8.111) the equation defining a Green's function $P(\omega)$ for an Hamiltonian H (see Sect. 8.3.1) and the solutions of (8.111) are the matrix elements of the Green's function

$$P(s,\omega|s_0) = \langle s|(i\omega[I]-[H])^{-1}|s_0\rangle. \tag{8.112}$$

Similarly, the effective medium is defined by an Hamiltonian H_{c} corresponding to a uniform rate $W_{\mathrm{c}}(\omega)$ which, so far, has not yet been specified. To determine this value, in CPA, we pick a particular pair (1,2) of neighbor sites in the effective medium, we replace $W_{\mathrm{c}}(\omega)$ on this bond by $W_{1 \to 2}$ and we require that this operation must have on the average no effect on the effective medium (Fig. 8.20). Let us define the Hamiltonian $H_{\mathrm{c}} + V$ of the effective medium in which $W_{\mathrm{c}}(\omega)$ is replaced by $W_{1 \to 2}$ on the bond (1,2). We impose the condition

$$\langle (i\omega[I]-[H_{\mathrm{c}}]-[V])^{-1} \rangle = (i\omega[I]-[H_{\mathrm{c}}])^{-1}, \tag{8.113}$$

where $\langle ... \rangle$ denotes the average over the distribution of $W_{1 \to 2}$. The Green's function of the effective medium can be easily obtained either analytically or

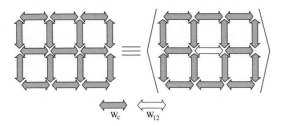

Fig. 8.20. Schematic representation of the principle of the Coherent Potential Approximation (adapted from [548])

numerically depending on the lattice and its dimensionality [548]. From this, the Green's function of the perturbed system (H_c+V) can be calculated using the Dyson's equation (8.31). Since V has only components on sites 1 and 2, one leads to a 2×2 matrix equation which must be solved self-consistently. After straightforward algebra [548], one obtains the following condition on $W_c(\omega)$

$$\left\langle \frac{W_c(\omega) - W_{12}}{1 - 2\left([P_c]_{11} - [P_c]_{12}\right)(W_c(\omega) - W_{12})} \right\rangle = 0, \qquad (8.114)$$

where $[P_c]_{11}$ and $[P_c]_{12}$ are matrix elements of the Green's function of the effective medium. Then, using (8.103) and (8.109), we obtain

$$\sigma(\omega) = \frac{ne^2}{kT} \, a^2 W_c(\omega). \qquad (8.115)$$

8.5.4 Example of a Network of Silicon Nanocrystals

As a representative example we discuss the case of a cubic array of silicon crystallites in which a fraction x of the bonds are broken. This system was studied in order to simulate the hopping conductivity in porous silicon [553, 554]. It is composed of spherical crystallites connected by cylindrical silicon bridges narrow enough for the coupling between the spheres to be small. Nanocrystals with a Gaussian distribution of band-gap energies centered around 2 eV and full width at half maximum of 0.1 eV have been considered. The diameter of the bridges is taken equal to half the diameter d of the smaller sphere and their length follows a uniform distribution comprised between 0.3 nm and $0.39d$ (beyond which the hopping rate becomes negligible). A carrier density of 0.3×10^{16} cm^{-3} is considered, which is typical of porous silicon. The hopping probabilities are calculated using a full tight binding description of the electronic structure and taking into account the electron–phonon coupling in the crystallites according to the theory described in Sect. 8.5.1 [554].

Figure 8.21 presents the temperature dependence of the dc conductivity for various values of x. It exhibits a strongly activated behavior, the activation energy increasing with the number of broken bonds up to the percolation threshold (predicted to be 2/3 in CPA for the bond percolation model). This comes from the fact that, at small x, the carriers easily find a path with

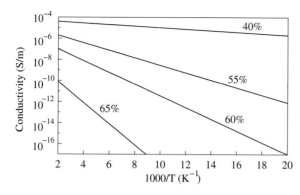

Fig. 8.21. Temperature dependence of the dc conductivity in a cubic array of Si crystallites for various fractions of broken bonds

small thermal barriers. On the contrary, when x increases there are less possible trajectories and the carrier must experience more and more jumps with weak probabilities, especially for large to small crystallites where the thermal barrier is large.

Note that hopping transport is also observed in networks of metal nanocrystals [481] and in complex molecules, as it seems to dominate the long distance ($>$ 20 Å) transport in DNA [555]. Also, recent works have concerned the transport properties of nanocrystal solids produced by colloidal chemistry [480, 556]. Long-range Coulomb interactions seem to play an important role in the case of undoped nanocrystals [556]. These effects are not described in the theory developed in this section: they require a more complex treatment of electron–electron interactions (e.g. see [557]).

A Matrix Elements of the Renormalizing Potential

To determine the excitonic states we need to calculate the matrix elements $\langle \psi_{vc}|V_{\text{eff}}|\psi_{v'c'}\rangle$ of the renormalizing potential within the subspace of all states corresponding to one electron–hole excitation. The matrix elements (1.102) then take the form

$$\langle \psi_{vc}|V_{\text{eff}}|\psi_{v'c'}\rangle = \sum_{\alpha,s} \frac{\langle \psi_{vc}|\sum_i V_s(r_i)|\psi_\alpha\rangle \langle \psi_\alpha|\sum_i V_s(r_i)|\psi_{v'c'}\rangle}{E - E_\alpha - \omega_s}. \quad (A.1)$$

$\sum_i V_s(r_i)$ is a sum of one-electron contributions. In a Slater determinant basis there will be only:

- diagonal contributions equal to $\sum_k n_k^i \langle k|V_s|k\rangle$ where the n_k^i are the occupation number for the determinant $|\psi_i\rangle$ under consideration,
- non diagonal elements equal to $\langle k|V_s|k'\rangle$ for determinants differing by one spin-orbital $u_k \neq u_{k'}$ when they are positioned at the same column in each determinant, all other columns being identical.

With this, starting from the left $|\psi_{vc}\rangle$ will be coupled only to the following excitations $|\psi_\alpha\rangle$ equal to

1. $|\psi_{vc}\rangle$,
2. $|\psi_{vc''}\rangle$ with $c'' \neq c$,
3. $|\psi_{v''c}\rangle$ with $v'' \neq v$,
4. $|0\rangle$ the ground state determinant,
5. $|\psi_{vc,v''c''}\rangle$ with $c'' \neq c$ and $v'' \neq v$.

At this stage it is important to make the following remark. One could think that including terms like 4 and 5 is not coherent since we are working within the manifold of the one electron–hole excitations. However as shown at the beginning the $|\psi_\alpha\rangle$ result from the compound states $|\psi_\alpha\rangle|s\rangle$ and it is necessary to include all of them to have a coherent representation of the renormalizing potential. We shall come back to this point later.

We can now calculate the contributions from the different $|\psi_\alpha\rangle$ 1 to 5 to the matrix element (A.1) and get:

1: $\dfrac{|\langle c|V_s|c\rangle - \langle v|V_s|v\rangle|^2}{E - E_{vc} - \omega_s}\delta_{cc'}\delta_{vv'}$

$+ (1-\delta_{cc'})\delta_{vv'}\dfrac{(\langle c|V_s|c\rangle - \langle v|V_s|v\rangle)\langle c|V_s|c'\rangle}{E - E_{vc} - \omega_s}$,

2: $\delta_{cc'}\delta_{vv'}\displaystyle\sum_{c''\neq c}\dfrac{|\langle c|V_s|c''\rangle|^2}{E - E_{vc''} - \omega_s} - (1-\delta_{cc'})(1-\delta_{vv'})\dfrac{\langle c|V_s|c'\rangle\langle v'|V_s|v\rangle}{E - E_{vc'} - \omega_s}$

$+ (1-\delta_{cc'})\delta_{vv'}\left[\dfrac{(\langle c'|V_s|c'\rangle - \langle v|V_s|v\rangle)\langle c|V_s|c'\rangle}{E - E_{vc'} - \omega_s}\right.$

$\left. + \displaystyle\sum_{c''\neq c,\neq c'}\dfrac{\langle c|V_s|c''\rangle\langle c''|V_s|c'\rangle}{E - E_{vc''} - \omega_s}\right]$,

3: $\delta_{cc'}\delta_{vv'}\displaystyle\sum_{v''\neq v}\dfrac{|\langle v|V_s|v''\rangle|^2}{E - E_{v''c} - \omega_s} - (1-\delta_{cc'})(1-\delta_{vv'})\dfrac{\langle v'|V_s|v\rangle\langle c|V_s|c'\rangle}{E - E_{v'c} - \omega_s}$

$+ (1-\delta_{vv'})\delta_{cc'}\left[-\dfrac{(\langle c|V_s|c\rangle - \langle v'|V_s|v'\rangle)\langle v'|V_s|v\rangle}{E - E_{v'c} - \omega_s}\right.$

$\left. + \displaystyle\sum_{v''\neq v,\neq v'}\dfrac{\langle v''|V_s|v\rangle\langle v'|V_s|v''\rangle}{E - E_{v''c} - \omega_s}\right]$,

4: $\dfrac{\langle c|V_s|v\rangle\langle v'|V_s|c'\rangle}{E - E_0 - \omega_s}$,

5: $\delta_{cc'}\delta_{vv'}\displaystyle\sum_{c''\neq c, v''\neq v}\dfrac{|\langle v''|V_s|c''\rangle|^2}{E - E_{vc,v''c''} - \omega_s}$

$- (1-\delta_{cc'})(1-\delta_{vv'})\dfrac{\langle v'|V_s|c'\rangle\langle c|V_s|v\rangle}{E - E_{vc,v'c'} - \omega_s}$

$- (1-\delta_{cc'})\delta_{vv'}\displaystyle\sum_{v''\neq v}\dfrac{\langle v''|V_s|c'\rangle\langle c|V_s|v''\rangle}{E - E_{vc,v''c'} - \omega_s}$

$- (1-\delta_{vv'})\delta_{cc'}\displaystyle\sum_{c''\neq c}\dfrac{\langle v'|V_s|c''\rangle\langle c''|V_s|v\rangle}{E - E_{vc,v'c''} - \omega_s}$. (A.2)

These expressions only include the contribution of one plasmon s and should be summed over s. It is also implicitely considered, as explained in Sect. 1.2.4 that

$$E_{vc} = E_0 + \varepsilon_c - \varepsilon_v ,$$
$$E_{vc,v'c'} = E_{vc} + \varepsilon_{c'} - \varepsilon_{v'} . \quad (A.3)$$

We now proceed to the calculation of the different coupling terms. We have, using (A.3) and regrouping terms

$c' = c\,, v' = v\,:$

$$\langle\psi_{vc}|V_{\text{eff}}|\psi_{vc}\rangle = \sum_{v'',c''} \frac{|\langle v''|V_s|c''\rangle|^2}{E - E_{vc} + \varepsilon_{v''} - \varepsilon_{c''} - \omega_s}$$

$$+ \sum_{c''} \frac{|\langle c|V_s|c''\rangle|^2}{E - E_{vc} + \varepsilon_c - \varepsilon_{c''} - \omega_s}$$

$$- \sum_{v''} \frac{|\langle c|V_s|v''\rangle|^2}{E - E_{vc} + \varepsilon_{v''} - \varepsilon_c - \omega_s}$$

$$- \sum_{c''} \frac{|\langle v|V_s|c''\rangle|^2}{E - E_{vc} + \varepsilon_v - \varepsilon_{c''} - \omega_s}$$

$$+ \sum_{v''} \frac{|\langle v|V_s|v''\rangle|^2}{E - E_{vc} + \varepsilon_{v''} - \varepsilon_v - \omega_s} + \frac{|\langle c|V_s|v\rangle|^2}{E - E_0 - \omega_s}$$

$$+ \frac{|\langle c|V_s|v\rangle|^2}{E - E_{vc} + \varepsilon_v - \varepsilon_c - \omega_s} - 2\frac{\langle c|V_s|c\rangle\langle v|V_s|v\rangle}{E - E_{vc} - \omega_s}\,. \quad (A.4)$$

$c' \neq c\,, v' = v\,:$

$$\langle\psi_{vc}|V_{\text{eff}}|\psi_{vc'}\rangle = \sum_{c''} \frac{\langle c|V_s|c''\rangle\langle c''|V_s|c'\rangle}{E - E_{vc'} + \varepsilon_{c'} - \varepsilon_{c''} - \omega_s}$$

$$- \sum_{v''} \frac{\langle c|V_s|v''\rangle\langle v''|V_s|c'\rangle}{E - E_{vc} + \varepsilon_{v''} - \varepsilon_{c'} - \omega_s}$$

$$+ \langle c|V_s|v\rangle\langle v|V_s|c'\rangle \left(\frac{1}{E - E_0 - \omega_s} + \frac{1}{E - E_{vc} + \varepsilon_v - \varepsilon_{c'} - \omega_s}\right)$$

$$- \langle c|V_s|c'\rangle\langle v|V_s|v\rangle \left(\frac{1}{E - E_{vc} - \omega_s} + \frac{1}{E - E_{vc'} - \omega_s}\right)\,. \quad (A.5)$$

$c' \neq c\,, v' \neq v\,:$

$$\langle\psi_{vc}|V_{\text{eff}}|\psi_{v'c'}\rangle$$

$$= \langle c|V_s|v\rangle\langle v'|V_s|c'\rangle \left(\frac{1}{E - E_0 - \omega_s} + \frac{1}{E - E_{vc} + \varepsilon_{v'} - \varepsilon_{c'} - \omega_s}\right)$$

$$- \langle c|V_s|c'\rangle\langle v'|V_s|v\rangle \left(\frac{1}{E - E_{vc'} - \omega_s} + \frac{1}{E - E_{v'c} - \omega_s}\right)\,. \quad (A.6)$$

Again these matrix elements have to be summed over s. The summations over v'' and c'' respectively correspond to occupied and unoccupied states of the N electron system. The interesting point is that the matrix elements contain terms which are completely similar to those obtained in GW. For instance the first term in $\langle\psi_{vc}|V_{\text{eff}}|\psi_{vc}\rangle$ is equal to $E_{\text{corr}}(N)$ at the condition of neglecting E_{vc} which corresponds to second order perturbation theory. The same is true for the second and third term which exactly give $\delta\varepsilon_c - \delta\varepsilon_v$ as defined by (1.109). If one looks now at $\langle\psi_{vc}|V_{\text{eff}}|\psi_{vc'}\rangle$ with $c' \neq c$ the first two terms give, at the condition that E_{vc} and $E_{vc'}$ are neglected, the correlation part $\langle c|\Sigma_{\text{corr}}|c'\rangle$ of the matrix element. Added to the Hartree–Fock contribution, this would give $\langle c|h + V_{\text{H}} + \Sigma|c'\rangle$ which must vanish if $|c\rangle$ and $|c'\rangle$ are eigenstates of this hamiltonian. By symmetry this is also true for $\langle \psi_{vc}|V_{\text{eff}}|\psi_{v'c}\rangle$. Finally the last four terms of all matrix elements are given quite generally by the expression of $\langle \psi_{vc}|V_{\text{eff}}|\psi_{v'c'}\rangle$ in (A.6) even if $c' = c, v' = v$.

A final term of importance in the discussion is $\langle \psi_0|V_{\text{eff}}|\psi_{vc}\rangle$ coupling the ground state. This should be close to zero if optimized single particle states are used. One can show along the previous lines that

$$\langle \psi_0 | V_{\text{eff}} | \psi_{vc}\rangle = \sum_s \left\{ \sum_{c'} \frac{\langle v|V_s|c'\rangle \langle c'|V_s|c\rangle}{E - E_{vc'} - \omega_s} - \sum_{v'} \frac{\langle v'|V_s|c\rangle \langle v|V_s|v'\rangle}{E - E_{v'c} - \omega_s} \right\}. \quad (A.7)$$

This can be expressed as

$$\langle \psi_0 | V_{\text{eff}} | \psi_{vc}\rangle = \sum_s \left\{ \sum_{c'} \frac{\langle v|V_s|c'\rangle \langle c'|V_s|c\rangle}{E - E_{vc} + \varepsilon_c - \varepsilon_{c'} - \omega_s} \right.$$
$$\left. - \sum_{v'} \frac{\langle v'|V_s|c\rangle \langle v|V_s|v'\rangle}{E - E_0 + \varepsilon_{v'} - \varepsilon_c - \omega_s} \right\}. \quad (A.8)$$

Again if one neglect $E - E_{vc}$ and $E - E_0$ in this expression one recovers the correlation part $\langle v|\Sigma_{\text{corr}}|c\rangle$. Added to the Hartree–Fock part, this gives $\langle v|h + V_{\text{H}} + \Sigma|c\rangle = 0$ for the full matrix element.

B Macroscopic Averages in Maxwell's Equations

In the macroscopic theory of dielectrics, macroscopic quantities are defined as suitable averages of the microscopic quantities. The aim of this appendix is to show that under such a procedure the Maxwell's equations remain invariant. More specifically, let us consider a microscopic quantity $f(\mathbf{r})$ and define the macroscopic quantity $F(\mathbf{r})$ as

$$F(\mathbf{r}) = \int g(\mathbf{r} - \mathbf{r}')f(\mathbf{r}')\mathrm{d}\mathbf{r}' , \qquad (\text{B.1})$$

where g is a weighting function which suppresses the short period oscillations contained in the microscopic function $f(\mathbf{r})$. For a bulk crystal, for instance, one could take the Fourier transform $g(\mathbf{q})$ as constant within the first Brillouin zone, zero outside.

Maxwell's equations include space derivative of the quantities and one must evaluate their average. For instance, we need to calculate:

$$A = \int g(\mathbf{r} - \mathbf{r}')\frac{\partial f(\mathbf{r}')}{\partial x'}\mathrm{d}\mathbf{r}' . \qquad (\text{B.2})$$

Making use of the fact that

$$\frac{\partial g(\mathbf{r} - \mathbf{r}')}{\partial x} = -\frac{\partial g(\mathbf{r} - \mathbf{r}')}{\partial x'} , \qquad (\text{B.3})$$

and, by integration by parts, we obtain

$$A = \int \frac{\partial g(\mathbf{r} - \mathbf{r}')}{\partial x}f(\mathbf{r}')\mathrm{d}\mathbf{r}' = \frac{\partial F(\mathbf{r})}{\partial x} . \qquad (\text{B.4})$$

Thus, the average of ∇f is equal to ∇F, and we can transpose the Maxwell's equations by changing all microscopic quantities to their macroscopic counterpart.

C Polarization Correction

We want to evaluate the polarization correction to the energy potential at r' induced by an excess electron at r when $r = r'$ or $r \approx r'$, i.e. $P_n(\cos\theta) \approx P_n(1) \approx 1$. Using (3.24), this is given by

$$\delta V = \frac{e^2}{4\pi\epsilon_0} \sum_{n=0}^{\infty} \frac{(\epsilon_{\text{in}} - \epsilon_{\text{out}})(n+1)r^n r'^n}{\epsilon_{\text{in}}[\epsilon_{\text{out}} + n(\epsilon_{\text{in}} + \epsilon_{\text{out}})]R^{2n+1}} \cdot \quad (C.1)$$

We rewrite it by defining $\eta = \epsilon_{\text{out}}/(\epsilon_{\text{in}} + \epsilon_{\text{out}})$. We thus need to evaluate

$$I = \sum_{n=0}^{\infty} \frac{n+1}{n+\eta} \frac{r^n r'^n}{R^{2n+1}} \quad (C.2)$$

as a function of η where $0 < \eta < 1$. It is convenient to separate the $n = 0$ term from the remaining sum. This gives

$$I = \frac{1}{\eta R} + \frac{J(\eta, x)}{R},$$

$$J(\eta, x) = \sum_{n=1}^{\infty} \frac{n+1}{n+\eta} x^n \text{ with } x = \frac{rr'}{R^2} \cdot \quad (C.3)$$

The main task is thus to evaluate J. The simplest case corresponds to $\eta \to 1$ which gives

$$J(1, x) = \sum_{n=1}^{\infty} x^n = \frac{x}{1-x} \cdot \quad (C.4)$$

The other limit is $\eta \to 0$

$$J(0, x) = \sum_{n=1}^{\infty} \left(1 + \frac{1}{n}\right) x^n = \frac{x}{1-x} + \sum_{n=1}^{\infty} \frac{x^n}{n},$$

$$= \frac{x}{1-x} - \ln(1-x) . \quad (C.5)$$

We can obtain $J(\eta, x)$ under integral form by rewriting (C.3)

$$J(\eta, x) = \sum_{n=1}^{\infty} x^n + (1-\eta) \sum_{n=1}^{\infty} \frac{x^n}{n+\eta}, \quad (C.6)$$

which can be transformed into

$$J(\eta, x) = \frac{x}{1-x} + (1-\eta)x^{-\eta} \int_0^x \sum_{n=1}^\infty u^{n+\eta-1} du , \quad (C.7)$$

and finally gives

$$J(\eta, x) = \frac{x}{1-x} + (1-\eta)x^{-\eta} \int_0^x \frac{u^\eta}{1-u} du . \quad (C.8)$$

To get a simple determination of the full $J(\eta, x)$, we can use a linear interpolation $J(\eta, x) = J(1, x) + (1-\eta)[J(0, x) - J(1, x)]$:

$$J(\eta, x) = \frac{x}{1-x} - (1-\eta)\ln(1-x) . \quad (C.9)$$

We have verified numerically that the difference between (C.9) and (C.8) in the whole range $0 \geq x \geq 1$ and $0 \geq \eta \geq 1$ is of the order of 5% in average, and is always smaller than 13%.

In Chap. 3, the determination of the electrostatic self-energy of a carrier in a spherical quantum dot requires to calculate the average of $J(\eta, x)$ for $r' = r$ (i.e., $x = r^2/R^2$) using the density of probability $|\phi(\mathbf{r})|^2$ in the effective mass approximation (Sect. 2.1.4):

$$|\phi(\mathbf{r})|^2 = \frac{1}{2\pi R}\left(\frac{\sin(\pi r/R)}{r}\right)^2 . \quad (C.10)$$

Using (C.9) and (C.10), the average is given by

$$\langle J(\eta) \rangle = \int J(\eta, x)|\phi(\mathbf{r})|^2 d\mathbf{r} = 0.557 + 0.376(1-\eta) , \quad (C.11)$$

where the coefficients have been obtained numerically.

References

1. D.R. Hartree: Proc. Cambridge Phil. Soc. **24**, 89 (1928)
2. F. Seitz: *The Modern Theory of Solids* (McGraw-Hill, New York 1940)
3. D. Pines: *Elementary Excitations in Solids* (Addison-Wesley, New York 1963)
4. J.C. Slater: Phys. Rev. **81**, 385 (1951) ; **82**, 538 (1951) ; **91**, 528 (1953).
5. V. Fock: Z. Physik **61**, 126 (1930)
6. D. Pines, P. Nozières: *The Theory of Quantum Liquids*, vol. I: Normal Fermi Liquids (Addison-Wesley, New York 1989)
7. J.C. Inkson: *Many-Body Theory of Solids, An Introduction*, (Plenum Press, New-York and London 1984)
8. D.M. Ceperley, B.J. Alder: Phys. Rev. Lett. **45**, 566 (1980) ; Y. Kwon, D.M. Ceperley, R.M. Martin: Phys. Rev. B **58**, 6800 (1998)
9. E.K.U. Gross, F.J. Dobson, M. Petersilka: *Density Functional Theory*, Topics in Current Chemistry, Vol. 180, ed. by R.F. Nakewajski (Springer, New York 1996)
10. R.G. Parr, W. Yang, Y. Weitao: *Density-Functional Theory of Atoms and Molecules*, International Series of Monograpohs on Chemistry, No. 16 (Oxford University Press, Oxford 1994)
11. L.H. Thomas: Proc. Cambridge Phil. Soc. **23**, 542 (1927) ; E. Fermi: Z. Physik **48**, 73 (1928)
12. J.C. Slater: *Quantum Theory of Molecules and Solids*, Vol. 1 (McGraw-Hill, New York 1963)
13. J.C. Slater: *Quantum Theory of Molecules and Solids*, Vol. 4 (McGraw-Hill, New York 1974)
14. P. Hohenberg, W. Kohn: Phys. Rev. **136**, 864 (1964)
15. W. Kohn, L.J. Sham: Phys. Rev. **140**, 1133 (1965)
16. R.O Jones, O. Gunnarsson: Reviews of Modern Physics **61**, 689 (1989)
17. J.P. Perdew, A. Zunger: Phys. Rev. B **23**, 5048 (1981)
18. D.C. Langreth, J.P. Perdew: Phys. Rev. B **21**, 5469 (1980)
19. D.C. Langreth, M.J. Mehl: Phys. Rev. B **28**, 1809 (1983)
20. J.P. Perdew, Y. Wang: Phys. Rev. B **33**, 8800 (1986)
21. J.P. Perdew: Phys. Rev. B **33**, 8822 (1986)
22. T. Koopmans: Physica **1**, 104 (1933)
23. O. Zakharov, A. Rubio, X. Blase, M.L. Cohen, S.G. Louie: Phys. Rev. B **50**, 10780 (1994)
24. X. Zhu, S.G. Louie: Phys. Rev. B **43**, 14142 (1991)
25. L.D. Landau: Zh. Eksp. Theor. Fiz. **30**, 1058 (1956) [Sov. Phys. - JETP **3**, 920 (1957)]
26. L. Hedin, B.I. Lundqvist: J. Phys. C **4**, 2064 (1971)

27. A. Fetter, J.D. Walecka: *Quantum Theory of Many Particle Systems* (McGraw-Hill Book Company, San Francisco 1971)
28. J. Lindhard: K. Dan. Vidensk. Selsk. Mat.- Fys. Medd. **28**, No. 8 (1954)
29. H. Ehrenreich, M.H. Cohen: Phys. Rev. **115**, 786 (1959)
30. S.L. Adler: Phys. Rev. **126**, 413 (1962)
31. N. Wiser: Phys. Rev. **129**, 62 (1963)
32. L. Hedin, S. Lundqvist: *Solid State Physics* **23**, ed. by H. Ehrenreich, F. Seitz, D. Turnbull (Academic Press, New York, London 1969) pp. 1
33. W.G. Aulbur, L. Jönsson, J.W. Wilkins: *Solid State Physics* **54**, ed. by H. Ehrenreich, F. Seitz, D. Turnbull (Academic Press, New York, London 1999) pp. 1
34. L. Hedin: Int. J. Quant. Chem. **56**, 445 (1995)
35. M. Lannoo, M. Schlüter, L.J. Sham: Phys. Rev. B **32**, 3890 (1985)
36. G. Onida, L. Reining, A. Rubio: Rev. Mod. Phys. **74**, 601 (2002)
37. L.J. Sham, T.M. Rice: Phys. Rev. **144**, 708 (1966)
38. G. Strinati: Phys. Rev. B **29**, 5718 (1984)
39. E. Runge, E.K.U. Gross: Phys. Rev. Lett. **52**, 997 (1984)
40. E.K.U Gross, W. Kohn: Phys. Rev. Lett. **55**, 2850 (1985) ; **57**, 923 (1986)
41. M. Petersilka, U.J. Grossmann, E.K.U. Gross: Phys. Rev. Lett. **76**, 1212 (1996)
42. A. Rubio, J.A. Alonso, X. Blase, L.C. Balbás, S.G. Louie: Phys. Rev. Lett. **77**, 247 (1996)
43. I. Vasiliev, S. Öğüt, J.R. Chelikowsky: Phys. Rev. B **65**, 115416 (2002)
44. S.J.A. van Gisbergen, J.G. Snijders, E.J. Baerends: Comp. Phys. Comm. **118**, 119 (1999)
45. M. Puranik, S. Umapathy, J.G. Snijders, J. Chandrasekhar: J. Chem. Phys. **115**, 6106 (2001)
46. R. Singh, B.M. Deb: Phys. Rep. **311**, 47 (1999)
47. K. Yabana, G.F. Bertsch: Phys. Rev. A **60**, 3809 (1999)
48. L. Reining, V. Olevano, A. Rubio, G. Onida: Phys. Rev. Lett. **88**, 066404 (2002)
49. F. Kootstra, P.L. de Boeij, J.G. Snijders: J. Phys. Chem. **112**, 6517 (2000)
50. V.I. Gavrilenko, F. Bechstedt: Phys. Rev. B **55**, 4343 (1997)
51. G.F. Bertsch, A. Schnell, K. Yabana: J. Chem. Phys. **115**, 4051 (2001)
52. S.G. Louie: Phys. Rev. B **22**, 1933 (1980)
53. R.C.Chaney, C. Lin, E.E. Lafon: Phys. Rev. B **3**, 459 (1971)
54. E.O. Kane: Phys. Rev. B **13**, 3478 (1976)
55. D.J. Chadi: Phys. Rev B **16**, 3572 (1977).
56. J.C. Slater, G.J. Koster: Phys. Rev. **94**, 1498 (1954)
57. W.A. Harrison: *Electronic Structure and the Properties of Solids, The Physics of the Chemical Bond* (Freeman, New York 1980)
58. P.O. Löwdin and H. Shull: Phys. Rev. **101**, 1730 (1956)
59. J. Van der Rest, P. Pécheur: J. Phys. Chem. Solids **45**, 563 (1984)
60. P. Vogl, H.P. Hjalmarson, J.D. Dow: J. Phys. Chem. Sol. **44**, 365 (1983)
61. K.C. Pandey, J.C. Philips, Solid Stat. Comm. **14**, 439 (1974)
62. J.R. Chelikowski, M.L. Cohen: Phys Rev. B **14**, 556 (1976)
63. C. Tserbak, H.M. Polatoglou, G. Theodorou: Phys. Rev. B **47**, 7104 (1993)
64. Y.M. Niquet, C. Delerue, G. Allan, M. Lannoo: Phys. Rev. B **62**, 5109 (2000)
65. L. Reining: private communication

66. J.M. Jancu, R. Scholz, F. Beltram, F. Bassani: Phys. Rev. B **57**,6493 (1998)
67. M. Lannoo: J. Phys. (Paris) **40**, 461 (1979)
68. G. Allan: *Electronic structure of crystal defects and disordered systems* (Les Editions de Physique, Paris 1981) pp. 37
69. D.C. Allan, E.J. Mele: Phys. Rev. Lett. **53**, 826 (1984)
70. O.L. Alerhand, E.J. Mele: Phys. Rev. Lett. **59**, 657 (1987); O.L. Alerhand, E.J. Mele: Phys. Rev. B **37**, 2536 (1988)
71. *Tight Binding Approach to Computational Materials Science*, Materials Research Society Symposium Proceedings Vol. 491, ed. by P. Turchi, A. Gonis, L. Colombo (Materials Research Society, Pittsburgh 1998).
72. J.A. Pople, D.L. Beveridge: *Approximate Molecular Orbital Theories* (McGraw-Hill, New York 1970)
73. M.L. Cohen, V. Heine, D. Weaire: *Solid State Physics* **24**, ed. by H. Ehrenreich, F. Seitz, D. Turnbull (Academic Press, New York, London 1970)
74. D.R. Hamann: Phys. Rev. Lett. **42**, 662 (1979)
75. M.L. Cohen, T.K. Bergstresser, Phys. Rev. **141**, 789 (1966)
76. L.-W. Wang, A. Zunger: J. Phys. Chem. **98**, 2158 (1994)
77. M. Lannoo, J. Bourgoin: *Point Defects in Semiconductors I*, Springer Series in Solid-State Sciences 22 (Springer-Verlag, Berlin 1981)
78. M. Cardona, F.H. Pollak: Phys. Rev. **142**, 530 (1966)
79. L.I. Schiff: *Quantum Mechanics*, 2nd ed. (McGraw-Hill, New York 1955)
80. E.O. Kane: J. Phys. Chem. Sol. **1**, 82 (1956)
81. E.O. Kane: J. Phys. Chem. Sol. **1**, 249 (1957)
82. C. Kittel, A.H. Mitchell: Phys. Rev. **96**, 1488 (1954)
83. G. Dresselhaus, A.F. Kip, C. Kittel: Phys. Rev. **98**, 368 (1955)
84. F. Bassani, G. Iadonisi, B. Preziosi: Rep. Prog. Phys. **37**, 1099 (1974)
85. M. Altarelli: *Heterojunctions and Semiconductor Superlattices*, ed. by G. Allan, G. Bastard, N. Boccara, M. Lannoo, M. Voos (Springer Verlag, Heidelberg 1986) pp. 12
86. G. Bastard: *Wave Mechanics Applied to Semiconductor Heterostructures* (Les Editions de Physique, Paris 1990)
87. J.M. Luttinger, W. Kohn: Phys. Rev. **97**, 869 (1955)
88. S.R. White, L.J. Sham: Phys. Rev. Lett. **47**, 879 (1981)
89. M. Altarelli: Phys. Rev. B **28**, 842 (1983)
90. W. Kohn: *Solid State Physics*, ed. by F. Seitz, D. Turnbull (Academic, New York 1957) pp. 258
91. T. Ando, A.B. Fowler, F. Stern: Rev. Mod. Phys. **54**, 437 (1982)
92. C. Weisbuch, B. Vinter: *Quantum Semiconductor Structures, Fundamentals and Applications* (Academic Press, New York 1991).
93. J.M. Luttinger: J. Math. Phys. **4**, 1154 (1963)
94. J. Voit: Rep. Prog. Phys. **58**, 977 (1995)
95. A. I Ekimov, A. A. Onushcenko, V. A. Tzekhomskii: Sov. Phys. Chem. Glass **6**, 511 (1980)
96. A. Henglein, B. Bunsenges: Phys. Chem. **88**, 301 (1982)
97. L. E. Brus: J. Chem. Phys. **79**, 5566 (1983)
98. L. Brus: Appl. Phys. A **53**, 465 (1991)
99. M.L. Steigerwald, L.E. Brus: Annu. Rev. Mater. Sci. **19**, 471 (1989)
100. A.P. Alivisatos: Science **271**, 933 (1996)
101. C.A. Coulson: Proc. R. Soc. London, Ser. A **169**, 413 (1939)

102. C.A. Coulson, H.C. Longuet-Higgins: Proc. R. Soc. London, Ser. A **192**, 16 (1947)
103. U. Kreibig, M. Vollmer: *Optical Properties of Metal Clusters*, Springer Series in Materials Science **25** (Springer, Berlin 1995)
104. G.A. Baraff, D. Gershoni: Phys. Rev. B **43**, 4011 (1991)
105. D. Gershoni, C.H. Henry, G.A. Baraff: IEEE Journal of Quantum Electronics **29**, 2433 (1993)
106. T.B. Bahder: Phys. Rev. B **41**, 11992 (1990)
107. Y.M. Niquet, C. Priester, C. Gourgon, H. Mariette, Phys. Rev. B **57**, 14850 (1998)
108. B. Grandidier, Y.M. Niquet, B. Legrand, J.P. Nys, C. Priester, D. Stiévenard, J.M. Gérard, V. Thierry-Mieg: Phys. Rev. Lett. **85**, 1068 (2000)
109. P.B. Allen, J.Q. Broughton, A.K. MacMahan: Phys. Rev. B **34**, 859 (1986)
110. C. Priester, G. Allan, M. Lannoo: Phys. Rev. B **38**, 9870 (1988)
111. A. Zunger, L.-W. Wang: Appl. Surf. Science **102**, 350 (1996)
112. H. Fu, L.-W. Wang, A. Zunger: Appl. Phys. Lett. **71**, 3433 (1997)
113. H. Fu, L.-W. Wang, A. Zunger: Appl. Phys. Lett. **73**, 1157 (1998)
114. H. Fu, L.-W. Wang, A. Zunger: Phys. Rev. B **57**, 9971 (1998)
115. D.M. Wood, A. Zunger: Phys. Rev. B **53**, 7949 (1996)
116. L.-W. Wang, J. Kim, A. Zunger: Phys. Rev. B **59**, 5678 (1999)
117. L.-W. Wang, A. Zunger: Phys. Rev. B **54**, 11417 (1996)
118. M.C. Payne, M.P. Teter, D.C. Allan, T.A. Arias, J.D. Joannopoulos: Rev. Mod. Phys. **64**, 1045 (1992)
119. G.H. Golub: *Matrix Computations*, third edition (Johns Hopkins University Press, Baltimore 1996)
120. C. Lanczos: J. Res. Natl. Bur. Std. US **45**, 255 (1950)
121. L.-W. Wang, A. Zunger: J. Chem. Phys. **100**, 2394 (1994)
122. L.-W. Wang, A. Zunger: Phys. Rev. B **59**, 15806 (1999)
123. F. Buda, J. Kohanoff, M. Parrinello: Phys. Rev. Lett. **69**, 1272 (1992)
124. J.W. Mintmire: J. Vac. Sci. Technol. A **11**, 1733 (1993)
125. A.J. Read, R.J. Needs, K.J. Nash, L.T. Canham, P.D.J Calcott, A. Qteish: Phys. Rev. Lett. **70**, 2050 (1993)
126. M. Hirao, T. Uda, Y. Murayama: Mater. Res. Soc. Symp. Proc. **283**, 425 (1993)
127. H.-Ch. Weissker, J. Furthmüller, F. Bechstedt: Phys. Rev. B **65**, 155328 (2002)
128. B. Delley, E.F. Steigmeier: Appl. Phys. Lett. **67**, 2370 (1995)
129. A.I. Ekimov, F. Hache, M.C. Schanne-Klein, D. Ricard, C. Flytzanis, I.A. Kudryavtsev, T.V. Yazeva, A.V. Rodina, Al.L. Efros: J. Opt. Soc. Am. B **10**, 100 (1993)
130. D.J. Norris, A. Sacra, C.B. Murray, M.G. Bawendi: Phys. Rev. Lett. **72**, 2612 (1994)
131. D.J. Norris, M.G. Bawendi: Phys. Rev. B **53**, 16338 (1996)
132. Y.M. Niquet, G. Allan, C. Delerue, M. Lannoo: Appl. Phys. Lett. **77**, 1182 (2000)
133. G. Allan, C. Delerue, M. Lannoo: Appl. Phys. Lett. **70**, 2437 (1997)
134. G. Allan, Y.M. Niquet, C. Delerue: Appl. Phys. Lett. **77**, 639 (2000)
135. C. Delerue, M. Lannoo, G. Allan: Phys. Rev. Lett. **76**, 3038 (1996)
136. S.Y. Ren, J.D. Dow: Phys. Rev. B **45**, 6492 (1992)
137. N.A. Hill, K.B. Whaley: Phys. Rev. Lett. **75**, 1130 (1995)

138. Y.M. Niquet, C. Delerue, G. Allan, M. Lannoo: Phys. Rev. B **65**, 165334 (2002)
139. Y.M. Niquet: Thèse de doctorat (Université de Lille 1, Villeneuve d'Ascq, 2001)
140. A. Germeau, A.L. Roest, D. Vanmaekelbergh, G. Allan, C. Delerue, E.A. Meulenkamp: Phys. Rev. Lett. **90**, 076803 (2003)
141. E. Bakkers, Z. Hens, A. Zunger, A. Franceschetti, L. Kouwenhoven, L. Gurevich, D. Vanmaekelbergh: Nano Lett. **1**, 551 (2001)
142. M. Shim, P. Guyot-Sionnest: Nature **407**, 981 (2000)
143. G. Allan, C. Delerue, Y.M. Niquet: Phys. Rev. B **63**, 205301 (2001)
144. G. Allan, C. Delerue: Phys. Rev. B **66**, 233303 (2002)
145. P.C. Sercel, K.J. Vahala: Phys. Rev. B **42**, 3690 (1990)
146. T. Richard, P. Lefébvre, H. Mathieu, J. Allègre: Phys. Rev. B **53** , 7287 (1996)
147. U. Banin, G. Cerullo, A. A. Guzelian, C. J. Bardeen, A. P. Alivisatos, C. V. Shank: Phys. Rev. B **55**, 7059 (1997)
148. J.M. Ziman: *Models of Disorder* (Cambridge University Press, Cambridge 1979)
149. G. Allan, C. Delerue, M. Lannoo: Phys. Rev. Lett. **78**, 3161 (1997)
150. F. Wooten, K. Winer, D. Weaire: Phys. Rev. Lett. **54**, 1392 (1985)
151. B.R. Djorjevic, M.F. Thorpe, F. Wooten: Phys. Rev. B **52**, 5685 (1995)
152. G. Allan, C. Delerue, M. Lannoo: Phys. Rev. B **61**, 10206 (2000)
153. P.N. Keating: Phys. Rev. **145**, 637 (1966)
154. M.J. Estes, G. Moddel: Appl. Phys. Lett. **68**, 1814 (1996)
155. L. Landau, E. Lifchitz: *Electrodynamique des milieux continus*, 2nd edn. (Mir, Moscow 1990)
156. E.M. Purcell: *Electricity and Magnetism, Berkeley Physics Course*, vol. 2 (Education Development Center, Inc., Newton, Massachusetts 1965)
157. C.J.F. Böttcher: *Theory of Electric Polarization*, 2nd edn., vol. 1 (Elsevier, Amsterdam 1973)
158. C. Kittel: *Introduction to Solid State Physics*, 5th edn. (John Wiley & Sons, Inc., New York 1976)
159. R. Courant and D. Hilbert: *Methods of Mathematical Physics*, Vol. 2 (Interscience Publishers, New York 1962)
160. L.V. Keldysh: Superlatt. Microstruct. **4**, 637 (1966)
161. L.V. Keldysh: Pis'ma Zh. Eksp. Teor. Fiz. **29**, 716 (1979) [JETP Lett. **29**, 658 (1979)]
162. M. Kumagai, T. Takagahara: Phys. Rev. B **40**, 12359 (1989)
163. L.C. Andreani, A. Pasquarello: Phys. Rev. B **42**, 8928 (1990)
164. N.D. Lang, W. Kohn: Phys. Rev. B **7**, 3541 (1973)
165. L.E. Brus: J. Chem. Phys. **80**, 4403 (1984)
166. M. Lannoo, C. Delerue, G. Allan: Phys. Rev. Lett. **74**, 3415 (1995)
167. G. Allan, C. Delerue, M. Lannoo, E. Martin: Phys. Rev. B **52**, 11982 (1995)
168. L. Banyai, P. Gilliot, Y.Z. Hu, S.W. Koch: Phys. Rev. B **45**, 14136 (1992)
169. F. Stern: *Heterojunctions and Semiconductor Superlattices*, Proceedings of the Les Houches Winterschool, ed. by G. Allan, G. Bastard, N. Boccara, M. Lannoo and M. Voos (Springer Verlag, Berlin 1965)
170. A. Kumar, S.E. Laux, F. Stern: Phys. Rev. B **42**, 5166 (1990).
171. E. Martin, C. Delerue, G. Allan, M. Lannoo: Phys. Rev. B **50**, 18258 (1994)
172. M. Lannoo: Phys. Rev. B **10**, 2544 (1974)

173. Y.M. Niquet, C. Delerue, M. Lannoo, G. Allan: Phys. Rev. B **64**, 113305 (2001)
174. M. Chamarro, M. Dib, V. Voliotis, A. Filoramo, P. Roussignol, T. Gacoin, J. P. Boilot, C. Delerue, G. Allan, and M. Lannoo: Phys. Rev. B **57**, 3729 (1998)
175. A. Franceschetti, A. Zunger: Phys. Rev. B **62**, 2614 (2000)
176. M. Rontani, F. Rossi, F. Manghi, E. Molinari: Phys. Rev. B **59**, 10165 (1999)
177. K. Hirose, N. S. Wingreen: Phys. Rev. B **59**, 4604 (1999)
178. M. Macucci, K. Hess, G.J. Iafrate: Phys. Rev. B **55**, 4879 (1997)
179. S. Nagaraja, P. Matagne, V.-Y. Thean, J.-P. Leburton, Y.-H. Kim, R.-M. Martin: Phys. Rev. B **56**, 15752 (1997)
180. B. Tanatar, D.M. Ceperley: Phys. Rev. B **39**, 5005 (1989)
181. P.E. van Camp, V.E. van Doren, J.T. Devreese: Phys. Rev. B **24**, 1096 (1981)
182. S. Baroni, R. Resta: Phys. Rev. B **33**, 7017 (1986)
183. P.M. Echenique, J.M. Pitarke, E.V. Chulkov, A. Rubio: Chemical Physics **251**, 1 (2000)
184. O. Pulci, G. Onida, R. Del Sole, L. Reining: Phys. Rev. Lett. **81**, 5347 (1998)
185. M. Rohlfing, S.G. Louie: Phys. Rev. Lett. **83**, 856 (1999)
186. G. Onida, L. Reining, R.W. Godby, R. Del Sole, and W. Andreoni: Phys. Rev. Lett. **75**, 818 (1995)
187. M. Rohlfing, S.G. Louie: Phys. Rev. Lett. **80**, 3320 (1998)
188. J.C. Grossman, M. Rohlfing, L. Mitas, S.G. Louie, M.L. Cohen: Phys. Rev. Lett. **86**, 472 (2001)
189. W. Hanke, L.J. Sham: Phys. Rev. Lett. **33**, 582 (1974)
190. C. Delerue, M. Lannoo, G. Allan: Phys. Rev. B **56**, 15306 (1997)
191. C. Krzeminski, C. Delerue, G. Allan: J. Phys. Chem. B **105**, 6321 (2001)
192. I. Vasiliev, S. Öğüt, J.R. Chelikowsky: Phys. Rev. Lett. **78**, 4805 (1997); L. Kronig, I. Vasiliev, J.R. Chelikowsy: Phys. Rev. B **62**, 9992 (2000)
193. A. Rubio, J.A. Alonso, X. Blase, S.G. Louie: Int. J. Mod. Phys. B **11**, 2727 (1997), and references therein
194. K. Yabana, G.F. Bertsch: Phys. Rev. B **54**, 4484 (1996)
195. A. Domps, P.-G. Reinhard, E. Suraud: Phys. Rev. Lett. **81**, 5524 (1998)
196. S.J.A. van Gisbergen, P.R.T. Schipper, O.V. Gritsenko, E.J. Baerends, J.G. Snijders, B. Champagne, B. Kirtman: Phys. Rev. Lett. **83**, 694 (1999)
197. M.S. Hybertsen, S.G. Louie: Phys. Rev. B **35**, 5585 (1987)
198. L.W. Wang, A. Zunger: Phys. Rev. Lett. **73**, 1039 (1994)
199. R. Tsu, D. Babic: Appl. Phys. Lett. **64**, 1806 (1994)
200. C. Delerue, M. Lannoo, G. Allan: Phys. Rev. B **68**, 115411 (2003)
201. M. von Laue: Ann. der Phys. **44**, 1197 (1914); C. Kittel: *Quantum Theory of Solids* (John Wiley & Sons, New York 1963)
202. *Semiconductor Quantum Dots: Physics, Spectroscopy and Applications*, ed. by Y. Masumoto, T. Takagahara (Springer, Berlin 2002)
203. R.S. Knox: *Theory of Excitons*, Solid State Phys. **5**, ed. by F. Seitz, D. Turnbull (Academic Press, New York 1963)
204. T. Takagahara, K. Takeda: Phys. Rev. B **53**, R4205 (1996)
205. A. Franceschetti, L.-W. Wang, H. Fu, A. Zunger: Phys. Rev. **58**, R13367 (1998)
206. L.X. Benedict: Phys. Rev. B **66**, 193105 (2002)
207. M. Chamarro, C. Gourdon, P. Lavallard, O. Lublinskaya, A.I. Ekimov: Phys. Rev. B **53**, 1336 (1996)

208. M. Nirmal, D.J. Norris, M. Kuno, M.G. Bawendi, Al.L. Efros, M. Rosen: Phys. Rev. Lett. **75**, 3728 (1995)
209. M. Dib, M. Chamarro, V. Voliotis, J.L. Fave, C. Guenaud, P. Roussignol, T. Gacoin, J.P. Boilot, C. Delerue, G. Allan, M. Lannoo: Phys. Stat. Sol. (b) **212**, 293 (1999)
210. J.B. Xia: Phys. Rev. B **40**, 8500 (1989)
211. G.B. Grigoryan, E.M. Kazarian, Al.L. Efros, T.V. Yazeva: Soviet Phys. - Solid State **32**, 1031 (1990)
212. N. Thang, G. Fishman: Phys. Rev. B **31**, 2404 (1985)
213. Al.L. Efros, M. Rosen, M. Kuno, M. Nirmal, D.J. Norris, M.G. Bawendi: Phys. Rev. B **54**, 4843 (1996)
214. B. Segall, D.T.F. Marple: *Physics and Chemistry of II-VI Compounds*, ed. by M. Aven, J.S. Presner (North-Holland Publ. Co., Amsterdam 1967)
215. S.V. Goupalov, E.L. Ivchenko: J. Cryst. Growth **184/185**, 393 (1998)
216. P.D.J. Calcott, K.J. Nash, L.T. Canham, M.J. Kane, D. Brumhead: J. Phys. Condens. Matter. **5**, L91 (1993)
217. T. Suemoto, K. Tanaka, A. Nakajima, T. Itakura: Phys. Rev. Lett. **70**, 3659 (1993)
218. X.L. Zheng, W. Wang, H.C. Chen: Appl. Phys. Lett. **60**, 986 (1992)
219. J.C. Vial, A. Bsiesy, G. Fishman, F. Gaspard, R. Hérino, M. Ligeon, F. Muller, R. Romestain, R.M. Macfarlane, *Microcrystalline Semiconductors - Material Science and Devices*, ed. by P.M. Fauchet, C.C. Tsai, L.T. Canham, I. Shimizu, Y. Aoyagi, Materials Research Society Symposia Proceedings **283** (Materials Research Society, Pittsburgh 1993)
220. G. Fishman, R. Romestain, J.C. Vial: J. Phys. **IV 3**, 355 (1993)
221. J.P. Proot, C. Delerue, G. Allan: Appl. Phys. Lett. **61**, 1948 (1992)
222. C. Delerue, G. Allan, M. Lannoo: Phys. Rev. B **48**, 11024 (1993)
223. M. Hybertsen, M. Needels: Phys. Rev. B **48**, 4608 (1993)
224. J. Shumway, A. Franceschetti, A. Zunger: Phys. Rev. B **63**, 155316 (2001)
225. M.S. Hybertsen, S.G. Louie: Phys. Rev. Lett. **55**, 1418 (1985)
226. M.S. Hybertsen, S.G. Louie: Phys. Rev. B **34**, 5390 (1986)
227. R.W. Godby, M. Schlüter, L.J. Sham: Phys. Rev. B **37**, 10159 (1988)
228. R.W. Godby, I.D. White: Phys. Rev. Lett. **80**, 3161 (1998)
229. A. Franceschetti, L.W. Wang, A. Zunger: Phys. Rev. Lett. **83**, 1269 (1999)
230. S. Öğüt, J.R. Chelikowsky, S.G. Louie: Phys. Rev. Lett. **80**, 3162 (1998)
231. L.J. Sham, M. Schlüter: Phys. Rev. Lett. **51**, 1888 (1983)
232. J.P. Perdew, M. Levy: Phys. Rev. Lett. **51**, 1884 (1983)
233. R.W. Godby, L.J. Sham: Phys. Rev. B **49**, 1849 (1994)
234. C. Delerue, M. Lannoo, G. Allan: Phys. Rev. Lett. **84**, 2457 (2000); Erratum: Phys. Rev. Lett. **89**, 249901 (2002)
235. C. Delerue, G. Allan, M. Lannoo: Phys. Rev. Lett. **90**, 076803 (2003)
236. J.C. Slater: *Quantum Theory of Atomic Structure* (McGraw-Hill, New York 1960)
237. M. Lannoo, G.A. Baraff, M. Schlüter: Phys. Rev. B **24**, 943 (1981)
238. F. Bechstedt, R. Del Sole: Phys. Rev. B **38**, 7710 (1988)
239. F. Gygi, A. Baldereschi: Phys. Rev. Lett. **62**, 2160 (1989)
240. I. Vasiliev, S. Öğüt, J.R. Chelikowsky: Phys. Rev. Lett. **82**, 1919 (1999)
241. I.V. Tokatly, O. Pankratov: Phys. Rev. Lett. **86**, 2078 (2001)
242. S. Albrecht, G. Onida, L. Reining: Phys. Rev. B 55, 10278 (1997)

243. A.R. Porter, M.D. Towler, R.J. Needs: Phys. Rev. B **64**, 035320 (2001)
244. L.T. Canham: Appl. Phys. Lett. **57**, 1046 (1990)
245. M. Ehbrecht, B. Kohn, F. Huisken, M.A. Laguna, V. Paillard: Phys. Rev. B **56**, 6958 (1997)
246. G. Ledoux, J. Gong, F. Huisken, O. Guillois, C. Reynaud: Appl. Phys. Lett. **80**, 4834 (2002)
247. G. Ledoux, O. Guillois, D. Porterat, C. Reynaud, F. Huisken, B. Kohn, V. Paillard: Phys. Rev. B **62**, 15942 (2000)
248. M.V. Wolkin, J. Jorne, P.M. Fauchet, G. Allan, C. Delerue: Phys. Rev. Lett. **82**, 197 (1999)
249. U. Banin, Y. Cao, D. Katz, O. Millo: Nature **400**, 542 (1999)
250. D.V. Averin, A.N. Korotkov, K.K. Likharev: Phys. Rev. B **44**, 6199 (1991)
251. H. van Houten, C.W.J. Beenakker, A.A.M. Staring: *Single-charge tunneling: Coulomb blockade phenomena in nanostructures*, ed. by H. Grabert, M. H. Devoret (Plenum Press, New York 1992)
252. A. Franceschetti, A. Williamson, A. Zunger: J. Phys. Chem. **104**, 3398 (2000)
253. R. Rinaldi, S. Antonaci, M. De Vittorio, R. Cingolani, U. Hohenester, E. Molinari, H. Lipsanen, J. Tulkki: Phys. Rev. B **62**, 1592 (2000)
254. M. Lomascolo, A. Vergine, T.K. Johal, R. Rinaldi, A. Passaseo, R. Cingolani, S. Patanè, M. Labardi, M. Allegrini, F. Troiani, E. Molinari: Phys. Rev. B **66**, 041302 (2002)
255. J.M. Gérard, B. Gayral: J. Lightwave Technol. **17**, 2089 (1999)
256. G. Mie: Ann. Phys. **25**, 377 (1908)
257. K.L. Kelly, E. Coronado, L.L. Zhao, G.C. Schatz: J. Phys. Chem. B **107**, 668 (2003)
258. D.S. Chemla, D.A.B. Miller: Optics Lett. **11**, 522 (1986)
259. K.M. Leung: Phys. Rev. A **33**, 2461 (1986)
260. S. Schmitt-Rink, D.A.B. Miller, D.S. Chemla: Phys. Rev. B **35**, 8113 (1987)
261. J. Tauc: *Optical Properties of Solids* (Academic Press, New York 1966)
262. L.M. Ramaniah, S.V. Nair, K.C. Rustagi: Phys. Rev. B **40**, 12423 (1989)
263. A. Messiah: *Mécanique Quantique* (Dunod, Paris 1959)
264. A. Yariv: *Quantum Electronics* (Wiley, New York 1989)
265. E. Rosencher, B. Vinter: *Optoélectronique* (Masson, Paris 1998)
266. L. Genzel, T.P. Martin: Surf. Sci. **34**, 33 (1973)
267. E.M. Purcell: Phys. Rev. **69**, 681 (1946)
268. *Microcavities and Photonic Bandgaps: Physics and Applications*, ed. by J. Rarity, C. Weisbuch (Kluwer, Dordrecht 1995)
269. P. Goy, J.-M. Raymond, M. Gross, S. Haroche: Phys. Rev. Lett. **50**, 1903 (1983)
270. R.G. Hulet, E.S. Hilfer, D. Kleppner: Phys. Rev. Lett. **55**, 2137 (1985)
271. J.M. Gérard, B. Sermage, B. Gayral, B. Legrand, E. Costard, V. Thierry-Mieg: Phys. Rev. Lett. **81**, 1110 (1998)
272. M. Bayer, T.L. Reinecke, F. Weidner, A. Larinov, A. McDonald, A. Forchel: Phys. Rev. Lett. **86**, 3168 (2001)
273. G.S. Solomon, M. Pelton, Y. Yamamoto: Phys. Rev. Lett. **86**, 3903 (2001)
274. B. Gayral, J.M. Gérard, A. Lemaître, C. Dupuis, L. Manin, J.L. Pelouard: Appl. Phys. Lett. **75**, 1908 (1999)
275. E. Yablonovitch: Phys. Rev. Lett. **58**, 2059 (1987)
276. J. Joannopoulos, R.D. Meade, J.N. Winn: *Photonic Crystals: Guiding the Flow of the Light* (Princeton University Press, Princeton 1995)

277. C. Reese, C. Becher, A. Imamoğlu, E. Hu, B.D. Gerardot, P.M. Petroff: Appl. Phys. Lett. **78**, 2279 (2001)
278. T. Yoshie, A. Scherer, H. Chen, D. Huffaker, D. Deppe: Appl. Phys. Lett. **79**, 114 (2001)
279. J. Petit, M. Lannoo, G. Allan: Solid State Comm. **60**, 861 (1986)
280. G. Dresselhaus, M.S. Dresselhaus: Phys. Rev. **160**, 649 (1967)
281. L. Brey, C. Tejedor: Solid State Comm. **48**, 403 (1983)
282. M. Graf, P. Vogl: Phys. Rev. B **51**, 4940 (1995)
283. T.G. Pedersen, K. Pedersen, T.B. Kriestensen: Phys. Rev. B **63**, 201101(R) (2001)
284. B.A. Foreman: Phys. Rev. B **66**, 165212 (2002)
285. A. Selloni, P. Marsella, R. Del Sole: Phys. Rev. B **33**, 8885 (1986)
286. K. Leung, K.B. Whaley: Phys. Rev. B **56**, 7455 (1997)
287. T.B. Boykin, P. Vogl: Phys. Rev. B **65**, 035202 (2002)
288. M. Stoneham: *Theory of Defects in Solids* (Clarendon, Oxford 1975)
289. J. Bourgoin, M. Lannoo: *Point Defects in Semiconductors II*, Springer Series in Solid-State Sciences **35** (Spinger-Verlag, Berlin 1983)
290. S. Baroni, S. de Gironcoli, A. Dal Corso, P. Giannozi: Rev. Mod. Phys. **73**, 515 (2001)
291. O. Madelung: *Solid-State Theory*, Springer Series in Solid State Sciences **2** (Springer-Verlag, Berlin 1981)
292. M. Born, K.H. Huang: *Dynamical Theory of Crystal Lattices* (Oxford University Press, Oxford 1954)
293. A.A. Maradudin, G.K. Morton, Weiss: *Theory of Lattice Dynamics in the Harmonic Approximation*, Solid State Physics, Advances and Applications, ed. by H. Ehrenreich, F. Seitz, D. Turnbull (Academic Press, New York, 1963)
294. H. Rucker, E. Molinari, P. Lugli: Phys. Rev. B **44**, 3463 (1991)
295. C. Delerue, G. Allan, M. Lannoo: Phys. Rev. B **64**, 193402 (2001)
296. H. Fu, V. Ozolins, A. Zunger: Phys. Rev. B **59**, 2881 (1999)
297. J. Zi, H. Büscher, C. Falter, W. Ludwig, K. Zhang, X. Xie: Appl. Phys. Lett. **69**, 200 (1996)
298. H. Lamb, Proc. Math. Soc. London **13**, 187 (1882)
299. A.E.H Love: *A Treatrise on the Mathematical Theory of Elasticity* (Dover, New York 1944)
300. T. Takagahara: Phys. Rev. Lett. **71**, 3577 (1993)
301. S. Nomura, T. Kobayashi: Solid State Commun. **82**, 335 (1992)
302. B. Champagnon, B. Andrianasolo, E. Duval: J. Chem. Phys. **94**, 5237 (1991)
303. M.C. Klein, F. Hache, D. Ricard, C. Flytzanis: Phys. Rev. B **42**, 11123 (1990)
304. J.S. Pan, H.B. Pan: Phys. Status Solidi B **148**, 129 (1988)
305. E. Roca, C. Trallero-Giner, M. Cardona: Phys. Rev. B **49**, 13704 (1994)
306. M.P. Chamberlain, C. Trallero-Giner, M. Cardona: Phys. Rev. B **51**, 1680 (1995)
307. R.W. Schoenlein, D.M. Mittleman, JJ. Shiang, A.P. Alivisatos, C.V. Shank: Phys. Rev. Lett. **70**, 1014 (1993)
308. J.L. Machol, F.W. Wise, R.C. Patel, D.B. Tanner: Phys. Rev. B **48**, 2819 (1993)
309. T.D. Krauss, F.W. Wise: Phys. Rev. Lett. **79**, 5102 (1997)
310. K. Huang, A. Rhys: Proc. Roy. Soc. (London) **A204**, 406 (1950)
311. E.U. Condon: Phys. Rev. **32**, 858 (1928)

312. J. Franck: Trans. Faraday Soc. **21**, 536 (1925)
313. J.J. Markham: Rev. Mod. Phys. **31**, 956 (1959)
314. M.H.L. Pryce: *Phonons in Perfect Lattices and in Lattices with Point Imperfections* (Oliver and Boyd, Edinburgh 1966)
315. N. Lorente, M. Persson: Phys. Rev. Lett. **85**, 2997 (2000)
316. H. Fröhlich, H. Pelzer, S. Zienau: Philos. Mag. **41**, 221 (1950)
317. R. Lassnig: Phys. Rev. B **30**, 7132 (1984)
318. N. Mori, T. Ando: Phys. Rev. B **40**, 6175 (1989)
319. M.H. Degani, O. Hipólito: Phys. Rev. B **35**, 7717 (1987)
320. J.C. Marini, B. Stebe, E. Kartheuser: Phys. Rev. B **50**, 14302 (1994)
321. R. Englman, R. Ruppin: J. Phys. C **1**, 614 (1968)
322. J.J. Licari, R. Evrard: Phys. Rev. B **15**, 2254 (1977)
323. A.A. Lucas, E. Kartheuser, R.G. Badro: Phys. Rev. B **2**, 2488 (1970)
324. K. Oshiro, K. Akai, M. Matsuura: Phys. Rev. B **58**, 7986 (1998)
325. S.N. Klimin, E.P. Pokalitov, V.M. Fomin: Phys. Status Solidi B **184**, 373 (1994)
326. M.H. Degani, H.A. Farias: Phys. Rev. B **42**, 11950 (1990)
327. L. Wendler, A.V. Chaplik, R. Haupt, O. Hipólito: J. Phys.: Condens. Matter **5**, 8031 (1993)
328. K.-D. Zhu, S.-W. Gu: Phys. Lett. A **58**, 435 (1992)
329. S. Nomura, T. Kobayashi: Phys. Rev. B **45**, 1305 (1992)
330. A.I. Ekimov, I.A. Kudravtsev, M.G. Ivanov, Al.L. Efros: J. Lumin. **46**, 83 (1990)
331. D.V. Melnikov, W. Beall Fowler: Phys. Rev. B **63**, 165302 (2001)
332. A. M. Alcade, G. Weber: Semicond. Sci. Technol. **15**, 1082 (2000)
333. V.M. Fomin, V.N. Gladilin, J.T. Devreese, E.P. Pokatilov, S.N. Balaban, S.N. Klimin: Phys. Rev. B **57**, 2415 (1998)
334. E.P. Pokatilov, S.N. Klimin, V.M. Fomin, J.T. Devreese, F.W. Wise: Phys. Rev. B **65**, 075316 (2002)
335. V.M. Fomin, V.N. Gladilin, S.N. Klimin, J.T. Devreese, P.M. Koenraad, J.H. Wolter: Phys. Rev. B **61**, R2436 (2000)
336. H.A. Jahn, E. Teller: Proc. R. Soc. **A161**, 220 (1937)
337. R. Dingle, W. Wiegmann, C.H. Henry: Phys. Rev. Lett. **33**, 827 (1974)
338. Y. Kayanuma: Phys. Rev. B **38**, 9797 (1988)
339. V.I. Klimov, A. Mihkailovsky, S. Xu, A. Malko, J.A. Hollingsworth, C. Leatherdale, M. Bawendi: Science **290**, 314 (2000)
340. L. Landau, E. Lifchitz: *Mécanique Quantique* (Mir, Moscow 1974)
341. V.I. Klimov, Ch.J. Schwarz, D.W. McBranch, C.A. Leatherdale, M.G. Bawendi: Phys. Rev. B **60**, R2177 (1999)
342. C. Wang, M. Shim, P. Guyot-Sionnest: Science **291**, 2390 (2001)
343. S. Sauvage, P. Boucaud, F.H. Julien, J.-M. Gérard, V. Thierry-Mieg: Appl. Phys. Lett. **71**, 2785 (1997)
344. M. Nirmal, B.O. Dabbousi, M.G. Bawendi, J.J. Macklin, J.K. Trautman, T.D. Harris, L.E. Brus: Nature **383**, 802 (1996)
345. M. Shim, P. Guyot-Sionnest: Phys. Rev. B **64**, 245342 (2001)
346. M. Bissiri, G. Baldassarri Höger von Högersthal, A.S. Bhatti, M. Capizzi, A. Frova, P. Frigeri, S. Franchi: Phys. Rev. B **62**, 4642 (2000)
347. A. García-Cristóbal, A.W.E. Minnaert, V.M. Fomin, J.T. Devreese, A.Yu. Silov, J.E.M. Haverkort, J.H. Wolter: Phys. Status Solidi B **215**, 331 (1999)

348. T.D. Krauss, F. Wise: Phys. Rev. B **55**, 9860 (1997)
349. A.P. Alivisatos, T.D. Harris, P.J. Carroll, M.L. Steigerwald, L.E. Brus: J. Chem. Phys. **90**, 3463 (1989)
350. G. Scamarcio, V. Spagnolo, G. Ventruti, M. Lugará, G.C. Righini: Phys. Rev. B **53**, R10489 (1996)
351. M. Nirmal, C.B. Murray, M.G. Bawendi: Phys. Rev. B **50**, 2293 (1994)
352. A. Lemaître, A.D. Ashmore, J.J. Finley, D.J. Mowbray, M.S. Skolnick, M. Hopkinson, T.F. Krauss: Phys. Rev. B **63**, 161309 (2001)
353. R. Heitz, I. Mukhametzhanov, O. Stier, A. Madhukar, D. Bimberg: Phys. Rev. Lett. **83**, 4654 (1999)
354. S. Kalliakos, X.B. Zhang, T. Taliercio, P. Lefebvre, B. Gil, N. Grandjean, B. Damilano, J. Massies: Appl. Phys. Lett. **80**, 428 (2002)
355. *Towards the First Silicon Laser*, ed. by L. Pavesi, S. Gaponenko, L. Dal Negro, NATO Science Series, II. Mathematics, Physics and Chemistry, Vol. 93 (Kluwer Academic Publishers, Dordrecht 2003)
356. G.D. Sanders, Y.-C. Chang: Phys. Rev. B **45**, 9202 (1992)
357. E.F. Steigmeier, B. Delley, H. Auderset: Physica Scripta **T45**, 305 (1992)
358. M. S. Hybertsen: Phys. Rev. Lett. **72**, 1514 (1994)
359. G. Bastard: Phys. Rev. B **24**, 4714 (1981)
360. C. Priester, G. Allan, M. Lannoo: Phys. Rev. B **28**, 7194 (1983)
361. G. Bastard, J. A. Brum, R. Ferreira: Solid State Phys. **44**, 229 (1991)
362. F.J. Ribeiro, A. Latge: Phys. Rev. B **50**, 4913 (1994)
363. J.L. Zhu: Phys. Rev. B **50**, 4497 (1994)
364. J.M. Ferreyra, C.R. Proeto: Phys. Rev. B **52**, R2309 (1995)
365. G.T. Einevoll, Y.C. Chang: Phys. Rev. B **40**, 9683 (1989)
366. C. Bose: J. Appl. Phys. **83**, 3089 (1998)
367. C. Bose: Physica E **4**, 180 (1999)
368. N. Porras-Montenegro, S.T. Perez-Merchancano: Phys. Rev. B **46**, 9780 (1992)
369. H.J. von Bardeleben, D. Stiévenard, A. Grosman, C. Ortega, J. Siejka: Phys. Rev. B **47**, 10899 (1993)
370. M.S. Brandt, M. Stutzmann: Appl. Phys. Lett. **61**, 2569 (1992)
371. E.H. Poindexter, P.J. Caplan: Prog. Surf. Science. **14**, 201 (1983)
372. P.J. Caplan, E.H. Poindexter, B.E. Deal, R.R. Razouk: J. Appl. Phys. **50**, 5847 (1979)
373. N.M. Johnson, D.K. Biegelsen, M.D. Moyer, S.T. Chang, E.H. Poindexter, P.J. Caplan: Appl. Phys. Lett. **43**, 563 (1983)
374. K.L. Brower: Appl. Phys. Lett. **43**, 1111 (1983)
375. B. Henderson: Appl. Phys. Lett. **44**, 228 (1984)
376. W.B. Jackson: Solid State Comm. **44**, 477 (1982)
377. R.A. Street, J. Zesch, M.J. Thomson: Appl. Phys. Lett. **43**, 672 (1983)
378. N.M. Johnson, W.B. Jackson, M.D. Moyer: Phys. Rev. B **31**, 1194 (1985)
379. Y. Bar-Yam, J.D. Joannopoulos: Phys. Rev. Lett. **56**, 2203 (1986)
380. W.A. Harrison: Surf. Sci. **55**, 1 (1976)
381. M. Lannoo, G. Allan: Phys. Rev. B **25**, 4089 (1982)
382. V.I. Klimov, D.W. McBranch, C.A. Leatherdale, M.G. Bawendi: Phys. Rev. B **60**, 13740 (1999)
383. M.G. Bawendi, P.J. Carroll, W.L. Wilson, L.E. Brus: J. Chem. Phys. **96**, 946 (1992)
384. W. Hoheisel, V.L. Colvin, C.S. Johnson, A.P. Alivisatos: J. Chem. Phys. **101**, 8455 (1994)

385. T.W. Roberti, N.J. Cherepy, J.Z. Zhang: J. Chem. Phys. **108**, 2143 (1998)
386. C.B. Murray, D.J. Norris, M.G. Bawendi: J. Am. Chem. Soc. **115**, 8706 (1993)
387. J.E. Bowen Katari, V.L. Colvin, A.P. Alivisatos: J. Phys. Chem. **98**, 4109 (1994)
388. S.A. Majetich, A.C. Carter: J. Phys. Chem. **97**, 8727 (1993)
389. S.A. Majetich, A.C. Carter, J. Belot, R.D. McCullough: J. Phys. Chem. **98**, 13705 (1994)
390. M. Kuno, J.K. Lee, B.O. Dabbousi, F.V. Mikulec, M.G. Bawendi: J. Chem. Phys. **106**, 9869 (1997)
391. M. Kuno, J.K. Lee, B.O. Dabbousi, F.V. Mikulec, M.G. Bawendi: Mater. Res. Soc. Symp. Proc. **452**, 347 (1996)
392. D. Schooss, A. Mews, A. Eychmüller, H. Weller: Phys. Rev. B **49**, 17072 (1994)
393. X. Peng, M.C. Schlamp, A.V. Kadavanich, A.P. Alivisatos: J. Am. Chem. Soc. **119**, 7079 (1997)
394. B.O. Dabbousi, R. Rodriguez-Viejo, F.V. Mikulek, J.R. Heine, H. Matoussi, R. Ober, K.F. Jensen, M.G. Bawendi: J. Phys. Chem. **101**, 9463 (1997)
395. M.A. Hines, P. Guyot-Sionnest: J. Phys. Chem. **100**, 468 (1996)
396. L.R. Becerra, C.B. Murray, R.G. Griffin, M.G. Bawendi: J. Chem. Phys. **100**, 3297 (1994)
397. S. Pokrant, K.B. Whaley: Eur. Phys. J. D **6**, 255 (1999)
398. K. Leung, K.B. Whaley: Phys. Rev. B **110**, 11012 (1999)
399. J. Taylor, T. Kippeny, S.J. Rosenthal: Journal of Cluster Science **12**, 571 (2001)
400. U. Winkler, D. Eich, Z.H. Chen, R. Fink, S.K. Kulkarni, E. Umbach: Chem. Phys. Lett. **306**, 95 (1999)
401. J. Nanda, B.A. Kuruvilla, D.D. Sarma: Phys. Rev. B **59**, 7473 (1999)
402. J. Nanda, D. Sarma: J. Appl. Phys. **90**, 2504 (2001)
403. H. Borchert, D.V. Talapin, C. McGinley, S. Adam, A. Lobo, A.R.B. de Castro, T. Möller, H. Weller: J. Chem. Phys. **119**, 1800 (2003)
404. C. McGinley, M. Riedler, T. Möller, H. Borchert, S. Haubold, M. Haase, H. Weller: Phys. Rev. B **65**, 245308 (2002)
405. H. Fu, A. Zunger: Phys. Rev. B **56**, 1496 (1997)
406. O.I. Mićić, A.J. Nozik, E. Lifshitz, T. Rajh, O.G. Poluektov, M.C. Thurnauer: J. Phys. Chem. **106**, 4390 (2002)
407. L. Langof, E. Ehrenfreund, E. Lifshitz, O.I. Mićić, A.J. Nozik: J. Phys. Chem. **106**, 1606 (2002)
408. D.J. Lockwood: Solid State Comm. **92**, 101 (1994)
409. G. Allan, C. Delerue, M. Lannoo: Phys. Rev. Lett. **76**, 2961 (1996)
410. G. Allan, C. Delerue, M. Lannoo: Phys. Rev. B **48**, 7951 (1993)
411. G. Vincent, D. Bois: Solid State Comm. **27**, 431 (1978)
412. B.V. Kamenev, A.G. Nassiopoulou: J. Appl. Phys. **90**, 5735 (2001)
413. M.H. Nayfeh, N. Rigakis, Z. Yamani: Phys. Rev. B **56**, 2079 (1997)
414. M.H. Nayfeh, N. Barry, J. Therrien, O. Akcadir, E. Gratton, G. Belomoin: Appl. Phys. Lett. **78**, 1131 (2001)
415. A. Ourmazd, D.W. Taylor, J.A. Rentschler, J. Bevk: Phys. Rev. Lett. **59**, 213 (1987)
416. F. Herman, R.V. Kasowski: J. Vac. Sci. Technol. **19**, 395 (1981)
417. A. Puzder, A.J. Williamson, J.C. Grossman, G. Galli: Phys. Rev. Lett. **88**, 097401 (2002)

418. D. Goguenheim, M. Lannoo: J. Appl. Phys. **68**, 1059 (1990)
419. D. Goguenheim, M. Lannoo: Phys. Rev. B **44**, 1724 (1991)
420. C.H. Henry, D.V. Lang: Phys. Rev. B **15**, 989 (1977)
421. S.M. Prokes, W.E. Carlos, V.M. Bermudez: Appl. Phys. Lett. **61**, 1447 (1992)
422. B.K. Meyer, D.M. Hofmann, W. Stadler, V. Petrova-Koch, F. Koch, P. Omling, P. Emanuelsson: Appl. Phys. Lett. **63**, 2120 (1993)
423. P.T. Landsberg: *Recombination in Semiconductors* (Cambridge University Press, Cambridge 1991)
424. *Properties of Silicon*, EMIS Datareviews Series No. 4, ed. by INSPEC (The Institution of Electrical Engineers, London 1988)
425. J. Shah: *Ultrafast Spectroscopy of Semiconductors and Semiconductor Nanostructures* (Springer-Verlag, Berlin 1999)
426. I. Mihalcescu, J.C. Vial, A. Bsiesy, F. Muller, R. Romestain, E. Martin, C. Delerue, M. Lannoo, G. Allan: Phys. Rev. B **51**, 17605 (1995)
427. C. Delerue, M. Lannoo, G. Allan, E. Martin, I. Mihalcescu, J.C. Vial, R. Romestain, F. Muller, R. Romestain, A. Bsiesy: Phys. Rev. Lett. **75**, 228 (1995)
428. Al.L. Efros: cond-mat/0204437
429. L. Tsybeskov, J.V. Vandyshev, P. Fauchet: Phys. Rev. B **49**, 7821 (1994)
430. J.C. Vial, A. Bsiesy, F. Gaspard, R. Hérino, M. Ligeon, F. Muller, R. Romestain, R.M. Macfarlane: Phys. Rev. B **45**, 14171 (1992)
431. A. Bsiesy, F. Muller, I. Mihalcescu, M. Ligeon, F. Gaspard, R. Hérino, R. Romestain, J.C. Vial: in *Light Emission from Silicon*, ed. by J.C. Vial, L.T. Canham, W. Lang. In: J. Lum. **57**, 29 (1993)
432. A. Bsiesy, F. Muller, M. Ligeon, F. Gaspard, R. Hérino, R. Romestain, J.C. Vial: Phys. Rev. Lett. **71**, 637 (1993)
433. D. Kovalev, H. Heckler, G. Polisski, F. Koch: Phys. Stat. Solidi **215**, 871 (1999)
434. U. Banin, M. Bruchez, A.P. Alivisatos, T. Ha, S. Weiss, D.S. Chemla: J. Chem. Phys. **110**, 1195 (1999)
435. M. Kuno, D.P. Fromm, H.F. Hamann, A. Gallagher, D.J. Nesbitt: J. Chem. Phys. **112**, 3117 (2000)
436. R.G. Neuhauser, K. Shimizu, W.K. Woo, S.A. Empedocles, M.G. Bawendi: Phys. Rev. Lett. **85**, 3301 (2000)
437. J. Tittel, W. Gohde, F. Koberling, A. Mews, A. Kornowski, H. Weller, A. Eychmuller, Th. Basche, B. Bunsenges: Phys. Chem. **101**, 1626 (1997)
438. F. Koberling, A. Mews, Th. Basche: Phys. Rev. B **60**, 1921 (1999)
439. K.T. Shimizu, R.G. Neuhauser, C.A. Leatherdale, S.A. Empedocles, W.J. Woo, M.G. Bawendi: Phys. Rev. B **63**, 5316 (2001)
440. M. Kuno, D.P. Fromm, H.F. Hamann, A. Gallagher, D.J. Nesbitt: J. Chem. Phys. **115**, 1028 (2001)
441. M. Kuno, D. P. Fromm, A. Gallagher, D. J. Nesbitt, O. I. Micic, A. J. Nozik: Nano Lett. **1**, 557 (2001)
442. H.D. Robinson, B.B. Goldberg: Phys. Rev. B **61**, R5086 (2000)
443. T. Aoki, Y. Nishikawa, M. Kuwata-Gonokami: Appl. Phys. Lett. **78**, 1065 (2001)
444. M. D. Mason, G. M. Credo, K. D. Weston, S. K. Buratto: Phys. Rev. Lett. **80**, 5405 (1998)
445. Al.L. Efros, M. Rosen: Phys. Rev. Lett. **78**, 1110 (1997)

446. D.I. Chepic, Al.L. Efros, A.I. Ekimov, M.G. Ivanov, I.A. Kudriavtsev, V.A. Kharchenco, T.V. Yazeva: J. of Luminescence. **47**, 113 (1990)
447. P. Roussignol, D. Ricard, K.C. Rustagi, C. Flytzanis: Optics Commun. **55**, 143 (1985)
448. U. Bockelmann, G. Bastard: Phys. Rev. B **42**, 8947 (1990)
449. H. Benisty, C.M. Sottomayor-Torres, C. Weisbuch: Phys. Rev. B **44**, 10945 (1991)
450. T.S. Sosnowski, T.B. Norris, H. Jiang, J. Singh, K. Kamath, P. Bhattacharya: Phys. Rev. B **57**, R9423 (1998)
451. H. Jiang, J. Singh: Physica(Amsterdam) **2E**, 720 (1998)
452. I. Vurgaftman, J. Singh: Appl. Phys. Lett. **64**, 232 (1994)
453. B. Ohnesarge, M. Albrecht, J. Oshinowo, A. Forchel, Y. Arakawa: Phys. Rev. B **54**, 11532 (1996)
454. T. Inoshita, H. Sakaki: Phys. Rev. B **46**, 7260 (1992)
455. J. Uramaya, T.B. Norris, J. Singh, P. Bhattacharya: Phys. Rev. Lett. **86**, 4930 (2001)
456. U. Bockelmann, T. Egler: Phys. Rev. B **46**, 15574 (1992)
457. Al.L. Efros, V.A. Kharchenko, M. Rosen: Solid State Comm. **93**, 281 (1995)
458. V. Klimov, D. McBranch: Phys. Rev. Lett. **80**, 4028 (1998)
459. P. Guyot-Sionnest, M. Shim, C. Matranga, M. Hines: Phys. Rev. B **60**, 2181 (1999)
460. V.I. Klimov, A.A. Mihkailovsky, C. Leatherdale, M.G. Bawendi: Phys. Rev. B **61**, R13349 (2000)
461. J. Bellessa, V. Voliotis, T. Guillet, D. Roditchev, R. Grousson, X.L. Wang, M. Ogura: Eur. Phys. J B **21**, 499 (2001)
462. G. Ferreira, G. Bastard: Appl. Phys. Lett. **74**, 2818 (1999)
463. D.F. Schroeter, D.J. Griffiths, P.C. Sercel: Phys. Rev. B **54**, 1486 (1996)
464. *Computational Electronics: New Challenges and Directions*, ed. by M.S. Lundstrom, R.W. Dutton, D.K. Ferry, K. Hess. In: IEEE Transactions on Electron Devices, Vol. **47** (2000).
465. R.P. Andres, J.D. Bielefeld, J.I. Henderson, D.B. Janes, V.R. Kolagunta, C.P. Kubiak, W.J. Wahoney, R.G. Osifchin: Science **273**, 1690 (1996)
466. L.A. Bumm, J.J. Arnold, M.T. Cygan, T.D. Dunbar, T.P. Burgin, L. JonesII, D.L. Allara, J.M. Tour, P.M. Weiss: Science **271**, 1705 (1996)
467. A. Stabel, P. Herwig, K. Müllen, J.P. Rabe: Angew. Chem. Int. Ed. Engl. **34**, 1609 (1995)
468. W. Tian, S. Datta, S. Hong, R. Reifenberger, J.I. Henderson, P. Kubiak: J. Chem. Phys. **109**, 2874 (1998)
469. J. Tans, M. H. Devoret, H. Dai, A. Hess, R. E. Smalley, L. Geerligs, C. Dekker: Nature (London) **386**, 474 (1997)
470. M. A. Reed, C. Zhou, C. J. Muller, T. P. Burgin, J. M. Tour: Science **278**, 252 (1997)
471. J. Chen, M. A. Reed, A. M. Rawlett, J. M. Tour: Science **286**, 1550 (1999)
472. C. Kergueris, J.-P. Bourgoin, S. Palacin, D. Esteve, C. Urbina, M. Magoga, C. Joachim: Phys. Rev. B **59**, 12 505 (1999)
473. O. Millo, D. Katz, Y. Levi, Y. W. Cao, U. Banin: J. Low. Temp. Phys. **118**, 365 (2000)
474. O. Millo, D. Katz, Y. W. Cao, U. Banin: Phys. Rev. B **61**, 16773 (2000)
475. B. Alperson, I. Rubinstein, G. Hodes, D. Porath, O. Millo: Appl. Phys. Lett. **75**, 1751 (1999)

476. E. P. A. M. Bakkers, D. Vanmaekelbergh: Phys. Rev. B **62**, R7743 (2000)
477. D. L. Klein, R. Roth, A. K. L. Lim, A. P. Alivisatos, P. L. McEuen: Nature **389**, 699 (1997)
478. L.P. Kouwenhoven, D.G. Austing, S. Tarucha: Rep. Prog. Phys. **64**, 701 (2001)
479. A.L. Roest, J.J. Kelly, D. Vanmaekelbergh: Phys. Rev. Lett. **89**, 036801 (2002)
480. D. Yu, C. Wang, P. Guyot-Sionnest: Science **300**, 1277 (2003)
481. R. Parthasarathy, X.-M. Lin, H.M. Jaeger: Phys. Rev. Lett. **87**, 186807 (2001)
482. N.D. Lang: Phys. Rev. B **52**, 5335 (1995)
483. K. Hirose, M. Tsukada: Phys. Rev. B **51**, 5278 (1995)
484. A. Nitzan: Annu. Rev. Phys. Chem. **52**, 681 (2001)
485. J. Taylor, H. Guo, J. Wang: Phys. Rev. B **63**, 245407 (2001)
486. P.S. Damle, A.W. Ghosh, S. Datta: Phys. Rev. B **64**, 201403 (2001)
487. M. Brandbyge, J.-L. Mozos, P. Ordejón, J. Taylor, K. Stokbro: Phys. Rev. B **65**, 165401 (2002)
488. T.N. Todorov: J. Phys.: Condens. Matter **14**, 3049 (2002)
489. E.G. Emberly, G. Kirczenow: Phys. Rev. B **58**, 10911 (1998)
490. A.R. Williams, P.J. Feibelman, N.D. Lang: Phys. Rev. B **26**, 5433 (1982)
491. W.A. Harrison: Phys. Rev. **123**, 85 (1961)
492. J. Bardeen: Phys. Rev. Lett. **6**, 57 (1961)
493. J. Tersoff, D.R. Hamann: Phys. Rev. Lett. **25**, 1998 (1983)
494. J. Tersoff: Phys. Rev. B **41**, 1235 (1990)
495. A. Barraud, P. Millie, I. Yakimenko: J. Chem. Phys. **105**, 6972 (1996)
496. C.W.J. Beenakker: Phys. Rev. B **44**, 1646 (1991)
497. R.I. Shekhter: Zh. Éksp. Teor. Fiz. **63**, 1410 (1972) [Sov. Phys. JETP **36**, 747 (1973)]; I.O. Kulik, R.I. Shekhter, *ibid* **68**, 623 (1975) [*ibid* **41**, 308 (1975)]
498. S. Datta: *Electronic Transport in Mesoscopic Systems* (Cambridge University Press, Cambridge 1997)
499. S. Datta: Superlattices and Microstructures **28**, 253 (2000)
500. *Single-charge tunneling : Coulomb blockade phenomena in nanostructures*, ed. H. Grabert, M. H. Devoret, (Plenum Press, New York 1992)
501. D.K. Ferry, S.M. Goodnick: *Transport in Nanostructures* (Cambridge University Press, Cambridge, England, 1997)
502. H. Haug, A.-P. Jauho: *Quantum Kinetics in Transport and Optics of Semiconductors* (Springer-Verlag, Berlin 1996)
503. L.V. Keldysh: Sov. Phys. JETP **20**, 1018 (1965)
504. L.P. Kadanoff, G. Baym: *Quantum Statistical Mechanics* (Benjamin/Cummings, New York 1962)
505. C. Caroli, R. Combescot, P. Nozières, D. Saint-James: J. Phys. C **4**, 916 (1971)
506. C. Caroli, R. Combescot, D. Lederer, P. Nozières, D. Saint-James: J. Phys. C **4**, 2598 (1971)
507. Y. Meier, N.S. Wingreen: Phys. Rev. Lett. **68**, 2512 (1992)
508. A.-P. Jauho, N.S. Wingreen, Y. Meier: Phys. Rev. B **50**, 5528 (1994)
509. T.N. Todorov, G.A.D. Briggs, A.P. Sutton: J. Phys. C **5**, 2389 (1993)
510. G. Doyen, D. Drakova, M. Scheffler: Phys. Rev. B **47**, 9778 (1993)
511. P. Sautet, C. Joachim: Chem. Phys. Lett. **185**, 23 (1991)
512. V. Mujica, M. Kemp, M.A. Ratner: J. Chem. Phys. **101**, 6849 (1994)
513. M. Büttiker: Phys. Rev. B **33**, 3020 (1986)
514. B.A. Lippmann: Phys. Rev. Lett. **15**, 11 (1965) ; *ibid* **16**, 135 (1965)
515. P. Ehrenfest: Z. Phys. **45**, 455 (1927)

516. E.N. Economou: *Green's Functions in Quantum Physics* (Springer-Verlag, Berlin 1979)
517. R. Landauer: IBM J. Res. Dev. **1**, 223 (1957)
518. R. Landauer: J. Phys.: Condens. Matter **1**, 8099 (1989)
519. M. Büttiker, Y. Imry, R. Landauer, S. Pinhas: Phys. Rev. B **31**, 6207 (1985)
520. D.S. Fisher, P.A. Lee: Phys. Rev. B **23**, 6851 (1981)
521. M. Lannoo, P. Friedel: *Atomic and Electronic Structure of Surfaces*, Springer Series in Surface Sciences 16 (Springer-Verlag, Berlin 1991)
522. M.-C. Desjonquères, D. Spanjaard: *Concepts in Surface Physics*, Springer Series in Surface Sciences 30 (Springer-Verlag, Berlin 1993)
523. D. Goldhaber-Gordon, H. Shtrikman, D. Mahalu, D. Abusch-Magder, U. Meirav, M.A. Kastner: Nature **391**, 156 (1998)
524. S.M. Cronenwett, T.H. Oosterkamp, L.P. Kouwenhoven: Science **281**, 540 (1998)
525. J. Schmid, J. Weis, K. Eberl, K. von Klitzing: Physica B **256-258**, 375 (1998)
526. F. Simmel, R. H. Blick, J. P. Kotthaus, W. Wegscheider, M. Bichler: Phys. Rev. Lett. **83**, 804 (1999)
527. J. Park, A.N. Pasupathy, J.I. Goldsmith, C. Chang, Y. Yaish, J.R. Petta, M. Rinkoski, J.P. Sethna, H.D. Abruña, P.L. McEuen, D.C. Ralph: Nature **417**, 722 (2002)
528. W. Liang, M.P. Shores, M. Bockrath, J.R. Long, H. Park: Nature **417**, 725 (2002)
529. P.L. Pernas, A. Martin-Rodero, F. Flores: Phys. Rev. B **41**, 8553 (1990)
530. P.S. Damle, A.W. Ghosh, S. Datta: Chem. Phys. **171**, 281 (2002)
531. N.D. Lang, P. Avouris: Nano Lett. **3**, 737 (2003)
532. V. Mujica, A. Roitberg, M. Ratner: J. Chem. Phys. **112**, 6834 (2000)
533. C. Krzeminski, C. Delerue, G. Allan, D. Vuillaume, R.M. Metzger: Phys. Rev. B **64**, 085405 (2001)
534. R.M. Metzger, B. Chen, U. Höpfner, M.V. Lakshmikantham, D. Vuillaume, T. Kawai, X. Wu, H. Tachibana, T.V. Hughes, H. Sakurai, J.W. Baldwin, C. Hosh, M.P. Cava, L. Brehmer, C.J. Ashwell: J. Am. Chem. Soc. **119**, 10455 (1997)
535. D. Vuillaume, B. Chen, R. M. Metzger: Langmuir **15**, 4011 (1999)
536. T. Dittrich, P. Hänggi, G.-L. Ingold, B. Kramer, G. Schön, W. Zwerger: *Quantum Transport and Dissipation* (Wiley-VCH, Weinheim 1997)
537. *Mesoscopic Electron Transport*, ed. by L.L. Sohn *et al.* (Kluwer, Dordrecht 1997)
538. G.-M. Rignanese, X.Blase, S.G. Louie: Phys. Rev. Lett. **87**, 206405 (2001)
539. S. Datta: Proceedings of the International Electron Devices Meeting (IEDM) (IEEE Press 2002)
540. A.A. Middleton, N.S. Wingreen: Phys. Rev. Lett. **71**, 3198 (1993)
541. R.H. Tredgold: Proc. Phys. Soc. **80**, 807 (1962)
542. C.M. Hurd: J. Phys.C: Solid State Phys. **18**, 6487 (1985)
543. H. Scher, M. Lax: Phys. Rev. B **7**, 4491 (1973)
544. H. Scher, M. Lax: Phys. Rev. B **7**, 4502 (1973)
545. T. Nakayama, K. Yakubo, R.L. Orbach: Rev. Mod. Phys. **66**, 381 (1994)
546. M. Lax: Phys. Rev. **109**, 1921 (1958)
547. S. Kirkpatrick: Rev. Mod. Phys. **45**, 74 (1973)
548. T. Odagaki, M. Lax: Phys. Rev. B **24**, 5284 (1981)

549. I. Webman: Phys. Rev. Lett. **47**, 1496 (1981)
550. S. Summerfield: Solid State Comm. **39**, 401 (1981)
551. N.F. Mott: *Metal-Insulator Transitions* (Taylor & Francis Ltd., London 1974)
552. N.F. Mott, E.A. Davis: *Electronic Processes in Non-Crystalline Materials*, 2nd ed. (Oxford University Press, Oxford 1979)
553. M. Ben-Chorin, F. Möller, F. Koch, W. Schirmacher, M. Eberhard: Phys. Rev. B **51**, 2199 (1995)
554. E. Lampin, C. Delerue, M. Lannoo, G. Allan: Phys. Rev. B **58**, 12044 (1998)
555. Y.A. Berlin, A.L. Burin, M.A. Ratner: Chem. Phys. **275**, 61 (2002)
556. N.Y. Morgan, C.A. Leatherdale, M. Drndić, M.V. Jarosz, M.A. Kastner, M. Bawendi: Phys. Rev. B **66**, 075339 (2002)
557. A. Efros, B. Shklovskii: J. Phys. C **8**, L49 (1975)

Index

$k \cdot p$ method   43, 57–60
a-Si   74
a-Si:H   74

Absorption coefficient   143, 147, 156
Acoustic phonons   169
Activation energy   270
Addition spectra, tunneling spectroscopy   243
Adiabatic local density approximation   31
AlAs nanocrystals   68
AlP nanocrystals   68
AlSb nanocrystals   68
Amorphous semiconductors   74
– dangling bonds   205
Anomalous diffusion   266
Artificial atoms   53, 182
Auger auto-ionization   225, 232
Auger recombination   225–233
– III–V and II–VI nanocrystals   232
– Si nanocrystals   230
Auger thermalization   233

Ballistic transport   255
Bandgap problem   14
Bardeen formalism   239
Bethe–Salpeter equation   27
Black-body   148
Born approximation   248
Born–Oppenheimer approximation   154
– non-adiabatic transitions   172
Breit–Wigner formula   251

Capture cross section   220
Carrier relaxation   228
CdS nanocrystals   114

CdSe nanocrystals   68
– dangling bonds   209
Channels of conduction   51
Charging energy   258
– quantum dots   100
– quantum wells   103
– tunneling spectroscopy   133
Coherent potential approximation   268
Coherent transport   245
Complex refractive index   143
Configuration coordinate diagram   154
– dangling bond   220
– inelastic tunneling   263
– self-trapped excitons   210
Configuration interaction   6
– excitons   112
– multi-excitons   138
– screening   120
Contact resistance   255
Core–shell systems   208
Correlation energy   24, 89
Correlation length   266
Coulomb blockade   232, 244, 258
– scanning tunneling spectroscopy   135
Coulomb integrals   37
– screened   88
Coulomb oscillations   244
Coupling matrices   240

Dangling bonds   219
– bulk silicon   205
– CdSe nanocrystals   208, 209
– compensation mechanism   199
– silicon nanocrystals   207
Decimation method   252
Deep impurities   199, 200

Deep level transient spectroscopy 205
Deformation potentials 170
Density functional theory 10
– dielectric response 90
– effective mass approximation 89
– quantum confinement 63
– quasi-particle gap 121
– surface polarization charges 127
– transport 259
Density matrix
– linear response 17
– optical properties 144
– transport 247
Density of states 50–52
– optical properties 178
Dielectric constant 143
– nanostructures 94–100
Dielectric function 16, 18
– bulk 93
– optical properties 144
– static 91
Dielectric properties 77–103
Dielectric screening
– quantum dots 96
– quantum wells 95
Diffusion 266
Diffusion coefficient 266
Diffusive transport 255
Dipolar momentum 145
Direct optical transitions 157
Disorder 74
– hopping conductivity 266
– nanocrystal arrays 263
Donor potential 197
Donors 195
Dynamical matrix 152
Dyson's equation 247

eeh Auger mechanism 225
Effective charges 162
Effective mass approximation 43
– dielectric effects 84
– excitons 108
– optical transitions 174
– quantum dots 52
– quantum wells 47
– quantum wires 51
Effective medium 270
ehh Auger mechanism 225

Eigenvalue problems 62
Einstein relationships
– conductivity 268
– spontaneous emission 148
Elastic scattering formalism 245
Elastic tunneling 263
Electron affinity 14, 106
Electron correlations 6
Electron gaz 6
– effective mass theory 89
Electron paramagnetic resonance 205, 220
Electron spin resonance 205
Electron states
– direct gap semiconductors 69
– indirect gap semiconductors 70
Electron–electron interactions
– configuration interaction 113
– effective mass theory 86
– GW 21
– linear response 90
– quantum dots 83, 121
– quantum wells 82
– transport 253
Electron–hole inelastic scattering 233
Electron–hole interaction 120
Electron–phonon coupling
– acoustic modes 169
– Auger recombination 227
– capture at point defects 220
– coupling parameters 161
– general theory 151
– hopping transport 264
– hot carrier relaxation 233
– interband transitions 184
– intraband transitions 185
– non-adiabatic transitions 172
– optical modes 162
– Si nanocrystals 189, 193
Electron–plasmon coupling 21, 121
Envelope function 47, 195
Equation of motion 152
Exchange interactions 5, 6
– CdS nanocrystals 114
– nanostructures 122
– oscillator strength 180
– screening of the 27, 113
– Si nanocrystals 116

Exchange–correlation hole   6, 15
Exchange–correlation potential   10, 12
– effective mass theory   89
– surface polarization charges   127
– transport   253
Excitation spectra, tunneling spectroscopy   243
Excitons   108–121, 129–133, 138–140
– binding energy   110
– bulk   108, 176
– general theory   26
– quantum dots   114, 129, 179
– quantum wells   109, 178
– Si nanocrystals   116
Extinction coefficient   143

Form factors   41
Fourier transform infrared spectroscopy   214
Fröhlich coupling   162
Franck–Condon principle   155
Franck–Condon shift   155
– Si dangling bond   220
– tunneling between nanocrystals   263

GaAs nanocrystals   68
GaP nanocrystals   68
GaSb nanocrystals   68
Ge nanocrystals   68
– optical properties   185
Green's functions   25, 246, 269
GW approximation   21
– nanostructures   122, 123
– surface defects   200
– transport   260

Harmonic oscillators   152
Harrison's rules   34
Hartree approximation   3
– dielectric properties   86
Hartree–Fock approximation   5
– electron–hole excitations   26
– excitons   112
– transport   261
Hohenberg–Kohn theorem   11
Hole states
– III–V and II–VI nanocrystals   73
– silicon nanocrystals   73
Hopping conductivity   266

Hot carrier relaxation   233
Huang–Rhys factor   155
– interband transitions   184
– intraband transitions   185
– LO phonons   168
– numerical calculation   161
– Si dangling bond   220
– Si nanocrystals   189
Hydrogen desorption   211
Hydrogenic impurities   195
Hyper-polarizability   93

Image charges   80, 107
Impurity binding energy   197
InAs nanocrystals   68
– electron–phonon coupling   169
– tunneling spectroscopy   135
Inelastic tunneling   263
Infrared emission   225
Infrared optical transitions   183
InP nanocrystals   68
– dangling bonds   209
InSb nanocrystals   68
Inter-valley couplings   71
Interband transitions   175–181
– bulk   175
– quantum dots   179, 184, 186
– quantum wells   177
Intraband transitions   181–184
– bulk   181
– quantum dots   182, 185, 191
– quantum wells   181
Ionization energy   14, 106
Ionization levels   200, 258

Jahn–Teller effect   173

Kadanoff–Baym formalism   245
Kane Hamiltonian   46, 58
Keldysh formalism   245
Kohn–Sham equation   12
– effective mass approximation   89
– surface polarization charges   127
– transport   253, 259
Kondo effect   253
Koopman's theorem   14

Landauer formula   249
Langmuir–Blodgett films   239, 257

Leads 236
Ligand molecules 208
Linear combination of atomic orbitals 32
Linear response theory 16, 90
– exact formulation 18
– tight binding 37
Lippmann–Schwinger equation 246
LO phonons 164
Local density approximation 10
– effective mass theory 89
– transport 259
Local pseudopotentials 41
Local-fields 93, 143, 149
Luttinger–Kohn Hamiltonian 46
Lyddane–Sachs–Teller relation 164

Macroscopic dielectric function 93
Macroscopic optical properties 143
Macroscopic theory of dielectrics 78–84
Master equations
– coherent potential approximation 268
– orthodox theory 241
Maxwell–Garnett theory 143
Mean-field theories 1–13
– Coulomb blockade 258
– transport 253
Metal–insulator–metal junctions 239
Metallic island 244
Molecular electronics 235
Moments, method of 265
Multi-excitons 138
Multi-phonon capture 219
Multi-phonon relaxation 233

Nano-device 236
Network of nanostructures 263
No-phonon optical transitions 186
Non-adiabatic transitions 172
Non-equilibrium density operator 254
Non-equilibrium Green's functions 245
Non-local pseudopotentials 41
Normal coordinates 152

Optical absorption 141, 175
Optical cavities 148, 149

Optical line-shape 145, 158
Optical matrix elements 150, 156
– $\boldsymbol{k}\cdot\boldsymbol{p}$ theory 175
Optical phonons 162
Orthodox theory 239, 258
Oscillator strength 147
– bulk 176
– intraband transitions in nanocrystals 183
– quantum dots 180
– quantum wells 178
Oxygen defects 210, 214

P_b center 205, 219
Pauli principle 3
Percolating systems 269
Phonon bottleneck 233
Phonon line-shape 156, 265
Phonon scattering 255
Phonon-assisted optical transitions 160, 189
Phonons 153
Photo-darkening effect 232
Photoluminescence 114, 148
– intermittency 232
– self-trapped excitons 210
– Si nanocrystals 189
Photonic crystals 150
Polarizability 16, 91
Polarization charges 80, 94, 107, 121
Polarization selection rules 178, 181
Polaron 162, 169
Porous silicon 219
– Auger recombination 230
– dangling bonds 199
– oxygen defects 214
– self-trapped excitons 210
– transport 270
Pseudopotentials 39
– nanostructures 61
– optical matrix elements 150

Quantum confinement 47–76
– amorphous semiconductors 74
– quantum dots 52–56, 67
– quantum wells 47
– quantum wires 51
Quantum dots 52, 83, 96, 100, 179, 182, 196

Quantum wells 47, 80, 95, 103, 109, 177, 181, 195
Quantum wires 51
Quasi-particles 14, 105, 121

Radiative recombination
– amorphous semiconductors 76
Random intermittency 232
Random-phase approximation 16, 31, 91
Rectifying characteristics 257
Recursion method 252
Reflectance 143
Relaxation energy 155
– acoustic phonons 169
– LO phonons 168
Reservoirs 236
Resonant tunneling 252

Saturation of the photoluminescence 230
Scanning tunneling microscope 133, 239
Scattering operator 248
Schrödinger–Poisson solvers 253
Selection rules 182
Self-consistent calculations 4
– non-equilibrium 253
Self-consistent potential 255
Self-energy 24, 106, 122
– effective mass theory 85
– surface contribution 79
– – quantum dots 100
– – quantum wells 103
– transport 237, 249
Self-energy corrections 126, 201
– transport 260
Self-interaction 4, 88
Self-trapped excitons 210–213
Semi-empirical methods 31–46
Sequential tunneling 252
Shell-filling spectroscopy 243
Shell-tunneling spectroscopy 243
Si dimers 212
Si nanocrystals 116
– Auger recombination 225, 230
– confinement energies 68, 132
– dangling bonds 207, 219
– electron states 70
– hole states 73
– optical properties 185
– oxygen defects 214
– transport 270
Si–SiO$_2$ interface 205, 214, 219
Si=O bonds 216
Single-electron transistor 244
Singlet 116
Slater determinant 2, 112
Slater's $X\alpha$ method 13
Slater's transition level 123, 204
SO modes 166
Spherical Bessel functions 53, 164
Spherical harmonics 53
Spherical potential well 53, 111
Spin–orbit coupling 114
Spontaneous emission 141, 149
Square potential well 49
Stimulated emission 141
Stokes shift
– oxygen defects 214
– self-trapped excitons 210
– surface defects 200
Structure factors 41
Superlattices 59
Surface passivation 214

Thermal activation 220
Thermalization 233
Thomas–Fermi approximation 10
Thomas–Reiche–Kuhn rule 147
Tight binding method 32
– dielectric effects 85, 92
– nanostructures 61
– optical matrix elements 151
Time-dependent density functional theory 29
TOPO 208
Topological disorder 266
Total energies, tight binding 36
Transition levels 241
Transport properties 235–271
Triplet 116
Tunneling 239, 263
Tunneling spectroscopy 133
Two-level model 116

Unrestricted calculations 261

Valence force-field   153
Variable range hopping   268
Voltage quenching   230
Voltage tunable electro-luminescence   232

Weak coupling limit   237
Wentzel–Kramers–Brillouin approximation   239

X-ray photo-electron spectroscopy   209

ZnO nanocrystals   68
– intraband transitions   183
ZnSe nanocrystals
– electron–phonon coupling   169

Druck: betz-druck GmbH, D-64291 Darmstadt
Verarbeitung: Buchbinderei Schäffer, D-67269 Grünstadt